Two Manuals Combined into One Book

Military First Aid

and

Ranger Survival

Department of Defense

Pentagon
Publishing

*FIELD MANUAL
NO. 4-25.11
NAVY TACTICAL
REFERENCE
PUBLICATION
NO. 4-02.1
AIR FORCE MANUAL
NO. 44-163(I)

HEADQUARTERS
DEPARTMENT OF THE ARMY,
THE NAVY, AND THE AIR FORCE
Washington, DC, 23 December 2002

FIRST AID

TABLE OF CONTENTS

*This publication supersedes FM 21-11, 27 October 1988

PREFACE

This manual meets the first aid training needs of individual service members. Because medical personnel will not always be readily available, the nonmedical service members must rely heavily on their own skills and knowledge of life-sustaining methods to survive on the integrated battlefield. This publication outlines both self-aid and aid to other service members (buddy aid). More importantly, it emphasizes prompt and effective action in sustaining life and preventing or minimizing further suffering and disability. First aid is the emergency care given to the sick, injured, or wounded before being treated by medical personnel. The term *first aid* can be defined as "urgent and immediate lifesaving and other measures, which can be performed for casualties by nonmedical personnel when medical personnel are not immediately available." Nonmedical service members have received basic first aid training and should remain skilled in the correct procedures for giving first aid. This manual is directed to *all* service members. The procedures discussed apply to all types of casualties and the measures described are for use by both male and female service members.

This publication is in consonance with the following North Atlantic Treaty Organization (NATO) International Standardization Agreements (STANAGs) and American, British. Canadian, and Australian Quadripartite Standardization Agreements (QSTAGs).

TITLE	STANAG	QSTAG
Medical Training in First Aid, Basic Hygiene and Emergency Care	2122	535
First Aid Kits and Emergency Medical Care Kits	2126	
Medical First Aid and Hygiene Training in NBC Operations	2358	
First Aid Material for Chemical Injuries	2871	

These agreements are available on request, using Department of Defense (DD) Form 1425 from the Standardization Documents Order Desk, 700 Robins Avenue, Building 4, Section D, Philadelphia, Pennsylvania 19111-5094.

Unless this publication states otherwise, masculine nouns and pronouns do not refer exclusively to men.

Use of trade or brand names in this publication is for illustrative purposes only and does not imply endorsement by the Department of Defense (DOD).

The proponent for this publication is the US Army Medical Department Center and School. Submit comments and recommendations for the improvement of this publication directly to the **Commander, US Army Medical Department Center and School, ATTN: MCCS-FCD-L, 1400 East Grayson Street, Fort Sam Houston, Texas 78234-5052**.

CHAPTER 1

FUNDAMENTAL CRITERIA FOR FIRST AID

"The fate of the wounded rests in the hands of the ones who apply the first dressing."

Nicholas Senn (1898) (49th President of the American Medical Association)

1-1. General

When a nonmedical service member comes upon an unconscious or injured service member, he must accurately evaluate the casualty to determine the first aid measures needed to prevent further injury or death. He should seek medical assistance as soon as possible, but he should not interrupt the performance of first aid measures. To interrupt the first aid measures may cause more harm than good to the casualty. Remember that in a chemical environment, the service member should not evaluate the casualty until the casualty has been masked. After performing first aid, the service member must proceed with the evaluation and continue to monitor the casualty for development of conditions which may require the performance of necessary basic lifesaving measures, such as clearing the airway, rescue breathing, preventing shock, and controlling bleeding. He should continue to monitor the casualty until relieved by medical personnel.

Service members may have to depend upon their first aid knowledge and skills to save themselves (self-aid) or other service members (buddy aid/ combat lifesaver). They may be able to save a life, prevent permanent disability, or reduce long periods of hospitalization by knowing **WHAT** to do, **WHAT NOT** to do, and **WHEN** to seek medical assistance.

NOTE

The prevalence of various body armor systems currently fielded to US service members, and those in development for future fielding, may present a temporary obstacle to effective evaluation of an injured service member. You may have to *carefully remove* the body armor from the injured service member to complete the evaluation or administer first aid. Begin by removing the outer-most hard or soft body armor components (open, unfasten or cut the closures, fasteners, or straps), then remove any successive layers in the same manner. Be sure to follow other notes, cautions and warnings regarding procedures in contaminated situations and when a broken back or neck is suspected. Continue to evaluate.

1-2. Terminology

To enhance the understanding of the material contained in this publication, the following terms are used—

- *Combat lifesaver.* This is a US Army program governed by Army Regulation (AR) 350-41. The combat lifesaver is a member of a nonmedical unit selected by the unit commander for additional training beyond basic first aid procedures (referred to as *enhanced first aid*). A minimum of one individual per squad, crew, team, or equivalent-sized unit should be trained. The primary duty of this individual does not change. The additional duty of combat lifesaver is to provide enhanced first aid for injuries based on his training before the trauma specialist (military occupational specialty [MOS] 91W) arrives. The combat lifesaver's training is normally provided by medical personnel assigned, attached, or in direct support (DS) of the unit. The senior medical person designated by the commander manages the training program.

- *Trauma Specialist (US Army) or Hospital Corpsman (HM).* A medical specialist trained in emergency medical treatment (EMT) procedures and assigned or attached in support of a combat or combat support unit or marine forces.

- *Casualty evacuation.* Casualty evacuation (CASEVAC) is a term used by nonmedical units to refer to the movement of casualties aboard nonmedical vehicles or aircraft. See also the term *transported* below. Refer to FM 8-10-6 for additional information.

> **CAUTION**
>
> Casualties transported in this manner do not receive en route medical care.

- *Enhanced first aid (US Army).* Enhanced first aid is administered by the combat lifesaver. It includes measures, which require an additional level of training above self-aid and buddy aid, such as the initiation of intravenous (IV) fluids.

- *Medical evacuation.* Medical evacuation is the timely, efficient movement of the wounded, injured, or ill service members from the battlefield and other locations to medical treatment facilities (MTFs). Medical personnel provide en route medical care during the evacuation. Once the casualty has entered the medical stream (trauma specialist, hospital corpsman, evacuation crew, or MTF), the role of first aid in the care of the casualty ceases and the casualty becomes the responsibility of the health service support (HSS) chain. Once he has entered the HSS chain he is referred to as a *patient*.

- *First aid measures.* Urgent and immediate lifesaving and other measures, which can be performed for casualties (or performed by the casualty himself) by nonmedical personnel when medical personnel are not immediately available.

- *Medical treatment.* Medical treatment is the care and management of wounded, injured, or ill service members by medically trained (MOS-trained) HM, and area of concentration (AOC) personnel. It may include EMT, advanced trauma management (ATM), and resuscitative and surgical intervention.

- *Medical treatment facility.* Any facility established for the purpose of providing medical treatment. This includes battalion aid stations, Level II facilities, dispensaries, clinics, and hospitals.

- *Self-aid/buddy aid.* Each individual service member is trained to be proficient in a variety of specific first aid procedures. This training enables the service member or a buddy to apply immediate first aid measures to alleviate a life-threatening situation.

- *Transported.* A casualty is moved to an MTF in a nonmedical vehicle without en route care provided by a medically-trained service member (such as a Trauma Specialist or HM). First aid measures should be continually performed while the casualty is being transported. If the casualty is acquired by a dedicated medical vehicle with a medically-trained crew, the role of first aid ceases and the casualty becomes the responsibility of the HSS chain, and is then referred to as a *patient*. This method of transporting a casualty is also referred to as *CASEVAC*.

1-3. Understanding Vital Body Functions for First Aid

In order for the service member to learn to perform first aid procedures, he must have a basic understanding of what the vital body functions are and what the result will be if they are damaged or not functioning.

a. Breathing Process. All humans must have oxygen to live. Through the breathing process, the lungs draw oxygen from the air and put it into the blood. The heart pumps the blood through the body to be used by the cells that require a constant supply of oxygen. Some cells are more dependent on a constant supply of oxygen than others. For example, cells of the brain may die within 4 to 6 minutes without oxygen. Once these cells die, they are lost forever since they do not regenerate. This could result in permanent brain damage, paralysis, or death.

b. Respiration. Respiration occurs when a person inhales (oxygen is taken into the body) and then exhales (carbon dioxide [CO_2] is expelled from the body). Respiration involves the—

- *Airway.* The airway consists of the nose, mouth, throat, voice box, and windpipe. It is the canal through which air passes to and from the lungs.

- *Lungs.* The lungs are two elastic organs made up of thousands of tiny air spaces and covered by an airtight membrane. The *bronchial tree* is a part of the lungs.

- *Rib cage.* The rib cage is formed by the muscle-connected ribs, which join the spine in back, and the breastbone in front. The top part of the rib cage is closed by the structure of the neck, and the bottom part is separated from the abdominal cavity by a large dome-shaped muscle called the *diaphragm* (Figure 1-1). The diaphragm and rib muscles, which are under the control of the respiratory center in the brain, automatically *contract* and *relax*. *Contraction* increases and *relaxation* decreases the size of the rib cage. When the rib cage increases and then decreases, the air pressure in the lungs is first less and then more than the atmospheric pressure, thus causing the air to rush into and out of the lungs to equalize the pressure. This cycle of inhaling and exhaling is repeated about 12 to 18 times per minute.

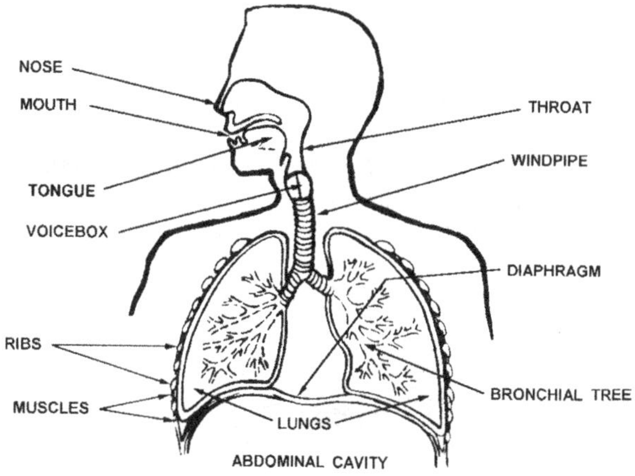

Figure 1-1. Airway, lungs, and rib cage.

c. Blood Circulation. The heart and the blood vessels (arteries, veins, and capillaries) circulate blood through the body tissues. The heart is divided into two separate halves, each acting as a pump. The left side pumps oxygenated blood (bright red) through the arteries into the capillaries; nutrients and oxygen pass from the blood through the walls of the capillaries into the cells. At the same time waste products and CO_2 enter the capillaries. From the capillaries the oxygen poor blood is carried through the veins to the right side of the heart and then into the lungs where it expels the CO_2 and picks up oxygen. Blood in the veins is dark red because of its low oxygen content. Blood does not flow through the veins in spurts as it does through the arteries. The entire system of the heart, blood vessels, and lymphatics is called the *circulatory system.*

(1) *Heartbeat.* The heart functions as a pump to circulate the blood continuously through the blood vessels to all parts of the body. It contracts, forcing the blood from its chambers; then it relaxes, permitting its chambers to refill with blood. The rhythmical cycle of contraction and relaxation is called the *heartbeat.* The normal heartbeat is from 60 to 80 beats per minute.

(2) *Pulse.* The heartbeat causes a rhythmical expansion and contraction of the arteries as it forces blood through them. This cycle of expansion and contraction can be felt (monitored) at various points in the body and is called the *pulse.* The common points for checking the pulse are at the—

- Side of the neck (*carotid*).
- Groin (*femoral*).
- Wrist (*radial*).
- Ankle (*posterior tibial*).

(*a*) *Carotid pulse.* To check the carotid pulse, feel for a pulse on the side of the casualty's neck closest to you. This is done by placing the tips of your first two fingers beside his Adam's apple (Figure 1-2).

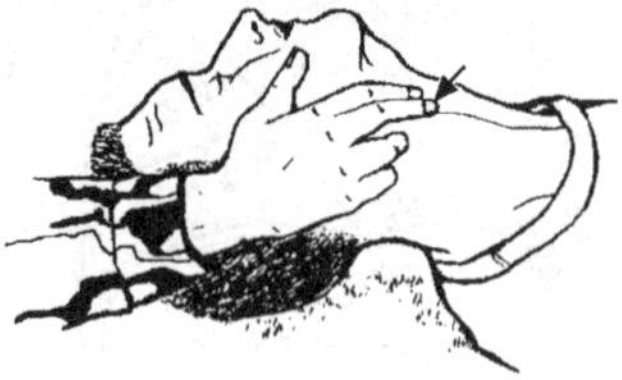

Figure 1-2. Carotid pulse.

(*b*) *Femoral pulse*. To check the femoral pulse, press the tips of your first two fingers into the middle of the groin (Figure 1-3).

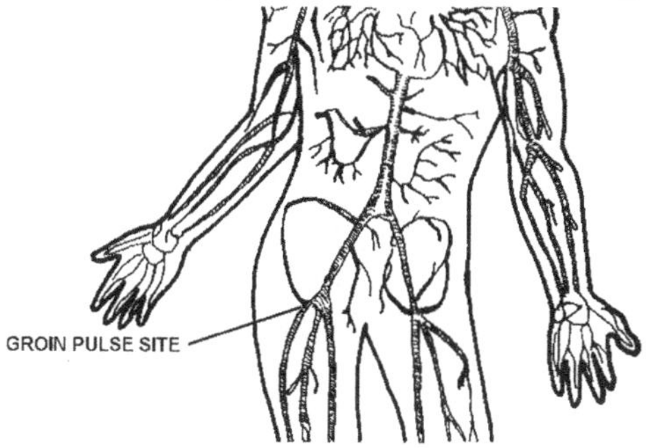

Figure 1-3. Femoral pulse.

(*c*) *Radial pulse*. To check the radial pulse, place your first two fingers on the thumb side of the casualty's wrist (Figure 1-4).

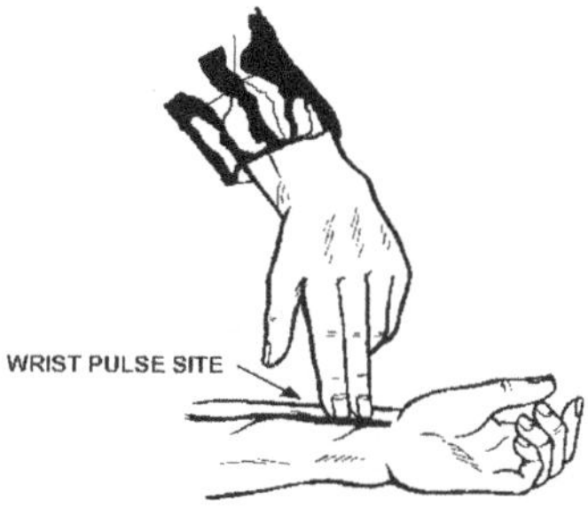

Figure 1-4. Radial pulse.

(*d*) *Posterior tibial pulse*. To check the posterior tibial pulse, place your first two fingers on the inside of the ankle (Figure 1-5).

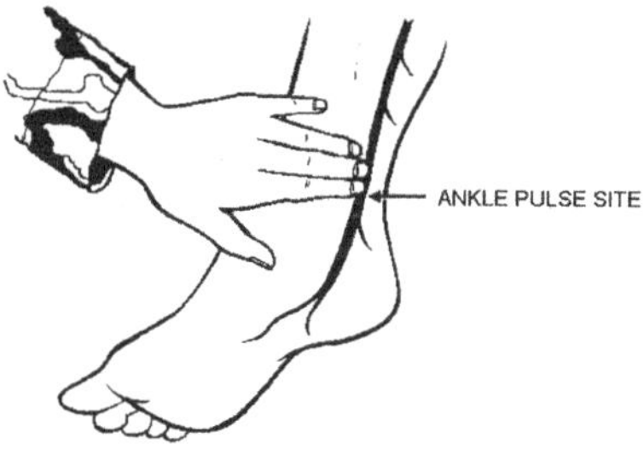

Figure 1-5. Posterior tibial pulse.

NOTE

DO NOT use your thumb to check a casualty's pulse because you may confuse the beat of your pulse with that of the casualty.

1-4. Adverse Conditions

a. Lack of Oxygen. Human life cannot exist without a continuous intake of oxygen. Lack of oxygen rapidly leads to death. First aid involves knowing how to open the airway and restore breathing.

b. Bleeding. Human life cannot continue without an adequate volume of blood circulating through the body to carry oxygen to the tissues. An important first aid measure is to stop the bleeding to prevent the loss of blood.

c. Shock. Shock means there is an inadequate blood flow to the vital tissues and organs. Shock that remains uncorrected may result in death even though the injury or condition causing the shock would not otherwise be fatal. Shock can result from many causes, such as loss of blood, loss of fluid from deep burns, pain, and reaction to the sight of a wound or blood. First aid includes preventing shock, since the casualty's chances of survival are much greater if he does not develop shock. Refer to paragraphs 2-21 through 2-24 for a further discussion of shock.

d. Infection. Recovery from a severe injury or a wound depends largely upon how well the injury or wound was initially protected. Infections result from the multiplication and growth (spread) of harmful microscopic organisms (sometimes referred to as germs). These harmful microscopic organisms are in the air, water, and soil, and on the skin and clothing. Some of these organisms will immediately invade (contaminate) a break in the skin or an open wound. The objective is to keep wounds clean and free of these organisms. A good working knowledge of basic first aid measures also includes knowing how to dress a wound to avoid infection or additional contamination.

1-5. Basics of First Aid

Most injured or ill service members are able to return to their units to fight or support primarily because they are given appropriate and timely first aid followed by the best medical care possible. Therefore, all service members must remember the basics.

- Check for **BREATHING**: Lack of oxygen intake (through a compromised airway or inadequate breathing) can lead to brain damage or death in very few minutes.

- Check for **BLEEDING**: Life cannot continue without an adequate volume of blood to carry oxygen to tissues.

- Check for **SHOCK**: Unless shock is prevented, first aid performed, and medical treatment provided, death may result even though the injury would not otherwise be fatal.

1-6. Evaluating a Casualty

a. The time may come when you must instantly apply your knowledge of first aid measures. This could occur during combat operations, in training situations, or while in a nonduty status. Any service member observing an unconscious and/or ill, injured, or wounded person must carefully and skillfully evaluate him to determine the first aid measures required to prevent further injury or death. He should seek help from medical personnel as soon as possible, but must not interrupt his evaluation of the casualty or fail to administer first aid measures. A second service member may be sent to find medical help. One of the cardinal principles for assisting a casualty is that you (the initial rescuer) must continue the evaluation and first aid measures, as the tactical situation permits, until another individual relieves you. If, during any part of the evaluation, the casualty exhibits the conditions (such as shock) for which the service member is checking, the service member must stop the evaluation and immediately administer first aid. In a chemical environment, the service member should not evaluate the casualty until both the individual and the casualty have been masked. If it is suspected that a nerve agent was used, administer the casualty's own nerve agent antidote autoinjector. After providing first aid, the service member must proceed with the evaluation and continue to monitor the casualty for further complications until relieved by medical personnel.

WARNING

Do not use your own nerve agent antidote autoinjector on the casualty.

NOTE

Remember, when evaluating and/or administering first aid to a casualty, you should seek medical aid as soon as possible. **DO NOT** stop first aid measures, but if the situation allows, send another service member to find medical aid.

b. To evaluate a casualty, perform the following steps:

(1) *Check the casualty for responsiveness.* This is done by gently shaking or tapping him while calmly asking, "Are you OK?" Watch for a response. If the casualty does not respond, go to step (2). If the casualty responds, continue with the evaluation.

(*a*) If the casualty is conscious, ask him where he feels different than usual or where it hurts. Ask him to identify the location of pain if he can, or to identify the area in which there is no feeling.

(*b*) If the casualty is conscious but is choking and cannot talk, stop the evaluation and begin first aid measures. Refer to paragraphs 2-10 and 2-11 for specific information on opening the airway.

WARNING

If a broken back or neck is suspected, do not move the casualty unless his life is in immediate danger (such as close to a burning vehicle). Movement may cause permanent paralysis or death.

(2) *Check for breathing.* (Refer to paragraph 2-6 for this procedure.)

(*a*) If the casualty is breathing, proceed to step (3).

(*b*) If the casualty is not breathing, stop the evaluation and begin first aid measures to attempt to ventilate the casualty. Attempt to open the airway, if an airway obstruction is apparent, clear the airway obstruction, then ventilate (see paragraphs 2-10 and 2-11).

(*c*) After successfully ventilating the casualty, proceed to step (3).

(3) *Check for pulse.* (Refer to paragraph 1-3*c*(2) for specific methods.) If a pulse is present and the casualty is breathing, proceed to step (4).

(*a*) If a pulse is present, but the casualty is still not breathing, start rescue breathing.

(*b*) If a pulse is not present, seek medical personnel for help.

(4) *Check for bleeding.* Look for spurts of blood or blood-soaked clothes. Also check for *both* entry and exit wounds. If the casualty is bleeding from an open wound, stop the evaluation and begin first aid procedures as follows for a—

(*a*) Wound of the arm or leg (refer to paragraphs 2-16 through 2-18 for information on putting on a field or pressure dressing).

(*b*) Partial or complete amputation, apply dressing (refer to paragraph 2-16 to 2-18) and then apply tourniquet if bleeding is not stopped (refer to paragraph 2-20 for information on putting on a tourniquet).

(*c*) Open head wound (refer to paragraph 3-10 for information on applying a dressing to an open head wound).

(*d*) Open chest wound (refer to paragraph 3-5 for information on applying a dressing to an open chest wound).

(*e*) Open abdominal wound (refer to paragraph 3-7 for information on applying a dressing to an open abdominal wound).

WARNING

In a chemically contaminated area, do not expose the wounds. Apply field dressing and then pressure dressing over wound area as needed.

(5) *Check for shock.* (Refer to paragraph 2-24 for first aid measures for shock.) If the signs and symptoms of shock are present, stop the evaluation, and begin first aid measures immediately. The following are the nine signs and symptoms of shock.

(*a*) Sweaty but cool skin (clammy skin).

(*b*) Paleness of skin. (In dark-skinned service members look for a grayish cast to the skin.)

(*c*) Restlessness or nervousness.

(*d*) Thirst.

(*e*) Loss of blood (bleeding).

(*f*) Confusion (does not seem aware of surroundings).

(*g*) Faster than normal breathing rate.

(*h*) Blotchy or bluish skin, especially around the mouth.

(*i*) Nausea or vomiting.

WARNING

Leg fractures must be splinted before elevating the legs as a first aid measure for shock.

(6) *Check for fractures*.

(*a*) Check for the following signs and symptoms of a back or neck injury and perform first aid procedures as necessary.

- Pain or tenderness of the back or neck area.
- Cuts or bruises on the back or neck area.
- Inability of a casualty to move or decreased sensation to extremities (paralysis or numbness).
  - Ask about ability to move (paralysis).
  - Touch the casualty's arms and legs and ask whether he can feel your hand (numbness).
- Unusual body or limb position.

(*b*) Immobilize any casualty suspected of having a back or neck injury by doing the following:

- Tell the casualty not to move.
- If a back injury is suspected, place padding (rolled or folded to conform to the shape of the arch) under the natural arch of the casualty's back. (For example, a blanket/poncho may be used as padding.)

WARNING

Do not move casualty to place padding.

- If a neck injury is suspected, immediately immobilize (manually) the head and neck. Place a roll of cloth under the casualty's neck, and put weighted boots (filled with dirt or sand) or rocks on both sides of his head.

(*c*) Check the casualty's arms and legs for open or closed fractures.

- Check for *open* fractures by looking for—
  - Bleeding.
  - Bones sticking through the skin.
  - Check for pulse.
- Check for *closed* fractures by looking for—
  - Swelling.
  - Discoloration.
  - Deformity.
  - Unusual body position.
  - Check for pulse.

(*d*) Stop the evaluation and begin first aid measures if a fracture to an arm or leg is suspected. Refer to Chapter 4 for information on splinting a suspected fracture.

(*e*) Check for signs/symptoms of fractures of other body areas (for example, shoulder or hip) and provide first aid as necessary.

(7) *Check for burns.* Look carefully for reddened, blistered, or charred skin; also check for singed clothing. If burns are found, stop the evaluation and begin first aid procedures. Refer to paragraph 3-9 for information on giving first aid for burns.

NOTE

Burns to the upper torso and face may cause respiratory complications. When evaluating the casualty, look for singed nose hair, soot around the nostrils, and listen for abnormal breath sounds or difficulty breathing.

(8) *Check for possible head injury.*

(*a*) Look for the following signs and symptoms:

- Unequal pupils.
- Fluid from the ear(s), nose, mouth, or injury site.
- Slurred speech.
- Confusion.
- Sleepiness.
- Loss of memory or consciousness.
- Staggering in walking.
- Headache.
- Dizziness.
- Nausea or vomiting.
- Paralysis.
- Convulsions or twitches.
- Bruising around the eyes and behind the ears.

(*b*) If a head injury is suspected, continue to watch for signs which would require performance of rescue breathing, first aid measures for shock, or control of bleeding; seek medical aid. Refer to paragraph 3-10 for information on first aid measures for head injuries.

CHAPTER 2

BASIC MEASURES FOR FIRST AID

2-1. General

Several conditions that require immediate attention are an inadequate airway, lack of breathing, and excessive loss of blood (circulation). A casualty without a clear airway or who is not breathing may die from lack of oxygen. Excessive loss of blood may lead to shock, and shock can lead to death; therefore, you must act immediately to control the loss of blood. All wounds are considered to be contaminated, since infection-producing organisms (germs) are always present on the skin and clothing, and in the soil, water, and air. Any missile or instrument (such as a bullet, shrapnel, knife, or bayonet) causing a wound pushes or carries the germs into that wound. Infection results as these organisms multiply. That a wound is contaminated does not lessen the importance of protecting it from further contamination. You must dress and bandage a wound as soon as possible to prevent further contamination.

NOTE

It is also important that you attend to any airway, breathing, or bleeding problems **IMMEDIATELY** because these problems, if left unattended, may become life threatening.

Section I. OPEN THE AIRWAY AND RESTORE BREATHING

2-2. Breathing Process

All humans must have oxygen to live. Through the breathing process, the lungs draw oxygen from the air and put it into the blood. The heart pumps the blood through the body to be used by the cells that require a constant supply of oxygen. Some cells are more dependent on a constant supply of oxygen than others. For example, cells of the brain may die within 4 to 6 minutes without oxygen. Once these cells die, they are lost forever since they do not regenerate. This could result in permanent brain damage, paralysis, or death.

2-3. Assessment of and Positioning the Casualty

a. **CHECK** for responsiveness (Figure 2-1A)—establish whether the casualty is conscious by gently shaking him and asking, "Are you OK?"

b. **CALL** for help (Figure 2-1B).

c. **POSITION** the unconscious casualty so that he is lying on his back and on a firm surface (Figure 2-1C).

> **WARNING**
>
> **If the casualty is lying on his chest (prone position), cautiously roll the casualty as a unit so that his body does not twist (which may further complicate a back, neck, or spinal injury).**

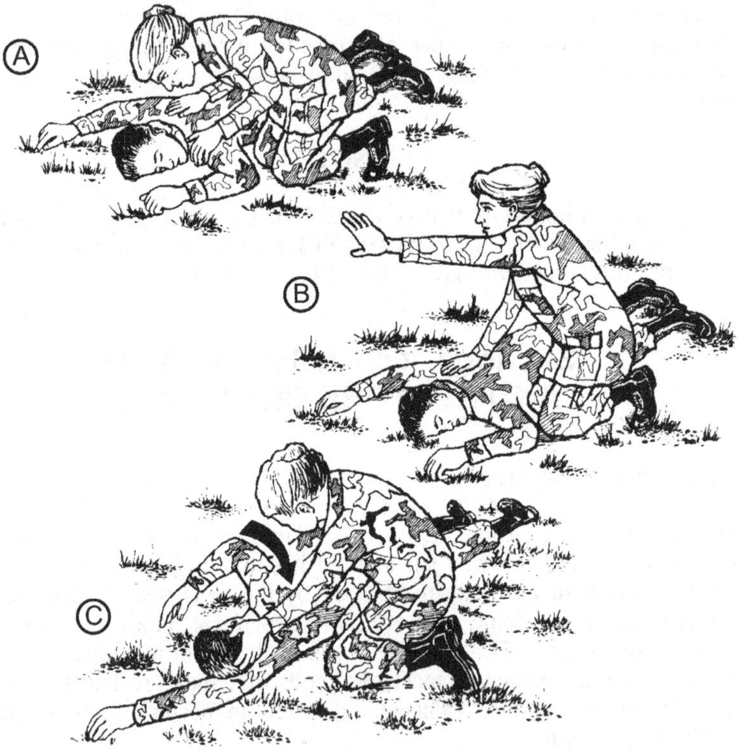

Figure 2-1. Assessment (Illustrated A—C).

(1) Straighten the casualty's legs. Take the casualty's arm that is nearest to you and move it so that it is straight and above his head. Repeat the procedure for the other arm.

(2) Kneel beside the casualty with your knees near his shoulders (leave space to roll his body) (Figure 2-1B). Place one hand behind his head and neck for support. With your other hand, grasp the casualty under his far arm (Figure 2-1C).

(3) Roll the casualty towards you using a steady, even pull. His head and neck should stay in line with his back.

(4) Return the casualty's arms to his side. Straighten his legs. Reposition yourself so that you are now kneeling at the level of the casualty's shoulders. However, if a neck injury is suspected and the jaw-thrust technique will be used, kneel at the casualty's head, looking towards his feet.

2-4. Opening the Airway of an Unconscious or Not Breathing Casualty

The tongue is the single most common cause of an airway obstruction (Figure 2-2). In most cases, simply using the head-tilt/chin-lift technique can clear the airway. This action pulls the tongue away from the air passage in the throat (Figure 2-3).

Figure 2-2. Airway blocked by tongue.

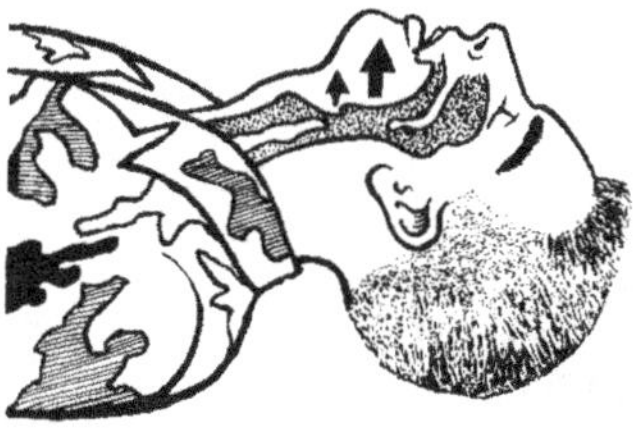

Figure 2-3. Airway opened by extending neck.

a. Call for help and then position the casualty. Move (roll) the casualty onto his back (Figure 2-1C). (Refer to paragraph 2-3*c* for information on positioning the casualty.)

NOTE

Perform finger sweep. If foreign material or vomitus is visible in the mouth, it should be removed, but do not spend an excessive amount of time doing so.

b. Open the airway using the jaw-thrust or head-tilt/chin-lift technique.

CAUTION

The head-tilt/chin-lift technique is an important procedure in opening the airway; however, use extreme care because excess force in performing this maneuver may cause further spinal injury. In a casualty with a suspected neck injury or severe head trauma, the safest approach to opening the airway is the jaw-thrust technique because in most cases it can be accomplished without extending the neck.

(1) *Perform the jaw-thrust technique.* The jaw-thrust may be accomplished by the rescuer grasping the angles of the casualty's lower jaw and lifting with both hands, one on each side, displacing the jaw forward and up (Figure 2-4). The rescuer's elbows should rest on the surface on which the casualty is lying. If the lips close, the lower lip can be retracted with the thumb. If mouth-to-mouth breathing is necessary, close the nostrils by placing your cheek tightly against them. The head should be carefully supported without tilting it backwards or turning it from side to side. If this is unsuccessful, the head should be tilted back very slightly. The jaw-thrust is the safest first approach to opening the airway of a casualty who has a suspected neck injury because in most cases it can be accomplished without extending the neck.

Figure 2-4. Jaw-thrust technique of opening airway.

(2) *Perform the head-tilt/chin-lift technique.* Place one hand on the casualty's forehead and apply firm, backward pressure with the palm to tilt the head back. Place the fingertips of the other hand under the bony part of the lower jaw and lift, bringing the chin forward. The thumb should not be used to lift the chin (Figure 2-5).

NOTE

The fingers should not press deeply into the soft tissue under the chin because the airway may be obstructed.

Figure 2-5. Head-tilt/chin-lift technique of opening airway.

(3) *Check for breathing (while maintaining an airway).* After establishing an open airway, it is important to maintain that airway in an open position. Often the act of just opening and maintaining the airway will allow the casualty to breathe properly. Once the rescuer uses one of the techniques to open the airway (jaw-thrust or head-tilt/chin-lift), he should maintain that head position to keep the airway open. Failure to maintain the open airway will prevent the casualty from receiving an adequate supply of oxygen. Therefore, while maintaining an open airway the rescuer should check for breathing by observing the casualty's chest and performing the following actions within 3 to 5 seconds:

(*a*) **LOOK** for the chest to rise and fall.

(*b*) **LISTEN** for air escaping during exhalation by placing your ear near the casualty's mouth.

(*c*) **FEEL** for the flow of air on your cheek (see Figure 2-6).

(*d*) **PERFORM** rescue breathing if the casualty does not resume breathing spontaneously.

NOTE

If the casualty resumes breathing, monitor and maintain the open airway. He should be transported to an MTF, as soon as practical.

2-5. Rescue Breathing (Artificial Respiration)

a. If the casualty does not promptly resume adequate spontaneous breathing after the airway is open, rescue breathing (artificial respiration) must be started. Be calm! Think and act quickly! The sooner you begin rescue breathing, the more likely you are to restore the casualty's breathing. If you are in doubt whether the casualty is breathing, give artificial respiration, since it can do no harm to a person who is breathing. If the casualty is breathing, you can feel and see his chest move. If the casualty is breathing, you can feel and hear air being expelled by putting your hand or ear close to his mouth and nose.

b. There are several methods of administering rescue breathing. The mouth-to-mouth method is preferred; however, it cannot be used in all situations. If the casualty has a severe jaw fracture or mouth wound or his jaws are tightly closed by spasms, use the mouth-to-nose method.

2-6. Preliminary Steps—All Rescue Breathing Methods

a. Establish unresponsiveness. Call for help. Turn or position the casualty.

b. Open the airway.

c. Check for breathing by placing your ear over the casualty's mouth and nose, and looking toward his chest.

(1) **LOOK** for rise and fall of the casualty's chest (Figure 2-6).

(2) **LISTEN** for sounds of breathing.

(3) **FEEL** for breath on the side of your face. If the chest does not rise and fall and no air is exhaled, then the casualty is not breathing.

(4) **PERFORM** rescue breathing if the casualty is not breathing.

NOTE

Although the rescuer may notice that the casualty is making respiratory efforts, the airway may still be obstructed and opening the airway may be all that is needed. If the casualty resumes breathing, the rescuer should continue to maintain an open airway.

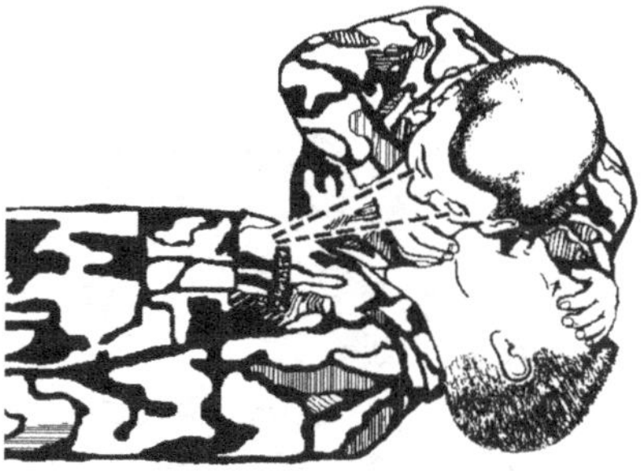

Figure 2-6. Check for breathing.

2-7. Mouth-to-Mouth Method

In this method of rescue breathing, you inflate the casualty's lungs with air from your lungs. This can be accomplished by blowing air into the person's mouth. The mouth-to-mouth rescue breathing method is performed as follows:

a. If the casualty is not breathing, place your hand on his forehead, and pinch his nostrils together with the thumb and index finger of this hand. Let this same hand exert pressure on his forehead to maintain the backward head tilt and maintain an open airway. With your other hand, keep your fingertips on the bony part of the lower jaw near the chin and lift (Figure 2-7).

Figure 2-7. Head tilt/chin lift.

NOTE

If you suspect the casualty has a neck injury and you are using the jaw-thrust technique, close the nostrils by placing your cheek tightly against them.

b. Take a deep breath and place your mouth (in an airtight seal) around the casualty's mouth (Figure 2-8). (If the injured person is small, cover both his nose and mouth with your mouth, sealing your lips against the skin of his face.)

Figure 2-8. Rescue breathing.

c. Blow two full breaths into the casualty's mouth (1 to 1 1/2 seconds per breath), taking a breath of fresh air each time before you blow. Watch out of the corner of your eye for the casualty's chest to rise. If the chest rises, sufficient air is getting into the casualty's lungs. Therefore, proceed as described in step (1). If the chest does not rise, do the following (*a*, *b*, and *c* below) and then attempt to ventilate again.

(1) Take corrective action immediately by reestablishing the airway. Make sure that air is not leaking from around your mouth or out of the casualty's pinched nose.

(2) Reattempt to ventilate.

(3) If the chest still does not rise, take the necessary action to open an obstructed airway (paragraph 2-10).

NOTE

If the initial attempt to ventilate the casualty is unsuccessful, reposition the casualty's head and repeat rescue breathing. Improper chin and head positioning is the most common cause of difficulty with ventilation. If the casualty cannot be ventilated after repositioning the head, proceed with foreign-body airway obstruction maneuvers (see paragraph 2-10).

(4) After giving two slow breaths, which cause the chest to rise, attempt to locate a pulse on the casualty. Feel for a pulse on the side of the casualty's neck closest to you by placing the first two fingers (index and middle fingers) of your hand on the groove beside the casualty's Adam's apple (carotid pulse) (Figure 2-9). (Your thumb should not be used for pulse taking because you may confuse your pulse beat with that of the casualty.) Maintain the airway by keeping your other hand on the casualty's forehead. Allow 5 to 10 seconds to determine if there is a pulse.

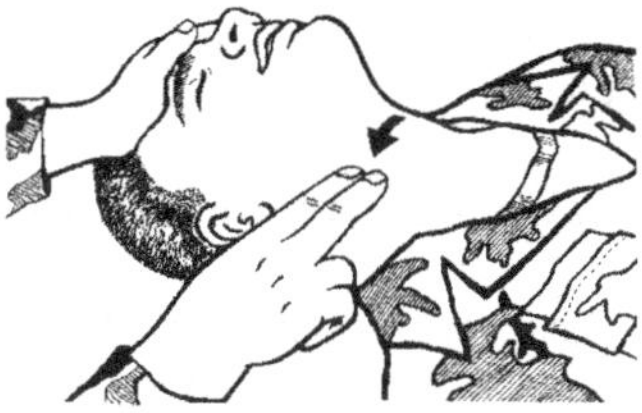

Figure 2-9. Placement of fingers to detect pulse.

(*a*) If signs of circulation are present and a pulse is found and the casualty is breathing—**STOP**; allow the casualty to breathe on his own. If possible, keep him warm and comfortable.

(*b*) If a pulse is found and the casualty is not breathing, continue rescue breathing.

(*c*) If a pulse is not found, seek medically trained personnel for help as soon as possible.

2-8. Mouth-to-Nose Method

Use this method if you cannot perform mouth-to-mouth rescue breathing because the casualty has a severe jaw fracture or mouth wound or his jaws are tightly closed by spasms. The mouth-to-nose method is performed in the same way as the mouth-to-mouth method except that you blow into the nose while you hold the lips closed with one hand at the chin. You then remove your mouth to allow the casualty to exhale passively. It may be necessary to separate the casualty's lips to allow the air to escape during exhalation.

2-9. Heartbeat

If a casualty's heart stops beating, you must immediately seek medical help. **SECONDS COUNT!** Stoppage of the heart is soon followed by cessation of respiration unless it has occurred first. Be calm! Think and act! When a casualty's heart has stopped, there is no pulse at all; the person is unconscious and limp, and the pupils of his eyes are open wide. When evaluating a casualty or when performing the preliminary steps of rescue breathing, feel for a pulse. If you DO NOT detect a pulse, seek medical help.

2-10. Airway Obstructions

In order for oxygen from the air to flow to and from the lungs, the upper airway must be unobstructed.

a. Upper airway obstructions often occur because—

(1) The casualty's tongue falls back into his throat while he is unconscious. The tongue *falls back* and *obstructs* the airway, it is not swallowed by the casualty.

NOTE

Ensure the correct positioning and maintenance of the open airway for an injured or unconscious casualty.

(2) Foreign bodies become lodged in the throat. These obstructions usually occur while eating. Choking on food (usually meat) is associated with—

- Attempting to swallow large pieces of poorly chewed food.
- Drinking alcohol.
- Slipping dentures.

(3) The contents of the stomach are regurgitated and may block the airway.

(4) Blood clots may form as a result of head and facial injuries.

b. Upper airway obstruction may cause either partial or complete airway blockage.

(1) *Partial airway obstruction.* The casualty may still have an air exchange. A good air exchange means that the casualty can cough

forcefully, though he may be wheezing between coughs. You, the rescuer, should not interfere, and should encourage the casualty to cough up the object obstructing his airway on his own. A poor air exchange may be indicated by weak coughing with a high pitched noise between coughs. Further, the casualty may show signs of shock (paragraph 1-6*b*[5]) indicating a need for oxygen. You should assist the casualty and treat him as though he had a complete obstruction.

(2) *Complete airway obstruction.* A complete obstruction (no air exchange) is indicated if the casualty cannot speak, breathe, or cough at all. He may be clutching his neck and moving erratically. In an unconscious casualty, a complete obstruction is also indicated if after opening his airway you cannot ventilate him.

2-11. Opening the Obstructed Airway—Conscious Casualty

Clearing a conscious casualty's airway obstruction can be performed with the casualty either standing or sitting and by following a relatively simple procedure.

WARNING

Once an obstructed airway occurs, the brain will develop an oxygen deficiency resulting in unconsciousness. Death will follow rapidly if breathing is not promptly restored.

a. Ask the casualty if he can speak or if he is choking. Check for the universal choking sign (Figure 2-10).

Figure 2-10. Universal sign of choking.

b. If the casualty can speak, encourage him to attempt to cough; the casualty still has a good air exchange. If he is able to speak or cough effectively, DO NOT interfere with his attempts to expel the obstruction.

c. Listen for high pitched sounds when the casualty breathes or coughs (poor air exchange). If there is poor air exchange or no breathing, CALL FOR HELP and immediately deliver manual thrusts (either an abdominal or chest thrust).

NOTE

The manual thrust with the hands centered between the waist and the rib cage is called an abdominal thrust (or Heimlich maneuver). The chest thrust (the hands are centered in the middle of the breastbone) is used only for an individual in the advanced stages of pregnancy, in the markedly obese casualty, or if there is a significant abdominal wound.

(1) Apply abdominal thrusts. This can be accomplished by using the following procedures:

(*a*) Stand behind the casualty and wrap your arms around his waist.

(*b*) Make a fist with one hand and grasp it with the other. The thumb side of your fist should be against the casualty's abdomen, in the midline and slightly above the casualty's navel, but well below the tip of the breastbone (Figure 2-11).

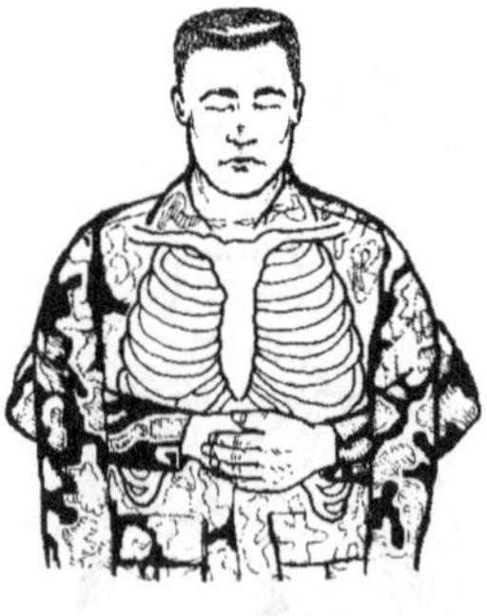

Figure 2-11. Anatomical view of abdominal thrust procedure.

(*c*) Press the fists into the abdomen with a quick backward and upward thrust (Figure 2-12).

Figure 2-12. Profile view of abdominal thrust.

(*d*) Each thrust should be a separate and distinct movement.

NOTE

Continue performing abdominal thrusts until the obstruction is expelled or the casualty becomes unresponsive.

(*e*) If the casualty becomes unresponsive, call for help as you proceed with steps to open the airway, and perform rescue breathing. (Refer to paragraph 2-7 for information on how to perform mouth-to-mouth resuscitation.)

(2) Apply chest thrusts. An alternate technique to the abdominal thrust is the chest thrust. This technique is useful when the casualty has an abdominal wound, when the casualty is pregnant, or when the casualty is so large that you cannot wrap your arms around the abdomen. To apply chest thrusts with casualty sitting or standing:

(*a*) Stand behind the casualty and wrap your arms around his chest with your arms under his armpits.

(*b*) Make a fist with one hand and place the thumb side of the fist in the middle of the breastbone (take care to avoid the tip of the breastbone and the margins of the ribs).

(*c*) Grasp the fist with the other hand and exert thrusts (Figure 2-13).

Figure 2-13. Profile view of chest thrust.

(*d*) Each thrust should be delivered slowly, distinctly, and with the intent of relieving the obstruction.

(*e*) Perform chest thrusts until the obstruction is expelled or the casualty becomes unresponsive.

(*f*) If the casualty becomes unresponsive, call for help as you proceed with steps to open the airway and perform rescue breathing.

2-12. Opening the Obstructed Airway—Casualty Lying Down or Unresponsive

The following procedures are used to expel an airway obstruction in a casualty who is lying down, who becomes unconscious, or who is found unconscious (the cause unknown):

- If a conscious casualty who is choking becomes unresponsive, call for help, open the airway, perform a finger sweep, and attempt rescue breathing (paragraphs 2-4 through 2-8). If you still cannot administer rescue breathing due to an airway blockage, then remove the airway obstruction using the procedures as in *b* below.

- If a casualty is unresponsive when you find him (the cause unknown), assess or evaluate the situation, call for help, position the casualty on his back, open the airway, establish breathlessness, and attempt to perform rescue breathing (paragraphs 2-4 through 2-8).

a. Open the airway and attempt rescue breathing (refer to paragraph 2-7 for information on how to perform mouth-to-mouth resuscitation).

b. If still unable to ventilate the casualty, perform 6 to 10 manual (abdominal or chest) thrusts.

(1) To perform the abdominal thrusts:

(*a*) Kneel astride the casualty's thighs (Figure 2-14).

Figure 2-14. Abdominal thrust on unresponsive casualty.

(*b*) Place the heel of one hand against the casualty's abdomen (in the midline slightly above the navel but well below the tip of the breastbone). Place your other hand on top of the first one. Point your fingers toward the casualty's head.

(*c*) Press into the casualty's abdomen with a quick, forward and upward thrust. You can use your body weight to perform the maneuver. Deliver each thrust quickly and distinctly.

(*d*) Repeat the sequence of abdominal thrusts, finger sweep, and rescue breathing (attempt to ventilate) as long as necessary to remove the object from the obstructed airway.

(*e*) If the casualty's chest rises, proceed to feeling for pulse.

(2) To perform chest thrusts:

(*a*) Place the unresponsive casualty on his back, face up, and open his mouth. Kneel close to the side of the casualty's body.

1. Locate the lower edge of the casualty's ribs with your fingers. Run the fingers up along the rib cage to the notch (Figure 2-15A).

2. Place the middle finger on the notch and the index finger next to the middle finger on the lower edge of the breastbone. Place the heel of the other hand on the lower half of the breastbone next to the two fingers (Figure 2-15B).

3. Remove the fingers from the notch and place that hand on top of the positioned hand on the breastbone, extending or interlocking the fingers (Figure 2-15C).

4. Straighten and lock your elbows with your shoulders directly above your hands without bending the elbows, rocking, or allowing the shoulders to sag. Apply enough pressure to depress the breastbone 1 1/2 to 2 inches, then release the pressure completely (Figure 2-15D). Do this 6 to 10 times. Each thrust should be delivered quickly and distinctly. See Figure 2-16 for another view of the breastbone being depressed.

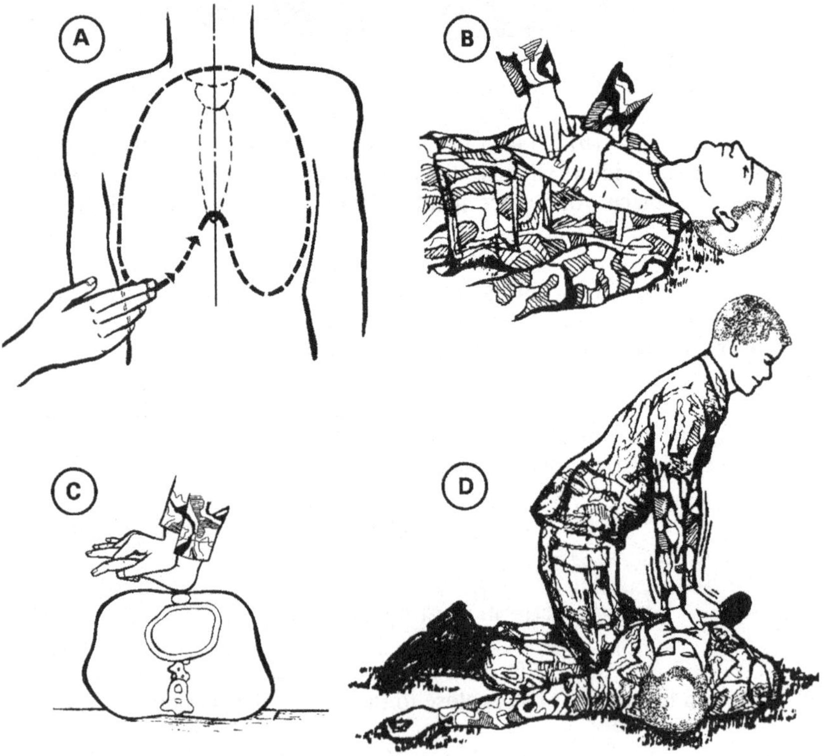

Figure 2-15. Hand placement for chest thrust (Illustrated A-D).

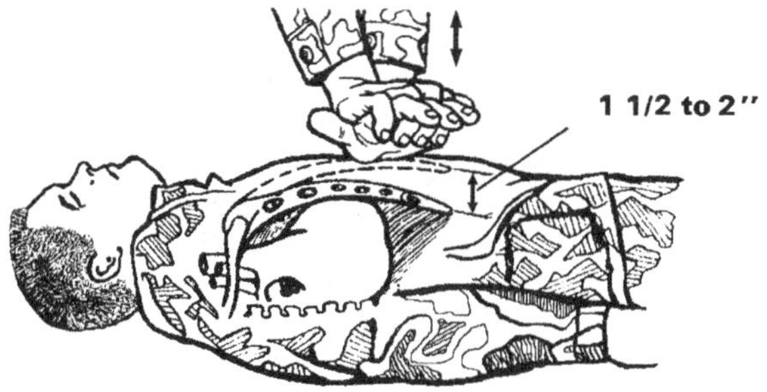

Figure 2-16. Breastbone depressed 1 1/2 to 2 inches.

(*b*) Repeat the sequence of chest thrust, finger sweep, and rescue breathing as long as necessary to clear the object from the obstructed airway. See paragraph (3) below.

(*c*) If the casualty's chest rises, proceed to feeling for his pulse.

(3) If you still cannot administer rescue breathing due to an airway obstruction, then remove the airway obstruction using the procedures in steps (*a*) and (*b*) below.

(*a*) Place the casualty on his back, face up, turn the unresponsive casualty as a unit, and call out for help.

(*b*) Perform finger sweep, keep casualty face up, use tongue-jaw lift to open mouth.

1. Open the casualty's mouth by grasping both his tongue and lower jaw between your thumb and fingers and lifting (tongue-jaw lift) (Figure 2-17). If you are unable to open his mouth, cross your fingers and thumb (crossed-finger method) and push his teeth apart (Figure 2-18) by pressing your thumb against his upper teeth and pressing your finger against his lower teeth.

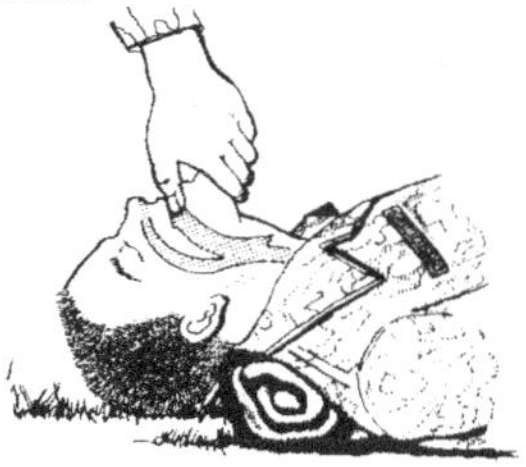

Figure 2-17. Opening casualty's mouth (tongue-jaw lift).

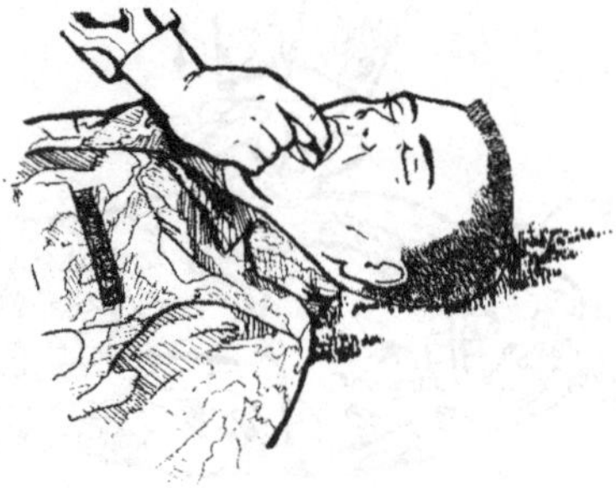

Figure 2-18. Opening casualty's mouth (crossed-finger method).

2. Insert the index finger of the other hand down along the inside of his cheek to the base of the tongue. Use a hooking motion from the side of the mouth toward the center to dislodge the foreign body (Figure 2-19).

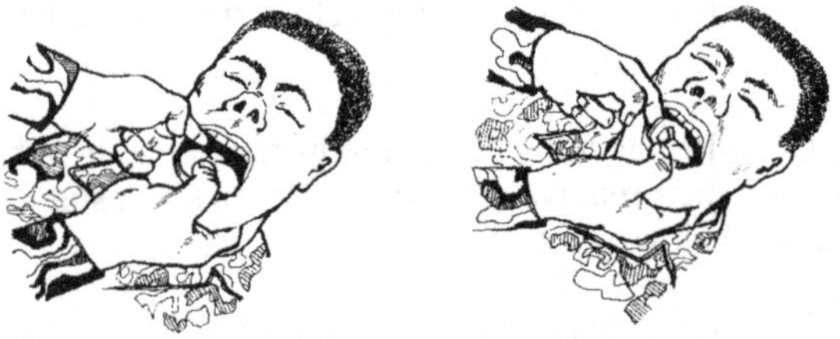

Figure 2-19. Using finger to dislodge a foreign body.

WARNING

Take care not to force the object deeper into the airway by pushing it with the finger.

Section II. STOP THE BLEEDING AND PROTECT THE WOUND

2-13. General

The longer a service member bleeds from a major wound, the less likely he will be able to survive his injuries. It is, therefore, important that the first aid provider promptly stop the external bleeding.

2-14. Clothing

In evaluating the casualty for location, type, and size of the wound or injury, cut or tear his clothing and carefully expose the entire area of the wound. This procedure is necessary to properly visualize injury and avoid further contamination. Clothing stuck to the wound should be left in place to avoid further injury. DO NOT touch the wound; keep it as clean as possible.

WARNING

DO NOT REMOVE protective clothing in a chemical environment. Apply dressings over the protective clothing.

2-15. Entrance and Exit Wounds

Before applying the dressing, carefully examine the casualty to determine if there is more than one wound. A missile may have entered at one point and exited at another point. The *EXIT* wound is usually *LARGER* than the entrance wound.

WARNING

The casualty should be continually monitored for development of conditions which may require the performance of necessary basic lifesaving measures, such as clearing the airway and mouth-to-mouth resuscitation. All open (or penetrating) wounds should be checked for a point of entry and exit and first aid measures applied accordingly.

WARNING

If the missile lodges in the body (fails to exit), DO NOT attempt to remove it or probe the wound. Apply a dressing. If there is an object extending from (impaled in) the wound, DO NOT remove the object. Apply a dressing around the object and use additional improvised bulky materials/dressings (use the cleanest material available) to build up the area around the object to stabilize the object and prevent further injury. Apply a supporting bandage over the bulky materials to hold them in place.

2-16. Field Dressing

a. Use the casualty's field dressing; remove it from the wrapper and grasp the tails of the dressing with both hands (Figure 2-20).

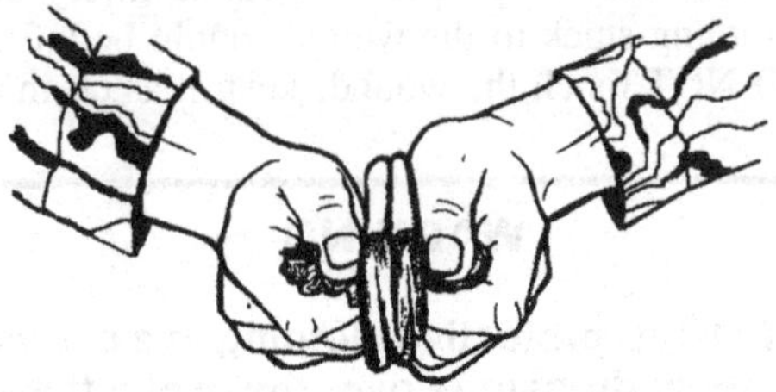

Figure 2-20. Grasping tails of dressing with both hands.

> **WARNING**
>
> **DO NOT touch the white (sterile) side of the dressing, and DO NOT allow it to come in contact with any surface other than the wound.**

b. Hold the dressing directly over the wound with the white side down. Pull the dressing open (Figure 2-21) and place it directly over the wound (Figure 2-22).

Figure 2-21. Pulling dressing open.

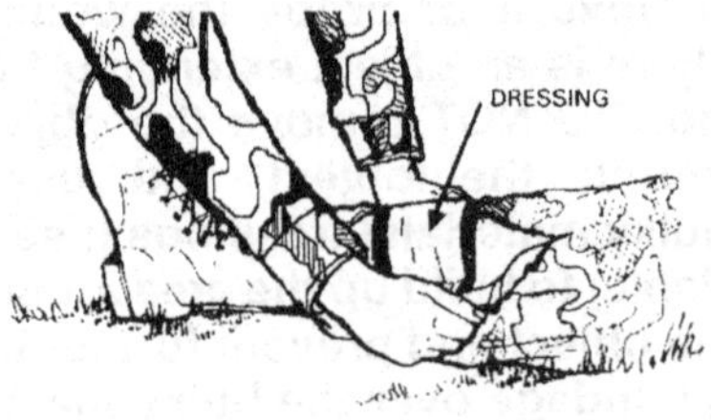

Figure 2-22. Placing dressing directly on wound.

c. Hold the dressing in place with one hand. Use the other hand to wrap one of the tails around the injured part, covering about one-half of the dressing (Figure 2-23). Leave enough of the tail for a knot. If the casualty is able, he may assist by holding the dressing in place.

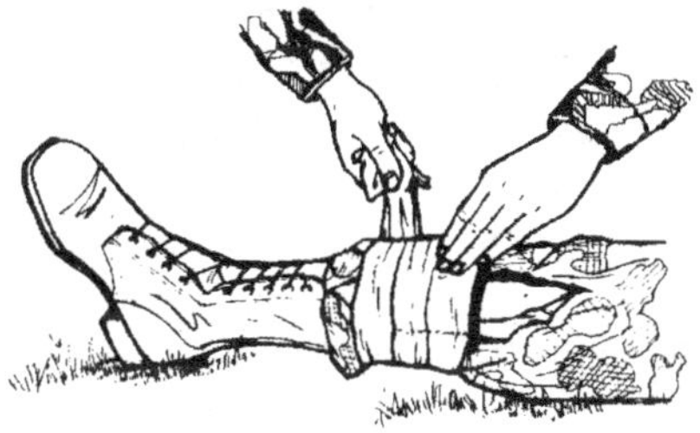

Figure 2-23. Wrapping tail of dressing around injured part.

d. Wrap the other tail in the opposite direction until the remainder of the dressing is covered. The tails should seal the sides of the dressing to keep foreign material from getting under it.

e. Tie the tails into a nonslip knot over the outer edge of the dressing (Figure 2-24). **DO NOT TIE THE KNOT OVER THE WOUND**. In order to allow blood to flow to the rest of an injured limb, tie the dressing firmly enough to prevent it from slipping but without causing a tourniquet-like effect; that is, the skin beyond the injury should not becomes cool, blue, or numb.

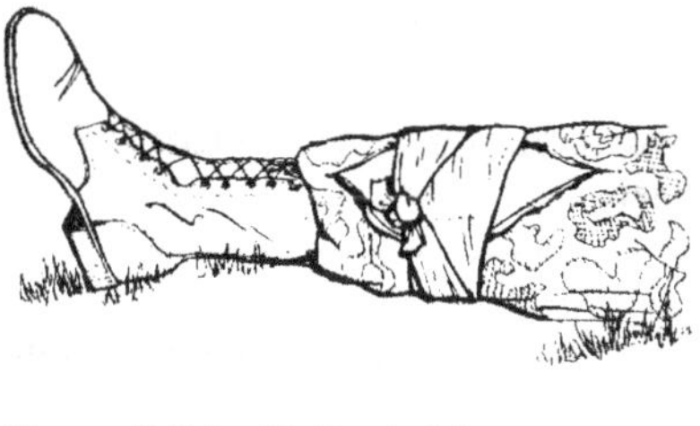

Figure 2-24. Tails tied into nonslip knot.

2-17. Manual Pressure

a. If bleeding continues after applying the sterile field dressing, direct manual pressure may be used to help control bleeding. Apply such pressure by placing a hand on the dressing and exerting firm pressure for 5 to 10 minutes (Figure 2-25). The casualty may be asked to do this himself if he is conscious and can follow instructions.

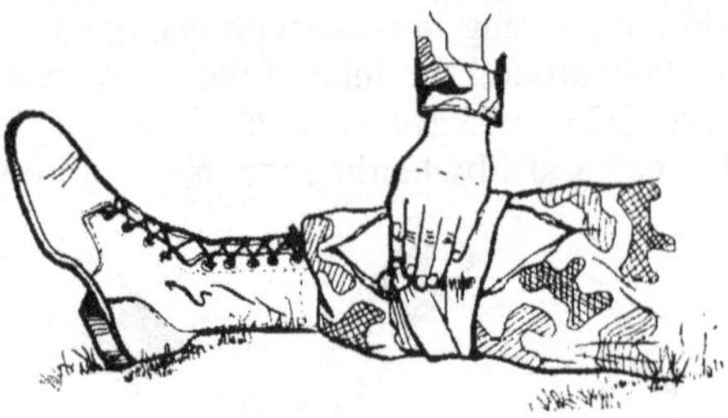

Figure 2-25. Direct manual pressure applied.

b. Elevate an injured limb slightly above the level of the heart to reduce the bleeding (Figure 2-26).

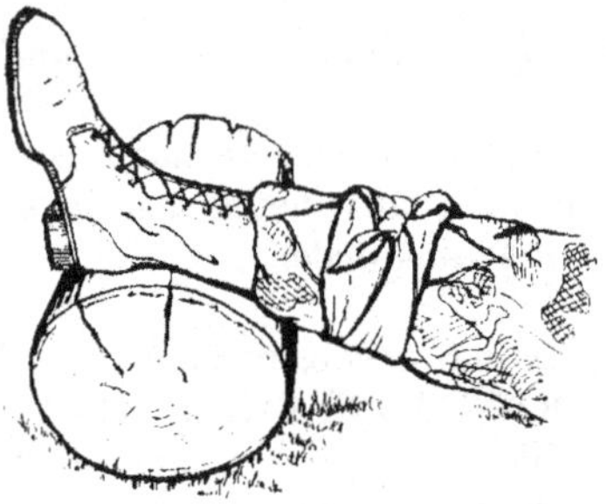

Figure 2-26. Injured limb elevated.

WARNING

DO NOT elevate a suspected fractured limb unless it has been properly splinted.

c. If the bleeding stops, check shock; administer first aid for shock as necessary. If the bleeding continues, apply a pressure dressing.

2-18. Pressure Dressing

Pressure dressings aid in blood clotting and compress the open blood vessel. If bleeding continues after the application of a field dressing, manual pressure, and elevation, then a pressure dressing must be applied as follows:

a. Place a wad of padding on top of the field dressing, directly over the wound (Figure 2-27). Keep the injured extremity elevated.

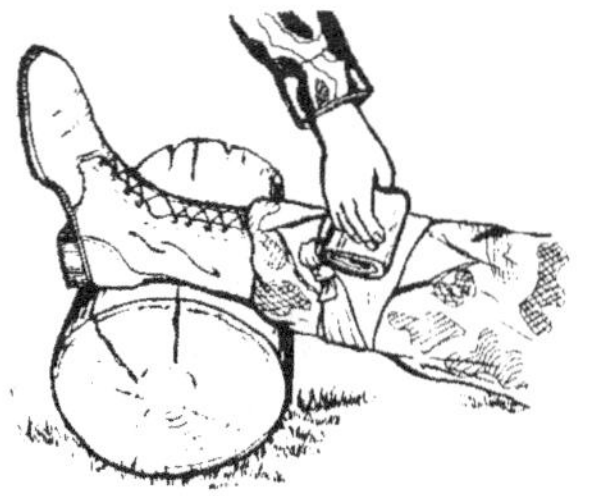

Figure 2-27. Wad of padding on top of field dressing.

NOTE

Improvised bandages may be made from strips of cloth. These strips may be made from T-shirts, socks, or other garments.

b. Place an improvised dressing (or cravat, if available) over the wad of padding (Figure 2-28). Wrap the ends tightly around the injured limb, covering the previously placed field dressing (Figure 2-29).

Figure 2-28. Improvised dressing over wad of padding

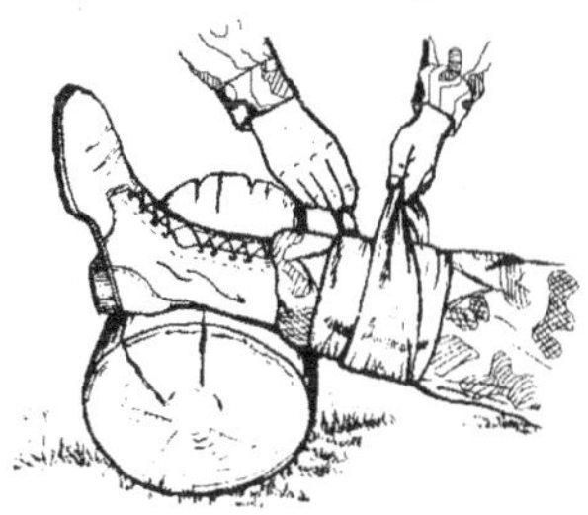

Figure 2-29. Ends of improvised dressing wrapped tightly around limb.

c. Tie the ends together in a nonslip knot, directly over the wound site (Figure 2-30). DO NOT tie so tightly that it has a tourniquet-like effect. If bleeding continues and all other measures have failed, or if the limb is severed, then apply a tourniquet. Use the tourniquet as a **LAST RESORT**. When the bleeding stops, check for shock; administer first aid for shock as necessary.

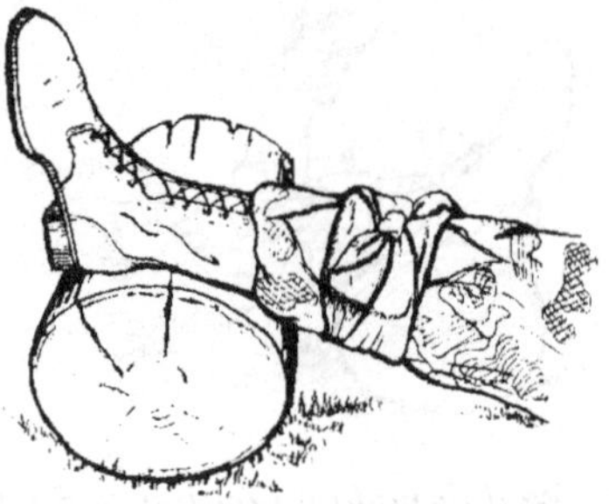

Figure 2-30. Ends of improvised dressing tied together in nonslip knot.

NOTE

Distal end of wounded extremities (fingers and toes) should be checked periodically for adequate circulation. The dressing must be loosened if the extremity becomes cool, blue, or numb.

NOTE

If bleeding continues and all other measures have failed (dressings and covering wound, applying direct manual pressure, elevating the limb above the heart level, and applying a pressure dressing while maintaining limb elevation) *then apply digital pressure* (see paragraph 2-19).

2-19. Digital Pressure

Digital pressure (often called "pressure points") is an alternative method to control bleeding. This method uses pressure from the fingers, thumbs, or hands to press at the site or point where a main artery supplying the wounded area lies near the skin surface or over bone (Figure 2-31). This pressure may help shut off or slow down the flow of blood from the heart to the wound and is used in combination with direct pressure and elevation. It may help in instances where bleeding is not easily controlled, where a pressure dressing has not yet been applied, or where pressure dressings are not readily available.

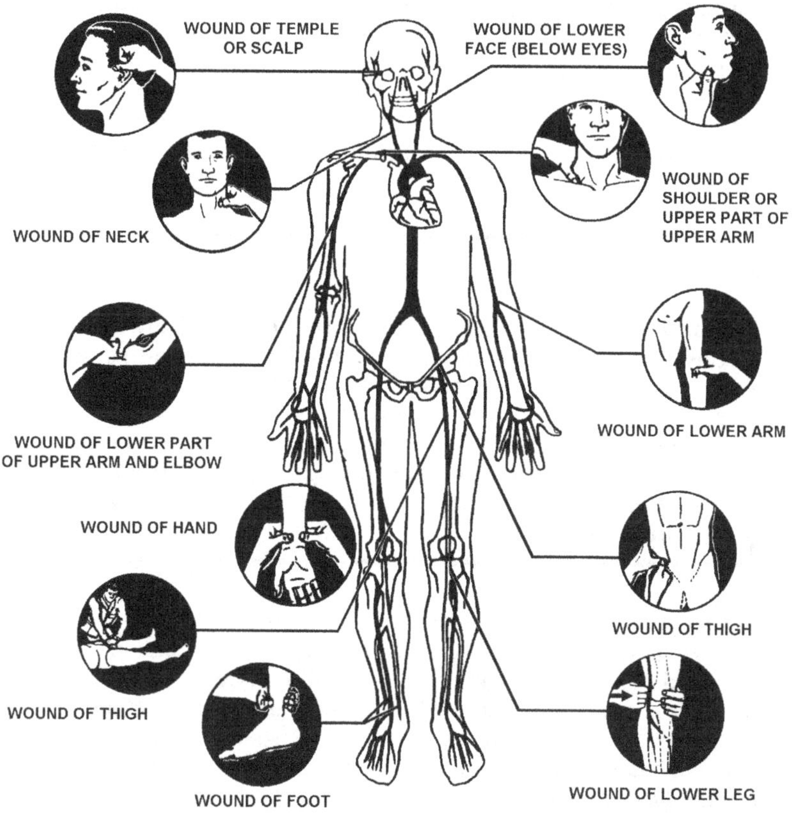

Figure 2-31. Digital pressure (pressure with fingers, thumbs or hands).

2-20. Tourniquet

DANGER

A tourniquet is only used on an arm or leg where there is a danger of the casualty losing his life (bleeding to death).

A tourniquet is a constricting band placed around an arm or leg to control bleeding. A service member whose arm or leg has been completely amputated may not be bleeding when first discovered, but a tourniquet should be applied anyway. This absence of bleeding is due to the body's normal defenses (contraction or clotting of blood vessels) as a result of the amputation, but after a period of time bleeding will start as the blood vessels relax or the clot may be knocked loose by moving the casualty. Bleeding from a major artery of the thigh, lower leg, or arm and bleeding from multiple arteries (which occurs in a traumatic amputation) may prove to be beyond control by manual pressure. If the pressure dressing (see paragraph 2-18, above) under firm hand pressure becomes soaked with blood and the wound continues to bleed, apply a tourniquet.

WARNING

Casualty should be continually monitored for development of conditions which may require the performance of necessary basic lifesaving measures, such as: clearing the airway, performing mouth-to-mouth resuscitation, preventing shock, and/or bleeding control. All open (or penetrating) wounds should be checked for a point of entry or exit and treated accordingly.

The tourniquet should not be used unless a pressure dressing has failed to stop the bleeding or an arm or leg has been cut off. On occasion, tourniquets have injured blood vessels and nerves. If left in place too long, a tourniquet can cause loss of an arm or leg. Once applied, it must stay in place, and the casualty must be taken to the nearest MTF as soon as possible. *DO NOT loosen or release a tourniquet after it has been applied as release could precipitate bleeding and potentially lead to shock.*

a. Improvising a Tourniquet. In the absence of a specially designed tourniquet, a tourniquet may be made from a strong, pliable material, such as gauze or muslin bandages, clothing, or cravats. An improvised tourniquet is used with a rigid stick-like object. To minimize skin damage, ensure that the improvised tourniquet is at least 2 inches wide.

WARNING

The tourniquet must be easily identified or easily seen.

WARNING

DO NOT use wire or shoestring for a tourniquet band.

b. Placing the Improvised Tourniquet.

(1) Place the tourniquet around the limb, between the wound and the body trunk (or between the wound and the heart). Never place it directly over a wound, a fracture, or joint. Tourniquets, for maximum effectiveness, should be placed on the upper arm or above the knee on the thigh (Figure 2-32).

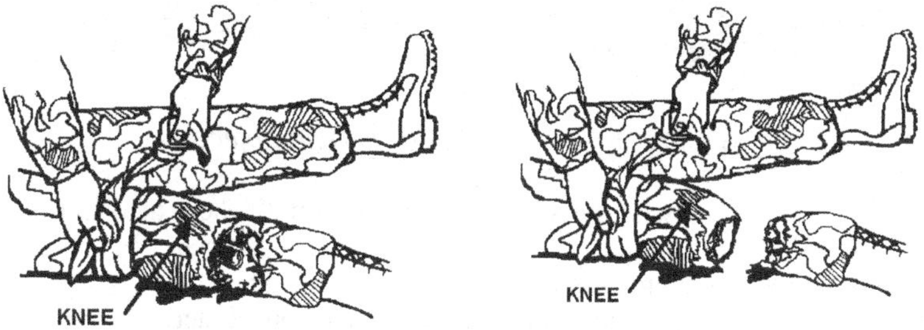

Figure 2-32. Tourniquet above knee.

(2) The tourniquet should be well-padded. If possible, place the tourniquet over the smoothed sleeve or trouser leg to prevent the skin from being pinched or twisted. If the tourniquet is long enough, wrap it around the limb several times, keeping the material as flat as possible. Damaging the skin may deprive the surgeon of skin required to cover an amputation. Protection of the skin also reduces pain.

c. Applying the Tourniquet.

(1) Tie a half-knot. (A half-knot is the same as the first part of tying a shoe lace.)

(2) Place a stick (or similar rigid object) on top of the half-knot (Figure 2-33).

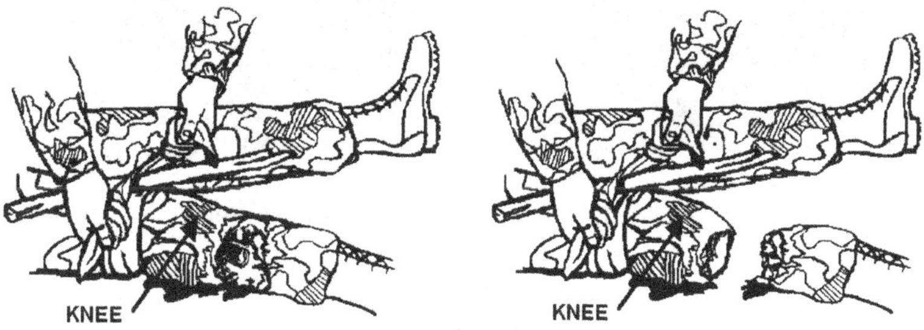

Figure 2-33. Rigid object on top of half-knot.

(3) Tie a full knot over the stick (Figure 2-34).

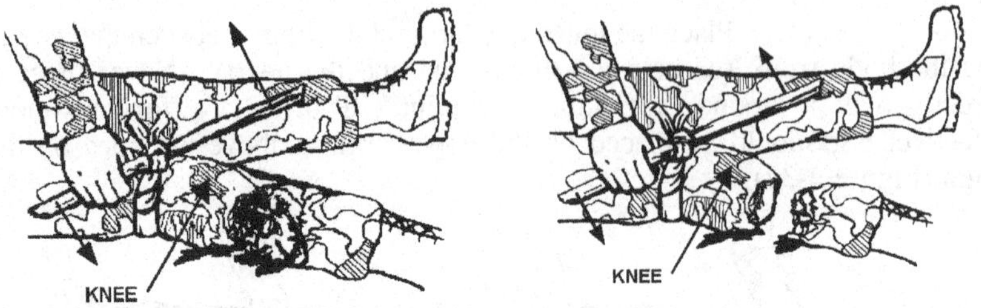

Figure 2-34. Full knot over rigid object.

(4) Twist the stick (Figure 2-35) until the tourniquet is tight around the limb and/or the bright red bleeding has stopped. In the case of amputation, dark oozing blood may continue for a short time. This is the blood trapped in the area between the wound and tourniquet.

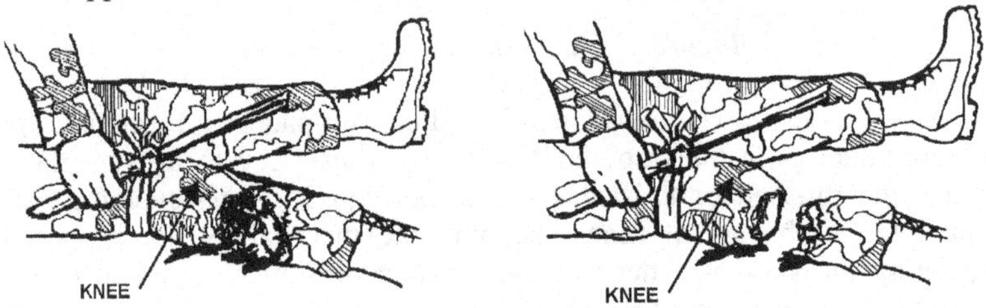

Figure 2-35. Stick twisted.

(5) Fasten the tourniquet to the limb by looping the free ends of the tourniquet over the ends of the stick. Then bring the ends around the limb to prevent the stick from loosening. Tie them together on the side of the limb (Figure 2-36).

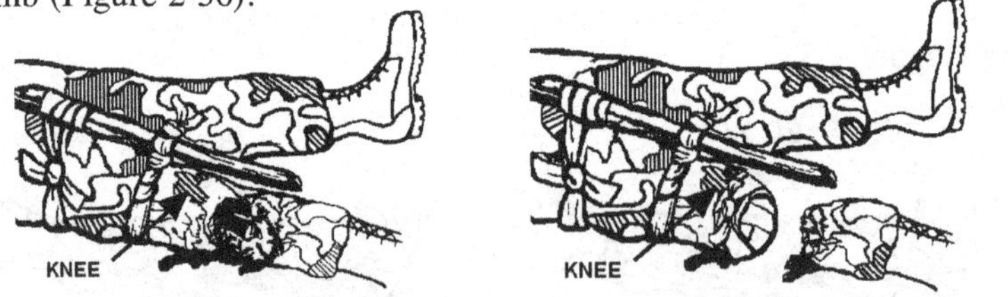

Figure 2-36. Tie free ends on side of limb.

NOTE

Other methods of securing the stick may be used as long as the stick does not unwind and no further injury results.

NOTE

If possible, save and transport any severed (amputated) limbs or body parts with (but out of sight of) the casualty.

(6) DO NOT cover the tourniquet—you should leave it in full view. If the limb is missing (total amputation), apply a dressing to the stump. All wounds should have a dressing to protect the wound from contamination.

(7) Mark the casualty's forehead with a "T" and the time to indicate a tourniquet has been applied. If necessary, use the casualty's blood to make this mark.

(8) Check and treat for shock.

(9) Seek medical aid.

CAUTION

Only appropriately skilled medical personnel may adjust or otherwise remove/release the tourniquet in the appropriate setting.

Section III. CHECK FOR SHOCK AND ADMINISTER FIRST AID MEASURES

2-21. General

The term *shock* has a variety of meanings. In medicine, it refers to a collapse of the body's cardiovascular system which includes an inadequate supply of blood to the body's tissues. Shock stuns and weakens the body. When the normal blood flow in the body is upset, death can result. Early recognition and proper first aid may save the casualty's life.

2-22. Causes and Effects

a. There are three basic mechanisms associated with shock. These are—

- The heart is damaged and fails to work as a pump.

- Blood loss (heavy bleeding) causes the volume of fluid within the vascular system to be insufficient.

- The blood vessels dilate (open wider) so that the blood within the system (even though it is a normal volume [the casualty is not bleeding or dehydrated]) is insufficient to provide adequate circulation within the body.

b. Shock may be the result of a number of conditions. These include—

- Dehydration.

- Allergic reaction to foods, drugs, insect stings, and snakebites.

- Significant loss of blood.

- Reaction to the sight of a wound, blood, or other traumatic scene.

- Traumatic injuries, such as—

  - Burns.

  - Gunshot or shrapnel wounds.

  - Crush injuries.

  - Blows to the body (which can cause broken bones or damage to internal organs).

  - Head injuries.

  - Penetrating wounds (such as from a knife, bayonet, or missile).

2-23. Signs and Symptoms of Shock

Examine the casualty to see if he has any of the following signs and symptoms:

- Sweaty but cool skin (clammy skin).

- Weak and rapid pulse.
- Paleness of skin (in dark-skinned individuals they may have a grayish look to their skin).
- Restlessness, nervousness.
- Thirst.
- Loss of blood (bleeding).
- Confusion (or loss of awareness).
- Faster-than-normal breathing rate.
- Blotchy or bluish skin (especially around the mouth and lips).
- Nausea and/or vomiting.

2-24. First Aid Measures for Shock

In the field, the first aid procedures administered for shock are identical to procedures that would be performed *to prevent shock.* When treating a casualty, assume that shock is present or will occur shortly. By waiting until actual signs and symptoms of shock are noticeable, the rescuer may jeopardize the casualty's life.

a. Position the Casualty. (DO NOT move the casualty or his limbs if suspected fractures have not been splinted. See Chapter 4 for details.)

(1) Move the casualty to cover, if cover is available and the situation permits.

(2) Lay the casualty on his back.

NOTE

A casualty in shock from a chest wound or one who is experiencing breathing difficulty, may breathe easier in a sitting position. If this is the case, allow him to sit upright, but monitor carefully in case his condition worsens.

(3) Elevate the casualty's feet higher than the level of his heart. Use a stable object (field pack or rolled up clothing) so that his feet will not slip off (Figure 2-37).

WARNING

DO NOT elevate legs if the casualty has an unsplinted broken leg, head injury, or abdominal injury.

Figure 2-37. Clothing loosened and feet elevated.

WARNING

Check casualty for leg fracture(s) and splint, if necessary, before elevating his feet. For a casualty with an abdominal wound, place his knees in an upright (flexed) position.

(4) Loosen clothing at the neck, waist, or wherever it may be binding.

CAUTION

DO NOT loosen or remove protective clothing in a chemical environment.

(5) Prevent chilling or overheating. The key is to maintain body temperature. In cold weather, place a blanket or other like item over him to keep him warm and under him to prevent chilling (Figure 2-38). However, if a tourniquet has been applied, leave it exposed (if possible). In hot weather, place the casualty in the shade and protect him from becoming chilled; however, avoid the excessive use of blankets or other coverings.

Figure 2-38. Body temperature maintained.

(6) Calm the casualty. Throughout the entire procedure of providing first aid for a casualty, the rescuer should reassure the casualty and keep him calm. This can be done by being authoritative (taking charge) and by showing self-confidence. Assure the casualty that you are there to help him.

(7) Seek medical aid.

b. Food and/or Drink. When providing first aid for shock, DO NOT give the casualty any food or drink. If you must leave the casualty or if he is unconscious, turn his head to the side to prevent him from choking if he vomits (Figure 2-39).

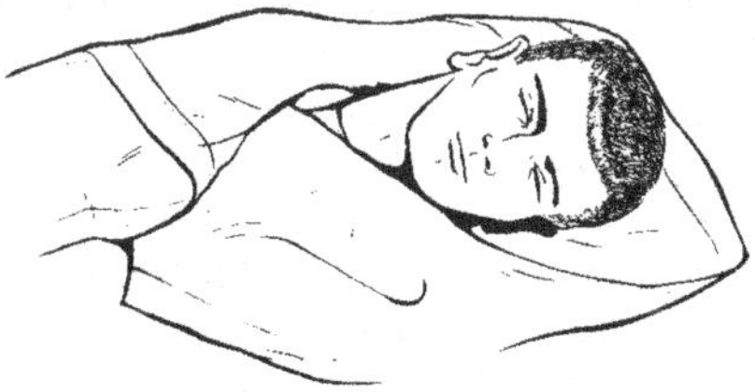

Figure 2-39. Casualty's head turned to side.

c. Evaluate Casualty. Continue to evaluate the casualty until medical personnel arrives or the casualty is transported to an MTF.

CHAPTER 3

FIRST AID FOR SPECIFIC INJURIES

3-1. General

Basic lifesaving steps are discussed in Chapters 1 and 2; they apply to first aid measures for all injuries. Some wounds and burns will require special precautions and procedures when applying these measures. This chapter discusses specific first aid procedures for wounds of the head, face, and neck; chest and stomach wounds; and burns. It also discusses the techniques for applying dressings and bandages to specific parts of the body.

3-2. Head, Neck, and Facial Injuries

a. Head Injuries.

(1) Head injuries range from minor abrasions or cuts on the scalp to severe brain injuries that may result in unconsciousness and sometimes death. Head injuries are classified as open or closed wounds. An open wound is one that is visible, has a break in the skin, and usually has evidence of bleeding. A closed wound may be visible (such as a depression in the skull) or the first aid provider may not be able to see any apparent injury (such as internal bleeding). Some head injuries result in unconsciousness; however, a service member may have a serious head wound and still be conscious. Casualties with head and neck injuries should be treated as though they also have a spinal injury. The casualty should not be moved until the head and neck is stabilized unless he is in immediate danger (such as close to a burning vehicle).

(2) Prompt first aid measures should be initiated for casualties with suspected head and neck injuries. The conscious casualty may be able to provide information on the extent of his injuries. However, as a result of the head injury, he may be confused and unable to provide accurate information. The signs and symptoms a first aid provider might observe are—

- Nausea and vomiting.
- Convulsions or twitches.
- Slurred speech.
- Confusion and loss of memory. (Does he know who he is? Does he know where he is? Does he know what day it is?)
- Recent unconsciousness.

- Dizziness.
- Drowsiness.
- Blurred vision, unequal pupils, or bruising (black eyes).
- Paralysis (partial or full).
- Complaint of headache.
- Bleeding or other fluid discharge from the scalp, nose, or ears.
- Deformity of the head (depression or swelling).
- Staggering while walking.

b. Neck Injuries. Neck injuries may result in heavy bleeding. Apply pressure above and below the injury, *but do not interfere with the breathing process*, and attempt to control the bleeding. Apply a dressing. Always evaluate the casualty for a possible neck fracture/spinal cord injury; if suspected, seek medical treatment immediately.

NOTE

> Establish and maintain the airway in cases of facial or neck injuries. If a neck fracture or spinal cord injury is suspected, immobilize the injury and, if necessary, perform basic life support measures.

c. Facial Injuries. Soft tissue injuries of the face and scalp are common. Abrasions (scrapes) of the skin cause no serious problems. Contusions (injury without a break in the skin) usually cause swelling. A contusion of the scalp looks and feels like a lump. Laceration (cut) and avulsion (torn away tissue) injuries are also common. Avulsions are frequently caused when a sharp blow separates the scalp from the skull beneath it. Because the face and scalp are richly supplied with blood vessels (arteries and veins), wounds of these areas usually bleed heavily.

3-3. General First Aid Measures

a. General Considerations. The casualty with a head injury (or suspected head injury) should be continually monitored for the development of conditions that *may require* basic lifesaving measures. After initiating first aid measures, request medical assistance and evacuation. If dedicated medical evacuation assets are not available, transport the casualty to an MTF as soon as the situation permits. The first aid provider should not attempt to remove a protruding object from the head or give the casualty anything to eat or drink. Further, the first aid provider should be prepared to—

- Clear the airway.
- Control bleeding (external).
- Administer first aid measures for shock.
- Keep the casualty warm.
- Protect the wound.

b. *Unconscious Casualty*. An unconscious casualty does not have control of all of his body's functions and may choke on his tongue, blood, vomitus, or other substances. (Refer to Figure 2-39.)

(1) *Breathing*. The brain requires a constant supply of oxygen. A bluish (or in an individual with dark skin—grayish) color of skin around the lips and nail beds indicates that the casualty is not receiving enough oxygen. Immediate action must be taken to clear the airway, to position the casualty on his side, or to initiate rescue breathing.

(2) *Bleeding*. Bleeding from a head injury usually comes from blood vessels within the scalp. Bleeding can also develop inside the skull or within the brain. In most instances visible bleeding from the head can be controlled by application of the field first aid dressing.

CAUTION

DO NOT attempt to put unnecessary pressure on the wound or attempt to push any brain matter back into the head (skull). **DO NOT** apply a pressure dressing.

c. *Concussion*. If an individual receives a heavy blow to the head or face, he may suffer a brain concussion (an injury to the brain that involves a temporary loss of some or all of the brain's ability to function). For example, the casualty may not breathe properly for a short period of time, or he may become confused and stagger when he attempts to walk. Symptoms of a concussion may only last for a short period of time. However, if a casualty is suspected of having suffered a concussion, he should be transported to an MTF as soon as conditions permit.

d. Convulsions. Convulsions (seizures/involuntary jerking) may occur even after a mild head injury. When a casualty is convulsing, protect him from hurting himself. Take the following measures:

(1) Ease him to the ground if he is standing or sitting.

(2) Support his head and neck.

(3) Maintain his airway.

(4) Protect him from further injury (such as hitting close-by objects).

NOTE

DO NOT forcefully hold the arms and legs if they are jerking because this can lead to broken bones. **DO NOT** force anything between the casualty's teeth—especially if they are tightly clenched because this may obstruct the casualty's airway. Maintain the casualty's airway if necessary.

e. Brain Damage. In *severe* head injuries where brain tissue is protruding, *leave the wound alone*; carefully place a loose moistened dressing (moistened with sterile normal saline if available) and also a first aid dressing over the tissue to protect it from further contamination. ***DO NOT** remove or disturb any foreign matter that may be in the wound.* Position the casualty so that his head is higher than his body. Keep him warm and *seek medical assistance immediately.*

NOTE

If there is an object extending from the wound, **DO NOT** remove the object. Improvise bulky dressings from the cleanest material available and place this material around the protruding object for support, then apply the field dressing.

3-4. Chest Wounds

Blunt trauma, bullet or missile wounds, stab wounds, or falls may cause chest injuries. These injuries can be serious and may cause death quickly if first aid is not administered in a timely manner. A casualty with a chest injury may complain of pain in the chest or shoulder area; he may have difficulty breathing. His chest may not rise normally when he breathes. The injury may cause the casualty to cough up blood and to have a rapid or a weak heartbeat. A casualty with an open chest wound has a punctured chest wall. The sucking sound heard when he breathes is caused by air leaking into his chest cavity. This particular type of wound is dangerous and will collapse the injured lung (Figure 3-1). Breathing becomes difficult for the casualty because the wound is open. The service members life may depend upon how quickly you apply an occlusive dressing over the wound (refer to paragraph 3-5).

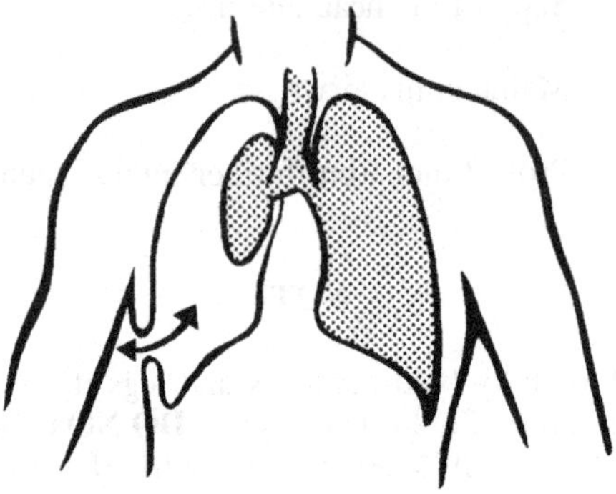

Figure 3-1. Collapsed lung.

3-5. First Aid for Chest Wounds

a. Evaluate the Casualty. Be prepared to perform first aid measures. These measures may include clearing the airway, rescue breathing, treatment for shock, and/or bleeding control.

b. Expose the Wound. If appropriate, cut or remove the casualty's clothing to expose the wound. Remember, **DO NOT** remove clothing that is stuck to the wound because additional injury may result. **DO NOT** attempt to clean the wound.

NOTE

Examine the casualty to see if there is an entry and exit wound. If there are two wounds (entry, exit), perform the same procedure for both wounds. Treat the more serious (heavier bleeding, larger) wound first. It may be necessary to improvise a dressing for the second wound by using strips of cloth, such as a torn T-shirt, or whatever material is available. Also, listen for sucking sounds to determine if the chest wall is punctured.

CAUTION

If there is an object impaled in the wound, **DO NOT** remove it. Apply a dressing around the object and use additional improvised bulky materials/dressings (use the cleanest materials available) to build up the area around the object. Apply a supporting bandage over the bulky materials to hold them in place.

CAUTION

DO NOT REMOVE protective clothing in a chemical environment. Apply dressings *over* the protective clothing.

c. *Open the Casualty's Field Dressing Plastic Wrapper.* In cases where there is a sucking chest wound, the plastic wrapper is used with the field dressing to create an occlusive dressing. If a plastic wrapper is not available, or if an additional wound needs to be treated; cellophane, foil, the casualty's poncho, or similar material may be used. The covering should be wide enough to extend 2 inches or more beyond the edges of the wound in all directions.

(1) Tear open one end of the casualty's plastic wrapper covering the field dressing. Be careful not to destroy the wrapper and **DO NOT** touch the inside of the wrapper.

(2) Remove the inner packet (field dressing).

(3) Complete tearing open the empty plastic wrapper using as much of the wrapper as possible to create a flat surface.

d. *Place the Wrapper Over the Wound.* Place the inside surface of the plastic wrapper directly over the wound *when the casualty exhales* and hold it in place (Figure 3-2). The casualty may hold the plastic wrapper in place if he is able.

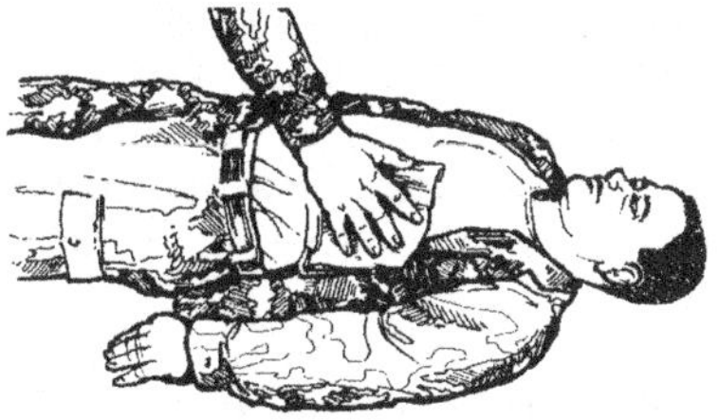

Figure 3-2. Open chest wound sealed with an occlusive dressing.

e. Apply the Dressing to the Wound.

(1) Use your free hand and shake open the field dressing (Figure 3-3).

Figure 3-3. Shaking open the field dressing.

(2) Place the white side of the dressing on the plastic wrapper covering the wound (Figure 3-4).

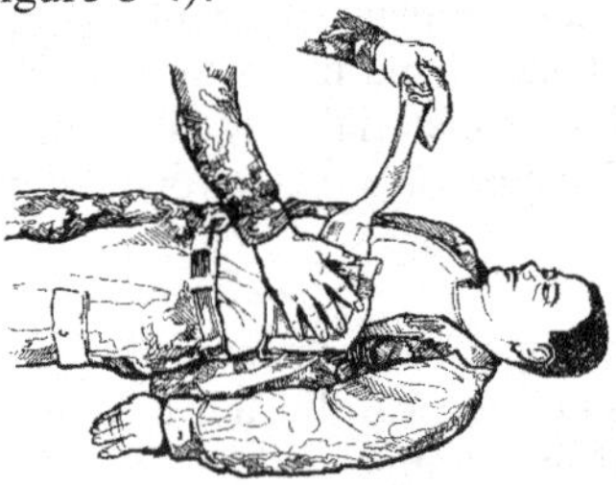

Figure 3-4. Field dressing placed on plastic wrapper.

NOTE

Use the casualty's field dressing, not your own.

(3) Have the casualty breathe normally.

(4) While maintaining pressure on the dressing, grasp one tail of the field dressing with the other hand and wrap it around the casualty's back. If tape is available, tape three sides of the plastic wrapper to the chest wall to provide occlusive type dressing. Leave one side untapped to provide emergency escape for air that may build up in the chest. If tape is not available, secure wrapper on three sides with field dressing leaving the fourth side as a flap.

(5) Wrap the other tail in the opposite direction, bringing both tails over the dressing (Figure 3-5).

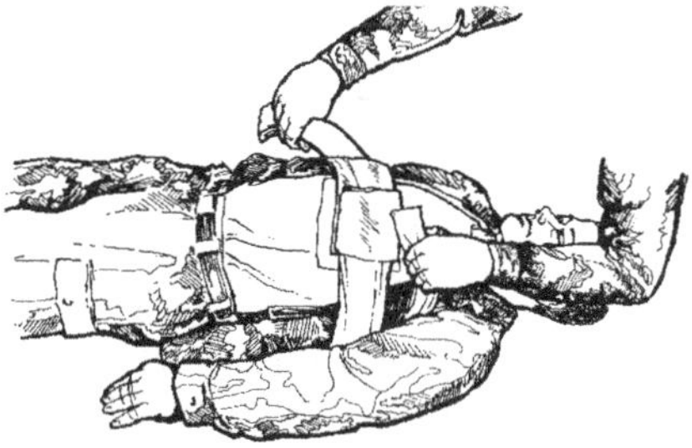

Figure 3-5. Tails of field dressing wrapped around casualty in opposite direction.

(6) Tie the tails into a square knot in the center of the dressing *after* the casualty exhales and *before* he inhales. This will aid in maintaining pressure on the bandage after it has been tied (Figure 3-6). Tie the dressing firmly enough to secure the dressing without interfering with the casualty's breathing.

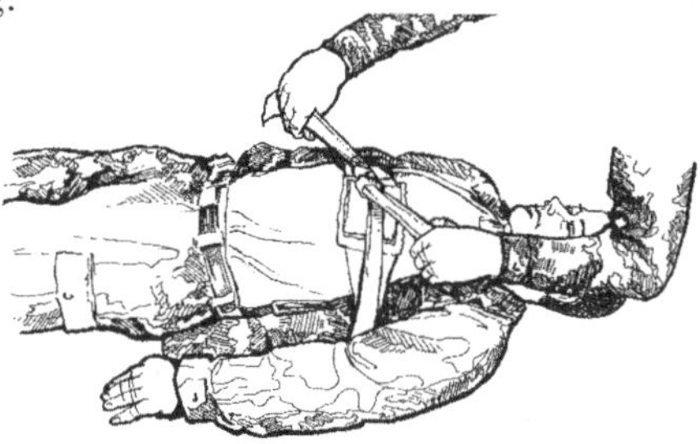

Figure 3-6. Tails of dressing tied into square knot over center of dressing.

NOTE

When practical, apply direct manual pressure over the dressing for 5 to 10 minutes to help control the bleeding.

f. Position the Casualty. Position the casualty on his injured side or in a sitting position, whichever makes breathing easier (Figure 3-7).

Figure 3-7. Casualty positioned (lying) on injured side.

g. Seek Medical Assistance. Contact medical personnel.

WARNING

If an occlusive dressing has been improperly placed, air may enter the chest cavity with no means of escape. This causes a life-threatening condition called *tension pneumothorax*. If the casualty's condition (for example, difficulty breathing, shortness of breath, restlessness, or blueness/grayness of the skin) worsens after placing the dressing, quickly lift or remove, and then replace the occlusive dressing.

3-6. Abdominal Wounds

The most serious abdominal wound is one in which an object penetrates the abdominal wall and pierces internal organs or large blood vessels. In these instances, bleeding may be severe and death can occur rapidly.

3-7. First Aid for Abdominal Wounds

a. Evaluate the Casualty. Be prepared to perform basic first aid measures. Always check for both entry and exit wounds. If there are two wounds (entry and exit), treat the wound that appears more serious first (for example, the heavier bleeding, protruding organs, larger wound, and so forth). It may be necessary to improvise dressings for the second wound by using strips of cloth, a T-shirt, or the cleanest material available.

b. Position the Casualty. Place and maintain the casualty on his back with his knees in an upright (flexed) position (Figure 3-8). The knees-up position helps relieve pain, assists in the treatment of shock, prevents further exposure of the bowel (intestines) or abdominal organs, and helps relieve abdominal pressure by allowing the abdominal muscles to relax.

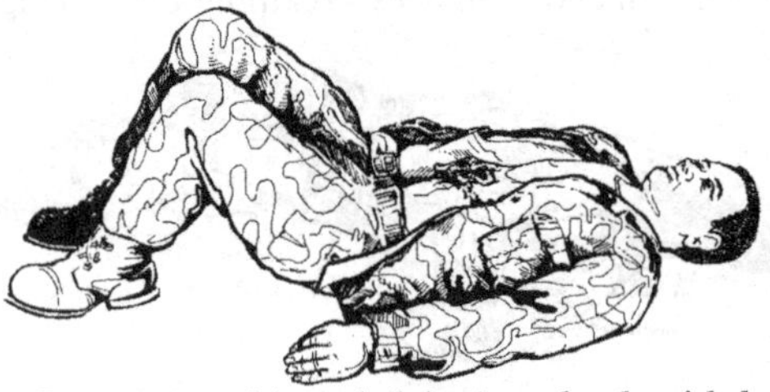

Figure 3-8. Casualty positioned (lying) on back with knees (flexed) up.

c. Expose the Wound.

(1) Remove the casualty's loose clothing to expose the wound. However, **DO NOT** attempt to remove clothing that is stuck to the wound; removing it may cause further injury.

CAUTION

DO NOT REMOVE protective clothing in a chemical environment. Apply dressings *over* the protective clothing.

(2) Gently pick up any organs that may be on the ground. Do this with a clean, dry dressing or with the cleanest available material. Place the organs on top of the casualty's abdomen (Figure 3-9).

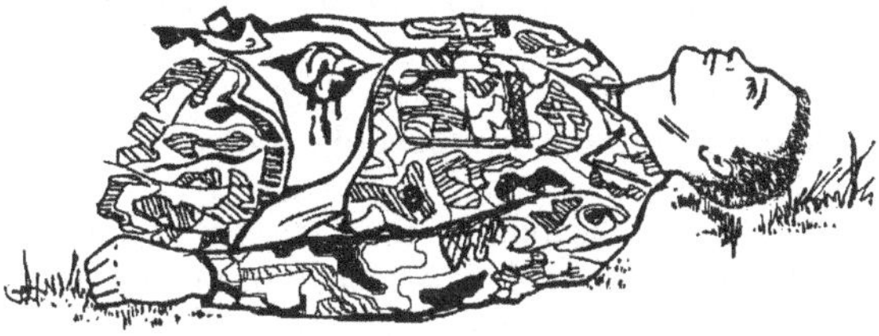

Figure 3-9. Protruding organs placed near wound.

NOTE

DO NOT probe, clean, or try to remove any foreign object from the abdomen. **DO NOT** touch with bare hands any exposed organs. **DO NOT** push organs back inside the body.

d. Apply the Field Dressing. Use the casualty's field dressing, not your own. If the field dressing is not large enough to cover the entire wound, the plastic wrapper from the dressing may be used to cover the wound first (placing the field dressing on top). Open the plastic wrapper carefully without touching the inner surface. If necessary, other improvised dressings may be made from clothing, blankets, or the cleanest materials available.

WARNING

If there is an object extending from the wound, DO NOT remove it. Place as much of the wrapper over the wound as possible without dislodging or moving the object. DO NOT place the wrapper over the object.

(1) Grasp the tails in both hands.

(2) Hold the dressing with the white side down directly over the wound. **DO NOT** touch the white (sterile) side of the dressing or allow anything except the wound to come in contact with it.

(3) Pull the dressing open and place it directly over the wound (Figure 3-10). If the casualty is able, he may hold the dressing in place.

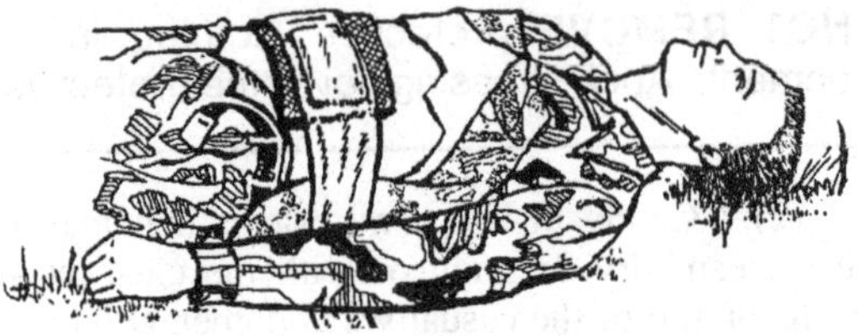

Figure 3-10. Dressing placed directly over the wound.

(4) Hold the dressing in place with one hand and use the other hand to wrap one of the tails around the body.

(5) Wrap the other tail in the opposite direction until the dressing is completely covered. Leave enough of the tail for a knot.

(6) Loosely tie the tails with a square knot at the casualty's side (Figure 3-11).

Figure 3-11. Dressing applied and tails tied with a square knot.

WARNING

When the dressing is applied, DO NOT put pressure on the wound or exposed internal parts, because pressure could cause further injury (vomiting, ruptured intestines, and so forth). Therefore, tie the dressing ties (tails) loosely at casualty's side, not directly over the dressing.

(7) Tie the dressing firmly enough to prevent slipping without applying pressure to the wound site (Figure 3-12).

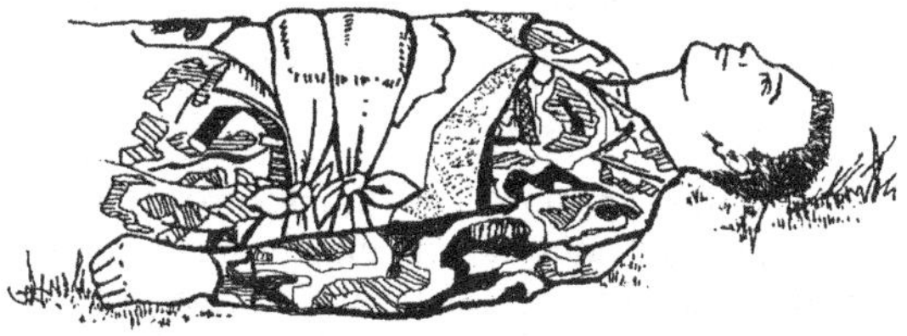

Figure 3-12. Field dressing covered with improvised material and loosely tied.

Field dressings can be covered with improvised reinforcement material (cravats, strips of torn T-shirt, or other cloth) for additional support and protection. Tie improvised bandage on the opposite side of the dressing ties firmly enough to prevent slipping but without applying additional pressure to the wound.

> **CAUTION**
>
> **DO NOT** give casualties with abdominal wounds food or water (moistening the lips is allowed).

e. Seek Medical Assistance. Notify medical personnel.

3-8. Burn Injuries

Burns often cause extreme pain, scarring, or even death. Before administering first aid, you must be able to recognize the type of burn. There are four types of burns:

- Thermal burns caused by fire, hot objects, hot liquids, and gases; or by nuclear blast or fireball.

- Electrical burns caused by electrical wires, current, or lightning.

- Chemical burns caused by contact with wet or dry chemicals or white phosphorus (WP)—from marking rounds and grenades.

- Laser burns (eye [ocular] injury).

3-9. First Aid for Burns

a. Eliminate the Source of the Burn. The source of the burn must be eliminated before any evaluation of the casualty can occur and first aid administered.

(1) Quickly remove the casualty from danger and cover the *thermal burn* with any large nonsynthetic material, such as a field jacket. If the casualty's clothing is still on fire, roll the casualty on the ground to smother (put out) the flames (Figure 3-13).

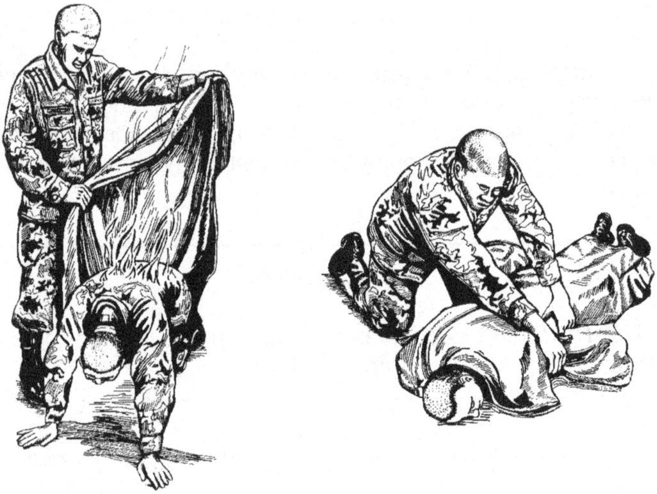

Figure 3-13. Casualty covered and rolled on ground.

> **CAUTION**
>
> Synthetic materials, such as nylon, may melt and cause further injury.

(2) Remove the *electrical burn* casualty from the electrical source by turning off the electrical current. **DO NOT** attempt to turn off the electricity if the source is not close by. Speed is critical, so **DO NOT** waste unnecessary time. If the electricity cannot be turned off, wrap any *nonconductive* material (*dry* rope, clothing, wood, and so forth) around the casualty's back and shoulders and drag the casualty away from the electrical source (Figure 3-14). **DO NOT** make body-to-body contact with the casualty or touch any wires because you could also become an electrical burn casualty.

Figure 3-14. Casualty removed from electrical source (using nonconductive material).

WARNING

High voltage electrical burns may cause temporary unconsciousness, difficulties in breathing, or difficulties with the heart (heartbeat).

(3) Remove the *chemical* from the *burned casualty*. Remove *liquid* chemicals by flushing with as much water as possible. Remove *dry* chemicals by brushing off loose particles (**DO NOT** use the bare surface of your hand because you could become a chemical burn casualty) and then flush with large amounts of water, if available. If large amounts of water are not available, then **NO** water should be applied because small amounts of water applied to a dry chemical burn may cause a chemical reaction. When WP strikes the skin, smother with a wet cloth or mud. Keep WP covered with a wet material to exclude air; this should help prevent the particles from burning.

(4) Remove the *laser burn* casualty from the source. When removing the casualty from the laser beam source, be careful not to enter the

beam or you may become a casualty. Never look directly at the beam source and if possible, wear appropriate eye protection.

NOTE

After the casualty is removed from the source of the burn, he should be evaluated for conditions requiring basic first aid measures.

b. *Expose the Burn.* Cut and gently lift away any clothing covering the burned area, without pulling clothing over the burns. Leave in place any clothing that is stuck to the burn. If the casualty's hands or wrists have been burned, remove jewelry if possible without causing further injury (rings, watches, and so forth) and place in his pockets. This prevents the necessity to cut off jewelry since swelling usually occurs as a result of a burn.

CAUTION

DO NOT lift or cut away clothing if in a chemical environment. Apply the dressing directly over the casualty's protective clothing. **DO NOT** attempt to decontaminate skin where blisters have formed.

c. *Apply a Field Dressing to the Burn.*

(1) Grasp the tails of the casualty's dressing in both hands.

(2) Hold the dressing directly over the wound with the white side down, pull the dressing open, and place it directly over the wound. **DO NOT** touch the white (sterile) side of the dressing or allow anything except the wound to come in contact with it. If the casualty is able, he may hold the dressing in place.

(3) Hold the dressing in place with one hand and use the other hand to wrap one of the tails around the limbs or the body.

(4) Wrap the other tail in the opposite direction until the dressing is completely covered.

(5) Tie the tails into a square knot over the outer edge of the dressing. The dressing should be applied lightly over the burn. Ensure that dressing is applied firmly enough to prevent it from slipping.

NOTE

Use the cleanest improvised dressing material available if a field dressing is not available or if it is not large enough for the entire wound.

d. Take the Following Precautions:

- **DO NOT** place the dressing over the face or genital area.
- **DO NOT** break the blisters.
- **DO NOT** apply grease or ointments to the burns.
- For electrical burns, check for both an entry and exit burn from the passage of electricity through the body. Exit burns may appear on any area of the body despite location of entry burn.
- For burns caused by wet or dry chemicals, flush the burns with large amounts of water and cover with a dry dressing.
- For burns caused by WP, flush the area with water, then cover with a wet material, dressing, or mud to exclude the air and keep the WP particles from burning.
- For laser burns, apply a field dressing.
- If the casualty is conscious and not nauseated, give him small amounts of water.

e. Seek Medical Assistance. Notify medical personnel.

3-10. Dressings and Bandages

a. Head Wounds.

(1) *Position the casualty.*

WARNING

DO NOT move the casualty if you suspect he has sustained a neck, spine, or head injury (which produces any signs or symptoms other than minor bleeding).

- If the casualty has a minor (superficial) scalp wound and is conscious:

  - Have the casualty sit up (unless other injuries prohibit or he is unable to).

  - If the casualty is lying down and is not accumulating fluids or drainage in his throat, elevate his head slightly.

  - If the casualty is bleeding from or into his mouth or throat, turn his head to the side or position him on his side so that the airway will be clear. Avoid putting pressure on the wound and place him on his uninjured side (Figure 3-15).

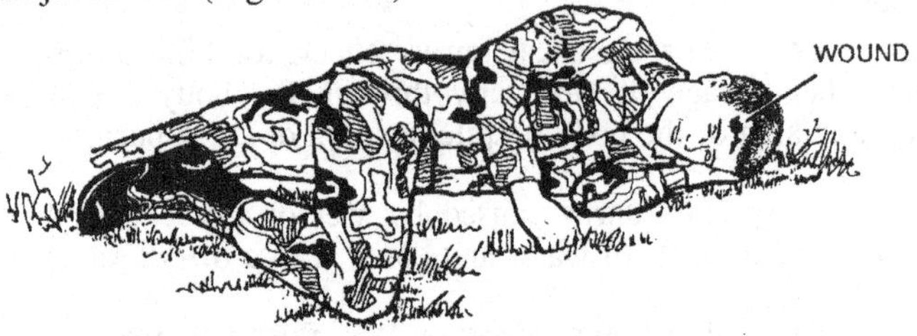

Figure 3-15. Casualty lying on side opposite injury.

- If the casualty is unconscious or has a severe head injury, then suspect and treat him as having a potential neck or spinal injury, *immobilize and* ***DO NOT*** *move the casualty.*

NOTE

If the casualty is choking or vomiting or is bleeding from or into his mouth (thus compromising his airway), position him on his uninjured side to allow for drainage and to help keep his airway clear.

WARNING

If it is necessary to turn a casualty with a suspected neck/spine injury; roll the casualty gently onto his side, keeping the head, neck, and body aligned while providing support for the head and neck. DO NOT roll the casualty by yourself but seek assistance. *Move him only if absolutely necessary*, otherwise keep the casualty immobilized to prevent further damage to the neck/spine.

(2) *Expose the wound.* Remove the casualty's helmet (if necessary). In a nuclear, biological, and chemical (NBC) environment, the first aid provider must leave the casualty as much protection (such as protective mask, mission-oriented protective posture [MOPP] overgarments) as possible. What items of protective equipment can be removed is dependent upon the casualty's injuries (where on the body and what type), the MOPP level, integrity of protective equipment (such as tears in the garment or mask seal), availability of chemical protective shelters, and the tactical situation.

WARNING

DO NOT attempt to clean the wound or remove a protruding object.

NOTE

Always use the casualty's field dressing, not your own.

(3) *Apply a dressing to a wound of the forehead or back of head.* To apply a dressing to a wound of the forehead or back of the head—

(*a*) Remove the dressing from the wrapper.

(*b*) Grasp the tails of the dressing in both hands.

(*c*) Hold the dressing (white side down) directly over the wound. **DO NOT** touch the white (sterile) side of the dressing or allow anything except the wound to come in contact with it.

(*d*) Place it directly over the wound.

(*e*) Hold it in place with one hand. If the casualty is able, he may assist.

(*f*) Wrap the first tail horizontally around the head; ensure the tail covers the dressing (Figure 3-16).

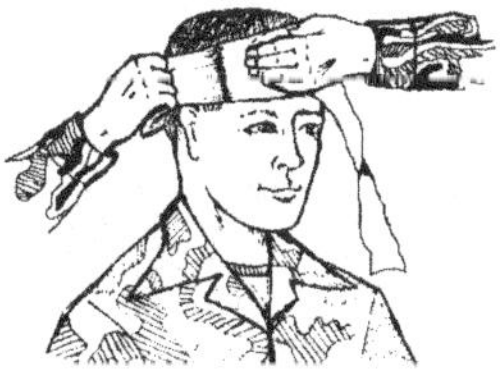

Figure 3-16. First tail of dressing wrapped horizontally around head.

(*g*) Hold the first tail in place and wrap the second tail in the opposite direction, covering the dressing (Figure 3-17).

Figure 3-17. Second tail wrapped in opposite direction.

(*h*) Tie a square knot and secure the tails at the side of the head, making sure they **DO NOT** cover the eyes or ears (Figure 3-18).

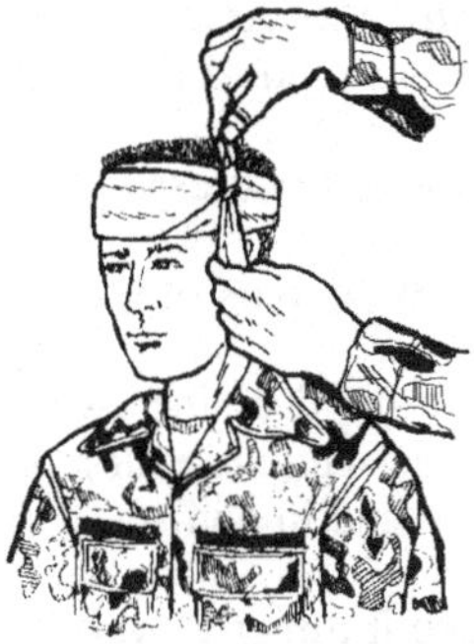

Figure 3-18. Tails tied in square knot at side of head.

(4) *Apply a dressing to a wound on top of the head.* To apply a dressing to a wound on top of the head—

(*a*) Remove the dressing from the wrapper.

(*b*) Grasp the tails of the dressing in both hands.

(*c*) Hold it (white side down) directly over the wound. **DO NOT** touch the white (sterile) side of the dressing or allow anything except the wound to come in contact with it.

(*d*) Place it over the wound (Figure 3-19).

Figure 3-19. Dressing placed over wound.

(*e*) Hold it in place with one hand. If the casualty is able, he may assist.

(*f*) Wrap one tail down under the chin (Figure 3-20), up in front of the ear, over the dressing, and in front of the other ear.

Figure 3-20. One tail of dressing wrapped under chin.

WARNING

Ensure the tails remain wide and close to the front of the chin to avoid choking the casualty.

(*g*) Wrap the remaining tail under the chin in the opposite direction and up the side of the face to meet the first tail (Figure 3-21).

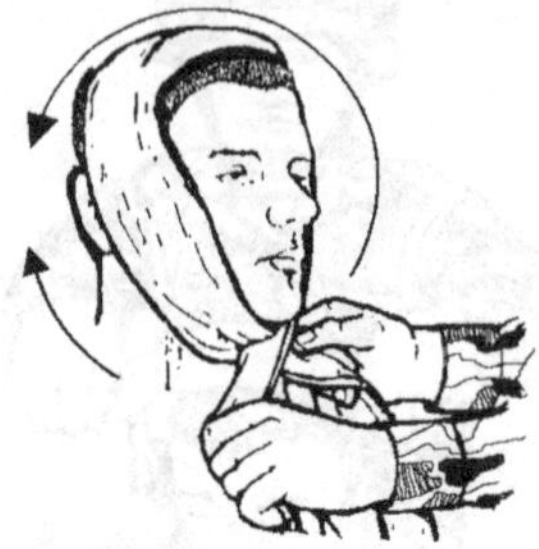

Figure 3-21. Remaining tail wrapped under chin in opposite direction.

(*h*) Cross the tails (Figure 3-22), bringing one around the forehead (above the eyebrows) and the other around the back of the head (at the base of the skull) to a point just above and in front of the opposite ear, and tie them using a square knot (Figure 3-23).

Figure 3-22. Tails of dressing crossed with one around forehead.

Figure 3-23. Tails tied in square knot (in front of and above ear).

(5) *Apply a triangular bandage to the head.* To apply a triangular bandage to the head—

(*a*) Turn the base (longest side) of the bandage up and center its base on the center of the forehead, letting the point (apex) fall on the back of the neck (Figure 3-24A).

(*b*) Take the ends behind the head and cross the ends over the apex.

(*c*) Take them over the forehead and tie them (Figure 3-24B).

(*d*) Tuck the apex behind the crossed part of the bandage or secure it with a safety pin, if available (Figure 3-24C).

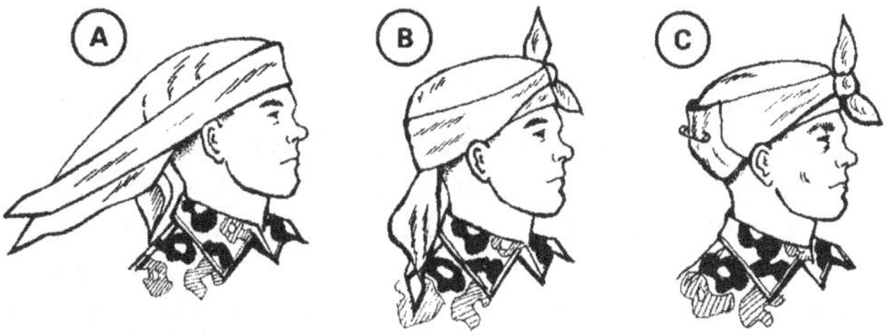

Figure 3-24. Triangular bandage applied to head (Illustrated A—C)

(6) *Apply a cravat bandage to the head.* To apply a cravat bandage to the head—

(*a*) Place the middle of the bandage over the dressing (Figure 3-25A).

(*b*) Cross the two ends of the bandage in opposite directions completely around the head (Figure 3-25B).

(*c*) Tie the ends over the dressing (Figure 3-25C).

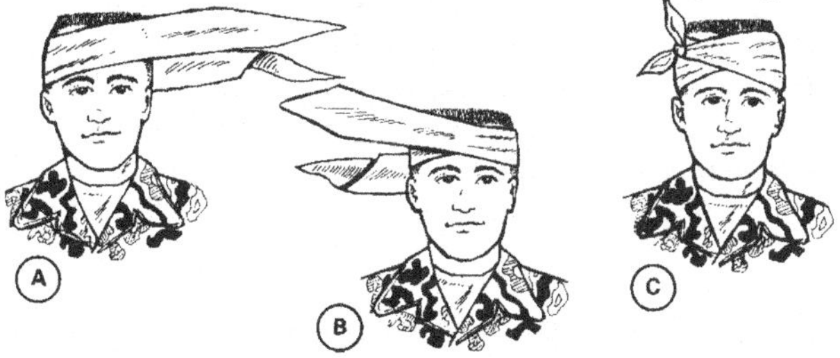

Figure 3-25. Cravat bandage applied to head (Illustrated A—C).

b. Eye Injuries. The eye is a vital sensory organ, and blindness is a severe physical handicap. Timely first aid of the eye may relieve pain and may also help to prevent shock, permanent eye injury, and possible loss of vision. Because the eye is very sensitive, any injury can be easily aggravated if it is improperly handled. Injuries of the eye may be quite severe. Cuts of the eyelids can appear to be very serious, but if the eyeball is not involved, a person's vision usually will not be damaged. However, lacerations (cuts) of the eyeball can cause permanent damage or loss of sight.

(1) *Lacerated/torn eyelids.* Lacerated eyelids may bleed heavily, but bleeding usually stops quickly. Cover the injured eye with a sterile dressing. **DO NOT** put pressure on the wound because you may injure the eyeball. Handle torn eyelids very carefully to prevent further injury. Place any detached pieces of the eyelid on a clean bandage or dressing and immediately send them with the casualty to the medical facility.

(2) *Lacerated eyeball (injury to the globe).* Lacerations or cuts to the eyeball may cause serious and permanent eye damage. Cover the injury with a loose sterile dressing. **DO NOT** put pressure on the eyeball because additional damage may occur. An important point to remember is that when one eyeball is injured, you should immobilize both eyes. This is done by applying a bandage to both eyes. Because the eyes move together, covering both will lessen the chances of further damage to the injured eye. (However, in hazardous surroundings, leave uninjured eye uncovered to enable casualty to see.)

CAUTION

DO NOT apply pressure when there is a possible laceration of the eyeball. The eyeball contains fluid. Pressure applied over the eye will force the fluid out, resulting in permanent injury. **APPLY PROTECTIVE DRESSING WITHOUT ADDED PRESSURE.**

(3) *Extruded eyeballs.* Service members may encounter casualties with severe eye injuries that include an extruded eyeball (eyeball out-of-socket). In such instances you should gently cover the extruded eye with a loose moistened dressing and also cover the unaffected eye. **DO NOT** bind or exert pressure on the injured eye while applying the dressing. Keep the casualty quiet, place him on his back, treat for shock, and evacuate him immediately.

(4) *Burns of the eyes.* Chemical burns, thermal (heat) burns, and light burns can affect the eyes.

(*a*) *Chemical burns.* Injuries from chemical burns require immediate first aid. Mainly acids or alkalies cause chemical burns. The first aid measures consist of flushing the eyes immediately with large amounts of water for at least 5 to 20 minutes, or as long as necessary to flush out the chemical and, once flushed, bandaging the eyes. If the burn is an acid burn, you should flush the eye for at least 5 to 10 minutes. If the burn is an alkali burn, you should flush the eye for at least 20 minutes. After the eye has been flushed evacuate the casualty immediately.

(*b*) *Thermal burns.* When an individual suffers burns of the face from a fire, the eyes will close quickly due to extreme heat. This reaction is a natural reflex to protect the eyeballs; however, the eyelids remain exposed and are frequently burned. If a casualty receives burns of the eyelids or face—

- **DO NOT** apply a dressing.
- **DO NOT** touch.
- **SEEK** medical assistance immediately.

(*c*) *Light burns.* Exposure to intense light can burn an individual. Infrared rays, eclipse light (if the casualty has looked directly at the sun), or laser burns cause injuries of the exposed eyeball. Ultraviolet rays from arc welding can cause a superficial burn to the surface of the eye. These injuries are generally not painful but may cause permanent damage to the eyes. Immediate first aid is usually not required. Loosely bandaging the eyes may make the casualty more comfortable and protect his eyes from further injury caused by exposure to other bright lights or sunlight.

CAUTION

With impaled objects or significant sized foreign bodies, both eyes are usually bandaged to help secure the foreign body in the injured eye. In a battlefield environment, leave the uninjured eye uncovered so that the casualty can see.

c. *Side-of-Head or Cheek Wound.* Facial injuries to the side of the head or the cheek may bleed profusely (Figure 3-26). Prompt action is necessary to ensure that the airway remains open and also to control the bleeding. It may be necessary to apply a dressing. To apply a dressing—

(1) Remove the dressing from its wrapper.

(2) Grasp the tails in both hands.

(3) Hold the dressing directly over the wound with the white side down and place it directly on the wound (Figure 3-27A). **DO NOT** touch the white (sterile) side of the dressing or allow anything except the wound to come in contact with it.

(4) Hold the dressing in place with one hand (the casualty may assist if able). Wrap the top tail over the top of the head and bring it down in front of the ear (on the side opposite the wound), under the chin (Figure 3-27B) and up over the dressing to a point just above the ear (on the wound side).

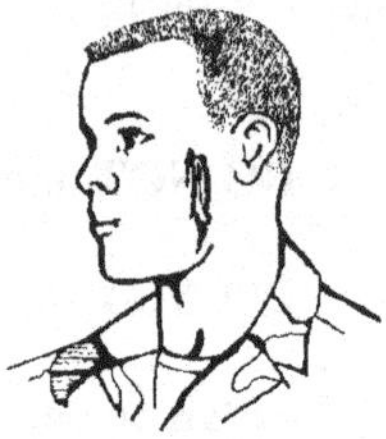

Figure 3-26. Side of head or cheek wound.

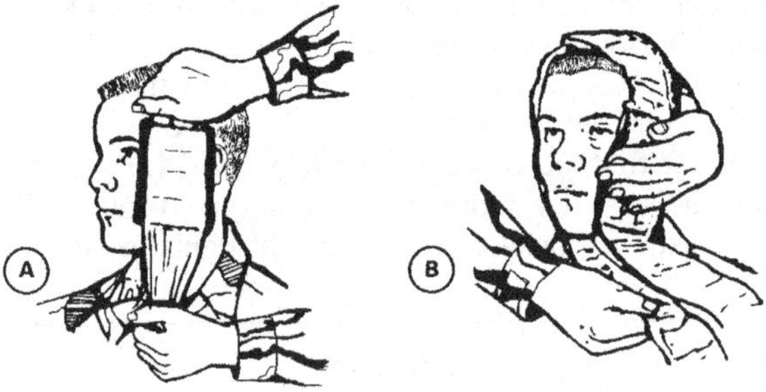

Figure 3-27. Dressing placed directly on wound. Top tail wrapped over top of head, down in front of ear, and under chin (Illustrated A—B).

NOTE

When possible, avoid covering the casualty's ear with the dressing, as this will decrease his ability to hear.

(5) Bring the second tail under the chin, up in front of the ear (on the side opposite the wound), and over the head to meet the other tail (on the wounded side) (Figure 3-28).

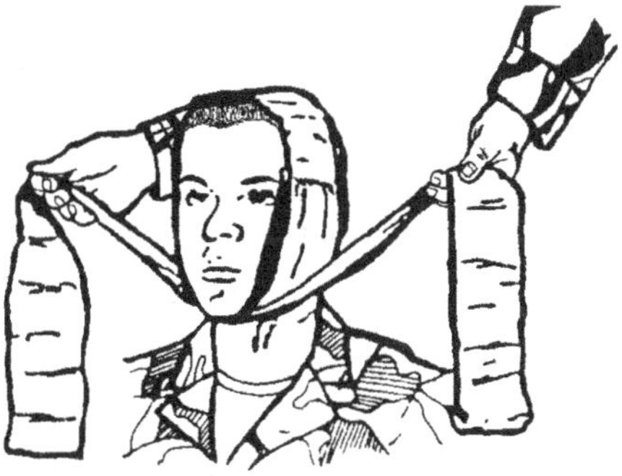

Figure 3-28. Bringing second tail under the chin.

(6) Cross the two tails (on the wound side) (Figure 3-29) and bring one end across the forehead (above the eyebrows) to a point just in front of the opposite ear (on the uninjured side).

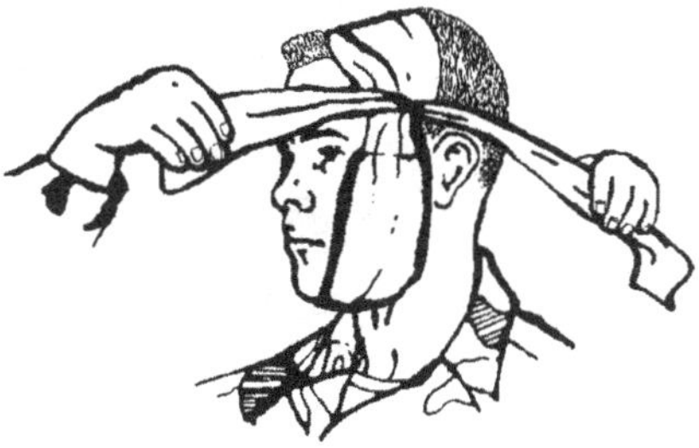

Figure 3-29. Crossing the tails on the side of the wound.

(7) Wrap the other tail around the back of the head (at the base of the skull), and tie the two ends just in front of the ear on the uninjured side with a square knot (Figure 3-30).

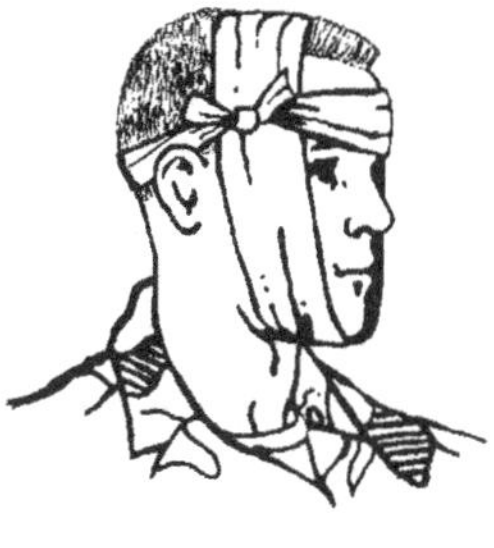

Figure 3-30. Tying the tails of the dressing in a square knot.

d. Ear Injuries. Lacerated (cut) or avulsed (torn) ear tissue may not, in itself, be a serious injury. Bleeding, or the drainage of fluids from the ear canal, however, may be a sign of a head injury, such as a skull fracture. **DO NOT** attempt to stop the flow from the inner ear canal nor put anything into the ear canal to block it. Instead, you should cover the ear lightly with a dressing. For minor cuts or wounds to the external ear, apply a cravat bandage as follows:

(1) Place the middle of the bandage over the ear (Figure 3-31A).

(2) Cross the ends, wrap them in opposite directions around the head, and tie them (Figures 3-31B and 3-31C).

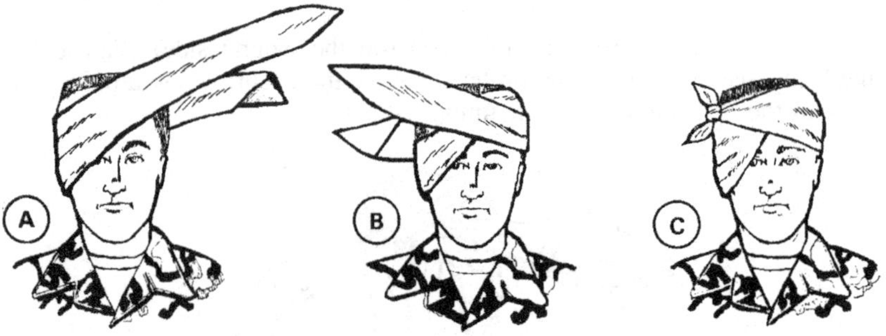

Figure 3-31. Applying cravat bandage to ear (Illustrated A—C).

(3) If possible, place some dressing material between the back of the ear and the side of the head to avoid crushing the ear against the head with the bandage.

e. Nose Injuries. Nose injuries generally produce bleeding. The bleeding may be controlled by placing an ice pack (if available) over the nose, or pinching the nostrils together. The bleeding may also be controlled by placing torn gauze (rolled) between the upper teeth and the lip.

CAUTION

DO NOT attempt to remove objects inhaled into the nose. An untrained person who removes such an object could worsen the casualty's condition and cause permanent injury.

f. Jaw Injuries. Before applying a bandage to a casualty's jaw, remove all loose or free-floating foreign material from the casualty's mouth.

If the casualty is unconscious, check for obstructions in the airway and remove if possible. If there is profuse bleeding in the oral cavity, the cavity may require loose packing with soft bandaging material (for example: Kerlix™ gauze) prior to applying a bandage. Care should be taken to avoid occluding the airway. When applying the bandage, allow the jaw enough freedom to permit passage of air and drainage from the mouth.

(1) *Apply bandages attached to field first aid dressing to the jaw.* After dressing the wound, apply the bandages using the same technique illustrated in Figure 3-32A—C.

NOTE

The dressing and bandaging procedure outlined for the jaw serves a twofold purpose. In addition to stopping the bleeding and protecting the wound, it also immobilizes a fractured jaw.

(2) *Apply a cravat bandage to the jaw.*

(*a*) Place the bandage under the chin and pull its ends upward. Adjust the bandage to make one end longer than the other (Figure 3-32A).

(*b*) Take the longer end over the top of the head to meet the short end at the temple and cross the ends over (Figure 3-32B).

(*c*) Take the ends in opposite directions to the other side of the head and tie them over the part of the bandage that was applied first (Figure 3-32C).

Figure 3-32. Applying a cravat bandage to jaw (Illustrated A—C).

NOTE

The cravat bandage technique is used to immobilize a fractured jaw or to maintain a sterile dressing that does not have tail bandages attached.

3-11. Shoulder Bandage

a. To apply bandages attached to the field first aid dressing—

(1) Take one bandage across the chest and the other across the back and under the arm opposite the injured shoulder.

(2) Tie the ends with a square knot (Figure 3-33).

Figure 3-33. Shoulder bandage.

b. To apply a cravat bandage to the shoulder or armpit—

(1) Make an extended cravat bandage by using two triangular bandages (Figure 3-34A); place the end of the first triangular bandage along the base of the second one (Figure 3-34B).

(2) Fold the two bandages into a single extended bandage (Figure 3-34C).

(3) Fold the extended bandage into a single cravat bandage (Figure 3-34D). After folding, secure the thicker part (overlap) with two or more safety pins (Figure 3-34E).

(4) Place the middle of the cravat bandage under the armpit so that the front end is longer than the back end and safety pins are on the outside (Figure 3-34F).

(5) Cross the ends on top of the shoulder (Figure 3-34G).

(6) Take one of the bandage ends across the back and under the arm on the opposite side and the other end across the chest. Tie the ends (Figure 3-34H).

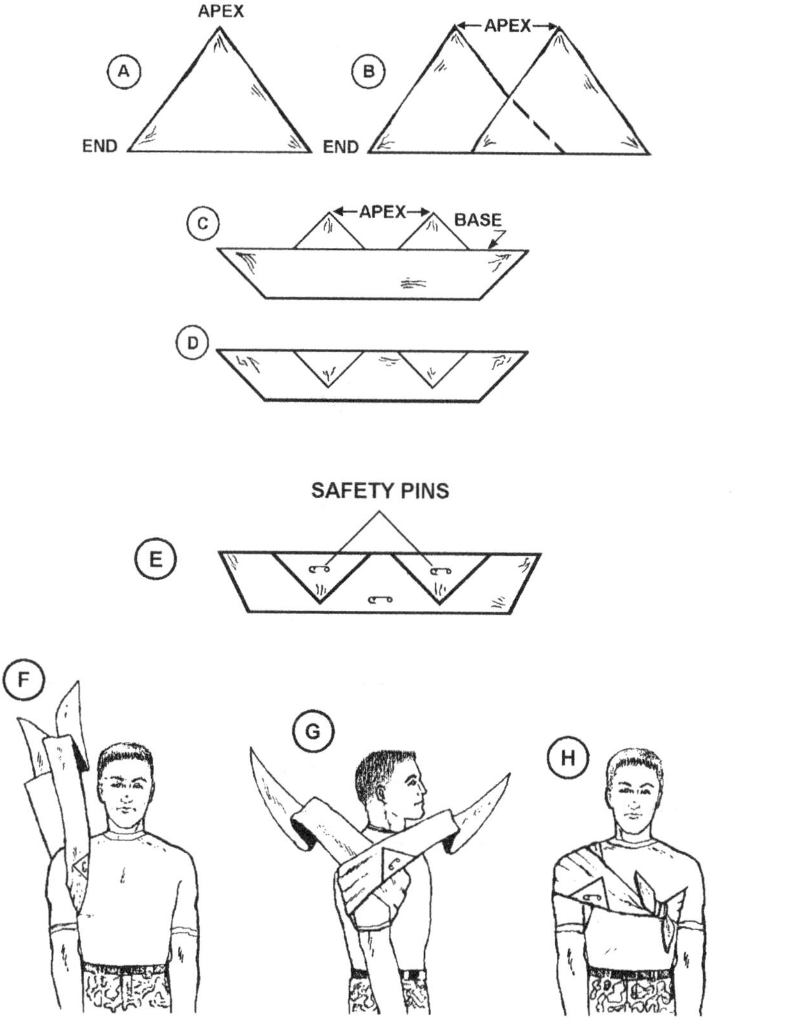

Figure 3-34. Extended cravat bandage applied to shoulder or armpit (Illustrated A—H).

Be sure to place sufficient wadding in the armpit. **DO NOT** tie the cravat bandage too tightly. Avoid compressing the major blood vessels in the armpit.

3-12. Elbow Bandage

To apply a cravat bandage to the elbow—

a. Bend the arm at the elbow and place the middle of the cravat at the point of the elbow bringing the ends upward (Figure 3-35A).

b. Bring the ends across, extending both downward (Figure 3-35B).

c. Take both ends around the arm and tie them with a square knot at the front of the elbow (Figure 3-35C).

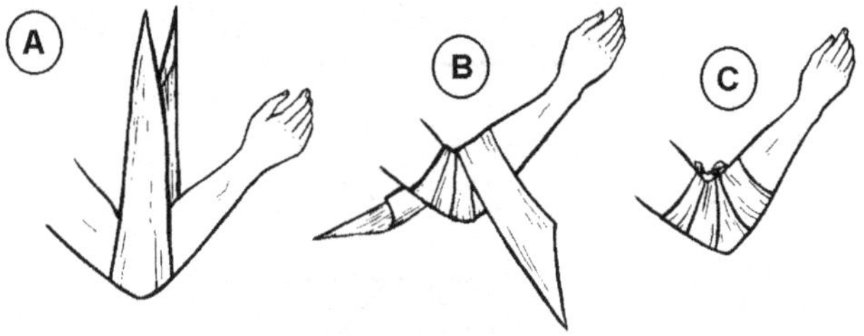

Figure 3-35. Elbow bandage (Illustrated A—C).

CAUTION

If an elbow fracture is suspected, **DO NOT** bend the elbow; bandage it in the position found.

3-13. Hand Bandage

a. To apply a triangular bandage to the hand—

(1) Place the hand in the middle of the triangular bandage with the wrist at the base of the bandage (Figure 3-36A). Ensure that the fingers are separated with absorbent material to prevent chafing and irritation of the skin.

(2) Place the apex over the fingers and tuck any excess material into the pleats on each side of the hand (Figure 3-36B).

(3) Cross the ends on top of the hand, take them around the wrist, and tie them (Figures 3-36C—E) with a square knot.

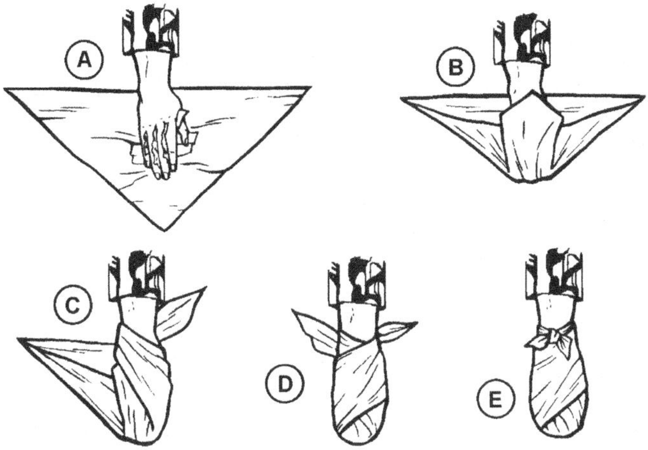

Figure 3-36. Triangular bandage applied to hand (Illustrated A—E).

b. To apply a cravat bandage to the palm of the hand—

(1) Lay the middle of the cravat over the palm of the hand with the ends hanging down on each side (Figure 3-37A).

(2) Take the end of the cravat at the little finger across the back of the hand, extending it upward over the base of the thumb; then bring it downward across the palm (Figure 3-37B).

(3) Take the thumb end across the back of the hand, over the palm, and through the hollow between the thumb and palm (Figure 3-37C).

(4) Take the ends to the back of the hand and cross them; then bring them up over the wrist and cross them again (Figure 3-37D).

(5) Bring both ends down and tie them with a square knot on top of the wrist (Figure 3-37E—F).

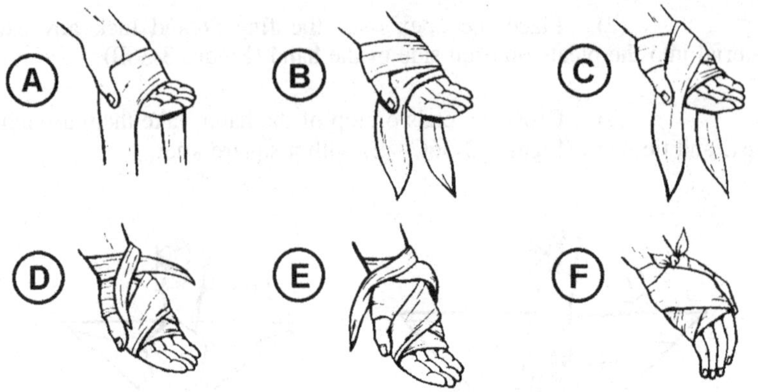

Figure 3-37. Cravat bandage applied to palm of hand (Illustrated A—F).

3-14. Leg (Upper and Lower) Bandage

To apply a cravat bandage to the leg—

a. Place the center of the cravat over the dressing (Figure 3-38A).

b. Take one end around and up the leg in a spiral motion and the other end around and down the leg in a spiral motion, overlapping part of each preceding turn (Figure 3-38B).

c. Bring both ends together and tie them (Figure 3-38C) with a square knot.

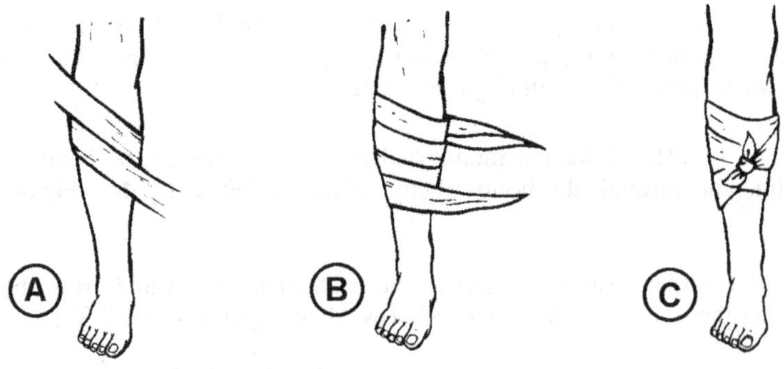

Figure 3-38. Cravat bandage applied to leg (Illustrated A—C).

3-15. Knee Bandage

To apply a cravat bandage to the knee as illustrated in Figure 3-39, use the same technique applied in bandaging the elbow.

CAUTION

If a fracture of the kneecap is suspected, **DO NOT** bend the knee; bandage it in the position found.

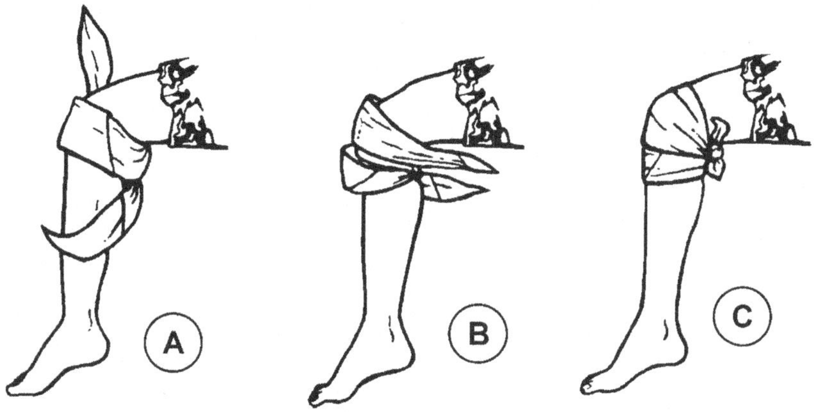

Figure 3-39. Cravat bandage applied to knee (Illustrated A—C).

3-16. Foot Bandage

To apply a triangular bandage to the foot—

a. Place the foot in the middle of the triangular bandage with the heel well forward of the base (Figure 3-40A). Ensure that the toes are separated by absorbent material to prevent chafing and irritation of the skin.

b. Place the apex over the top of the foot and tuck any excess material into the pleats on each side of the foot (Figure 3-40B).

c. Cross the ends on top of the foot, take them around the ankle, and tie them at the front of the ankle (Figure 3-40C—E).

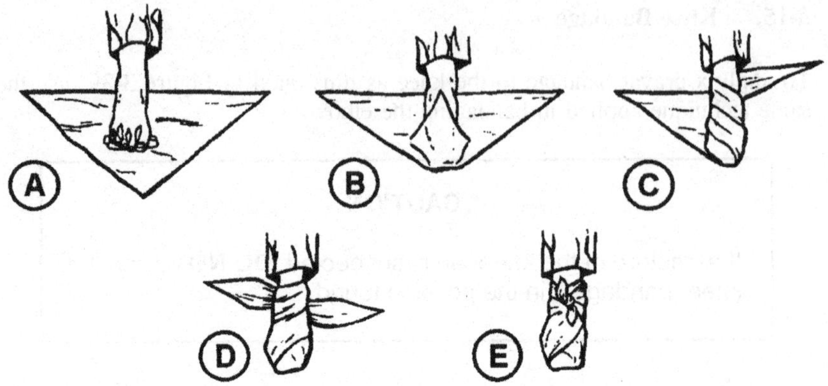

Figure 3-40. Triangular bandage applied to foot (Illustrated A—E).

CHAPTER 4

FIRST AID FOR FRACTURES

4-1. General

A fracture is any break in the continuity of a bone. Fractures can cause total disability or in some cases death by severing vital organs and/or arteries. On the other hand, they can most often be treated so there is a complete recovery. The potential for recovery depends greatly upon the first aid the individual receives before he is moved. First aid includes immobilizing the fractured part in addition to applying lifesaving measures when necessary. The basic splinting principle is to immobilize the joints above and below the fracture.

4-2. Kinds of Fractures

Figure 4-1 depicts types of fractures.

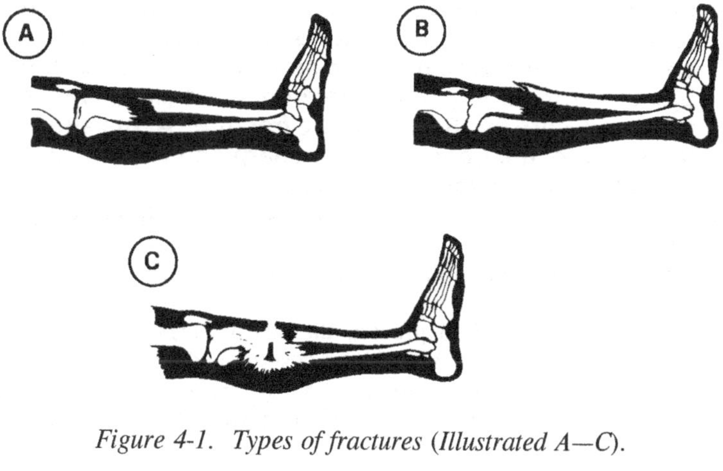

Figure 4-1. Types of fractures (Illustrated A—C).

a. Closed Fracture (Figure 4-1A). A closed fracture is a broken bone that does not break the overlying skin. The tissue beneath the skin may be damaged. A *dislocation* is when a joint, such as a knee, ankle, or shoulder, is not in the proper position. A *sprain* is when the connecting tissues of the joints have been torn. Dislocations and sprains (swelling, possible deformity, and discoloration) should be treated as closed fractures.

b. Open Fracture (Figure 4-1B and 4-1C). An open fracture is a broken bone that breaks (pierces) the overlying skin. The broken bone may come through the skin or a missile such as a bullet or shell fragment may go through the flesh and break the bone.

NOTE

An open fracture is contaminated and subject to infection.

4-3. Signs and Symptoms of Fractures

Indications of a fracture are deformity, tenderness, swelling, pain, inability to move the injured part, protruding bone, bleeding, or discolored skin at the injury site. A sharp pain when the service member attempts to move the part is also a sign of a fracture.

WARNING

DO NOT encourage the casualty to move the injured part in order to identify a fracture since such movement could cause further damage to surrounding tissues and promote shock. If you are not sure whether a bone is fractured, care for the injury as a fracture. At the site of the fracture, the bone ends are sharp and could cause vessel (artery and/or vein) damage.

4-4. Purposes of Immobilizing Fractures

A fracture is immobilized to prevent the sharp edges of the bone from moving and cutting tissue, muscle, blood vessels, and nerves. This reduces pain and helps prevent or control shock. In a closed fracture, immobilization keeps bone fragments from causing an open wound, which can become contaminated and subject to infection.

4-5. Splints, Padding, Bandages, Slings, and Swathes

a. Splints. Splints may be improvised from such items as boards, poles, sticks, tree limbs, or cardboard. If nothing is available for a splint, the chest wall can be used to immobilize a fractured arm and the uninjured leg can be used to immobilize (to some extent) the fractured leg.

b. Padding. Padding may be improvised from such items as a jacket, blanket, poncho, shelter half, or leafy vegetation.

c. Bandages. Bandages may be improvised from belts, rifle slings, kerchiefs, or strips torn from clothing or blankets. Narrow materials such as wire or cord should not be used to secure a splint in place. The application of wire and/or narrow material to an extremity could cause tissue damage and a tourniquet effect.

d. Slings. A sling is a bandage suspended from the neck to support an upper extremity. If a bandage is not available, a sling can be improvised by using the tail of a coat or shirt or pieces of cloth torn from such items as clothing and blankets. The triangular bandage is ideal for this purpose. Remember that the casualty's hand should be higher than his elbow, and the fingers should be showing at all times. The sling should be applied so that the supporting pressure is on the uninjured side.

e. Swathes. Swathes are any bands (pieces of cloth or load bearing equipment [LBE]) that are used to further immobilize a splinted fracture. Triangular and cravat bandages are often used and are called *swathe bandages*. The purpose of the swathe is to immobilize; therefore, the swathe bandage is placed above and/or below the fracture—not over it.

4-6. Procedures for Splinting Suspected Fractures

Before beginning first aid procedures for a fracture, gather whatever splinting materials are available. Ensure that splints are long enough to immobilize the joint above and below the suspected fracture. If possible, use at least four ties (two above and two below the fracture) to secure the splints. The ties should be square knots and should be tied away from the body on the splint. Distal pulses of the affected extremity should be checked before and after the application of the splint.

a. Evaluate the Casualty. Be prepared to perform any necessary lifesaving measures. Monitor the casualty for development of conditions that may require you to perform necessary lifesaving measures.

WARNING

Unless there is immediate life-threatening danger, such as a fire or an explosion, DO NOT move the casualty with a suspected back or neck injury. Improper movement may cause permanent paralysis or death.

> **WARNING**
>
> **In a chemical environment, DO NOT remove any protective clothing. Apply the dressings and splints over the garments.**

b. Locate the Site of the Suspected Fracture.

(1) Ask the casualty for the location of the injury.

- Does he have any pain?
- Where is it tender?
- Can he move the extremity?

NOTE

With the presence of an obvious deformity, do not make the casualty move extremity.

(2) Look for an unnatural position of the extremity.

(3) Look for a bone sticking out (protruding).

c. Prepare the Casualty for Splinting the Suspected Fracture.

(1) Reassure the casualty. Tell him that you will be providing first aid for him and that medical help is on the way.

(2) Loosen any tight or binding clothing.

(3) Remove all jewelry from the injured part and place it in the casualty's pocket. Tell the casualty you are doing this because if the jewelry is not removed and swelling occurs later, he may not be able to get it off and further bodily injury could result.

(4) Boots should not be removed from the casualty unless they are needed to stabilize a neck injury or there is actual bleeding from the foot.

d. Gather Splinting Materials. If standard splinting materials (splints, padding, and cravats) are not available, gather improvised materials. If splinting material is not available and the suspected fracture **CANNOT** be splinted, then swathes, or a combination of swathes and slings can be used to immobilize the extremity.

e. Pad the Splints. Pad the splints where they touch any bony part of the body, such as the elbow, wrist, knee, ankle, crotch, or armpit areas. Padding prevents excessive pressure on the area, which could lead to circulation problems.

f. Check the Circulation Below the Site of the Injury.

(1) Note any pale, white, or bluish-gray color of the skin, which may indicate impaired circulation. Circulation can also be checked by depressing the toe or fingernail beds and observing how quickly the color returns. A slower return of color to the injured side when compared with the uninjured side indicates a problem with circulation. The fingernail bed is the method to use to check the circulation in a dark-skinned casualty.

(2) Check the temperature of the injured extremity. Use your hand to compare the temperature of the injured side with the uninjured side. The body area below the injury may be colder to the touch indicating poor circulation.

(3) Question the casualty about the presence of numbness, tightness, cold, or tingling sensations.

WARNING

Casualties with fractures of the extremities may show impaired circulation, such as numbness, tingling, cold or pale to bluish skin tone. These casualties should be evacuated by medical personnel and treated as soon as possible. Prompt medical treatment may prevent possible loss of the limb.

WARNING

If it is an open fracture and the bone is protruding from the skin, *DO NOT ATTEMPT TO PUSH THE BONE BACK UNDER THE SKIN.* Apply a field dressing over the wound to protect the area.

g. Apply the Splint in Place.

(1) Splint the fracture in the position found. **DO NOT** attempt to reposition or straighten the injury. If it is an open fracture, stop the bleeding and protect the wound. Cover all wounds with field dressings before applying a splint. Remember to use the casualty's field dressing, not your own.

(2) Place one splint on each side of the fracture. Make sure that the splints reach, if possible, beyond the joints above and below the fracture.

(3) Tie the splints. Secure each splint in place above and below the fracture site with improvised (or actual) cravats. Improvised cravats, such as strips of cloth, belts, or whatever else you have, may be used. With minimal motion to the injured areas, place and tie the splints with the bandages. Push cravats through and under the natural body curvatures, and then gently position improvised cravats and tie in place. Use square knots. Tie all knots on the splint away from the casualty (Figure 4-2). **DO NOT** tie cravats directly over the suspected fracture site.

Figure 4-2. Square knots tied away from casualty.

h. Check the Splint for Tightness.

(1) **CHECK** to be sure that bandages are tight enough to securely hold splinting materials in place, but not so tight that circulation is impaired.

(2) **RECHECK** the circulation after application of the splint. Check the skin color and temperature. This is to ensure that the bandages holding the splint in place have not been tied too tightly. A fingertip check can be made by inserting the tip of the finger between the bandaged knot and the skin.

(3) **MAKE** any necessary adjustment without allowing the splint to become ineffective.

i. Apply a Sling. An improvised sling may be made from any available nonstretching piece of cloth, such as a battle dress uniform (BDU) shirt or trousers, poncho, or shelter half. Slings may also be improvised using the tail of a coat, belt, or a piece of cloth. Figure 4-3 depicts a shirttail used for support. A trousers belt or LBE may also be used for support (Figure 4-4). A sling should place the supporting pressure on the casualty's uninjured side. The supported arm should have the hand positioned slightly higher than the elbow showing the fingers.

Figure 4-3. Shirttail used for support.

Figure 4-4. Belt used for support.

(1) Insert the splinted arm in the center of the sling (Figure 4-5).

Figure 4-5. Arm inserted in center of improvised sling.

(2) Bring the ends of the sling up and tie them at the side (or hollow) of the neck on the uninjured side (Figure 4-6).

Figure 4-6. Ends of improvised sling tied to side of neck.

(3) Twist and tuck the corner of the sling at the elbow (Figure 4-7).

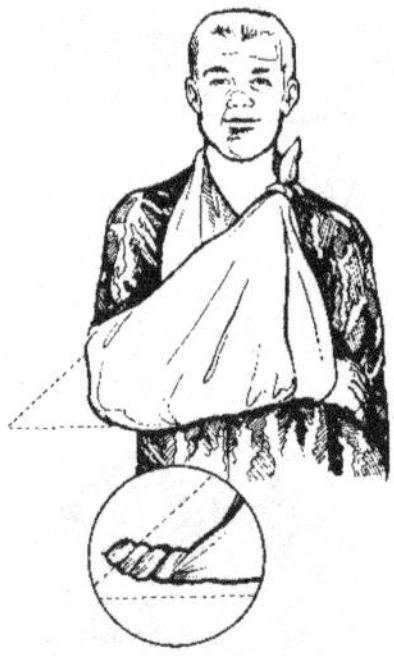

Figure 4-7. Corner of sling twisted and tucked at elbow.

j. Apply a Swathe. You may use any large piece of cloth, service member's belt, or pistol belt, to improvise a swathe.

WARNING

The swathe should not be placed directly on top of the injury, but positioned either above or below the fracture site.

(1) Apply swathes to the injured arm by wrapping the swathe over the injured arm, around the casualty's back, and under the arm on the uninjured side. Tie the ends on the uninjured side (Figure 4-8).

Figure 4-8. Arm immobilized with strip of clothing.

(2) A swathe is applied to an injured leg by wrapping the swathe around both legs and securing it on the uninjured side.

k. Seek Medical Assistance. Notify medical personnel, watch closely for development of life-threatening conditions and/or impaired circulation to the injured extremity. (Refer to Chapter 1 for additional information on life-threatening conditions.)

4-7. Upper Extremity Fractures

Figures 4-9 through 4-17 show how to apply slings, splints, and cravats (swathes) to immobilize and support fractures of the upper extremities. *Although the padding is not visible in some of the illustrations, it is always preferable to apply padding along the injured part for the length of the splint and especially where it touches any bony parts of the body.*

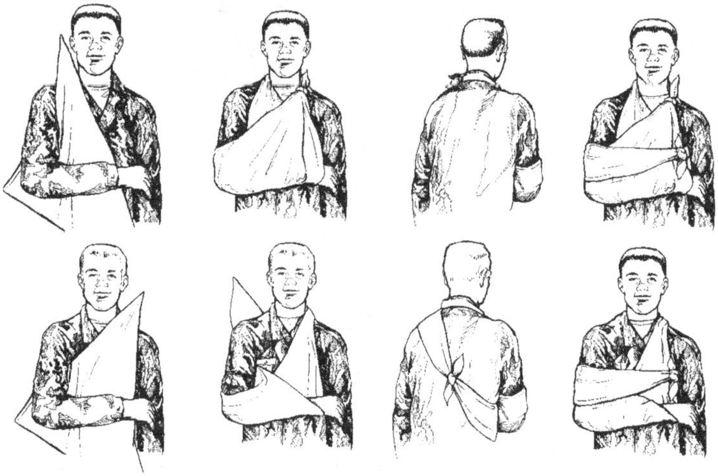

Figure 4-9. Application of triangular bandage to form sling (two methods).

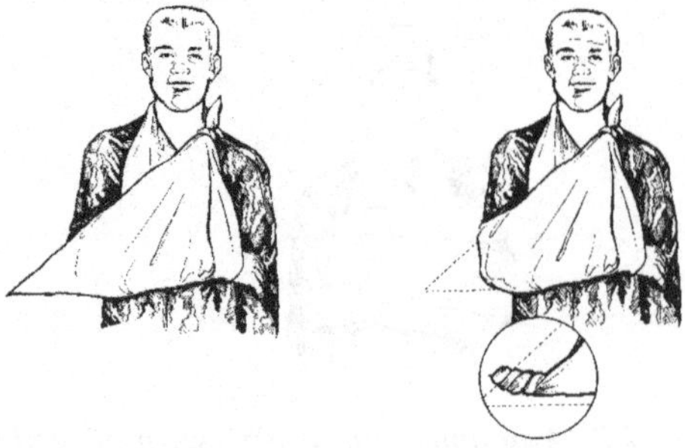

Figure 4-10. Completing sling sequence by twisting and tucking the corner of the sling at the elbow.

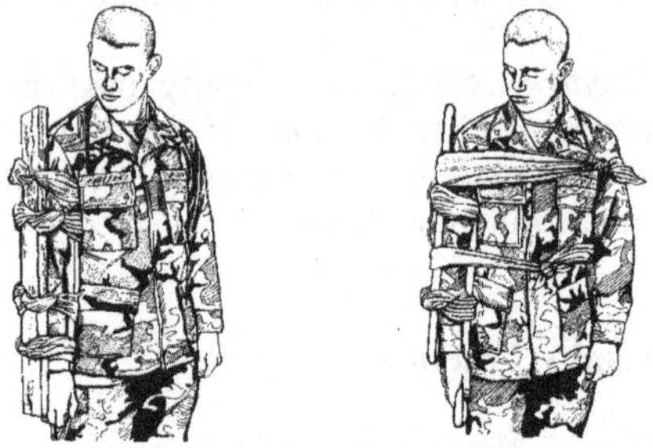

Figure 4-11. Board splints applied to fractured elbow when elbow is not bent (two methods).

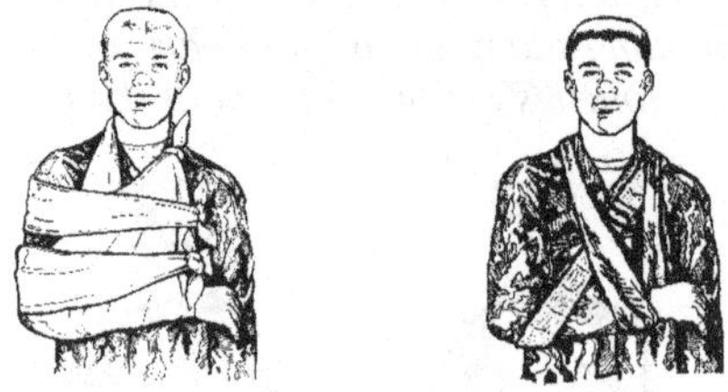

Figure 4-12. Chest wall used as splint for upper arm fracture when no splint is available.

Figure 4-13. Chest wall, sling, and cravat used to immobilize fractured elbow when elbow is bent.

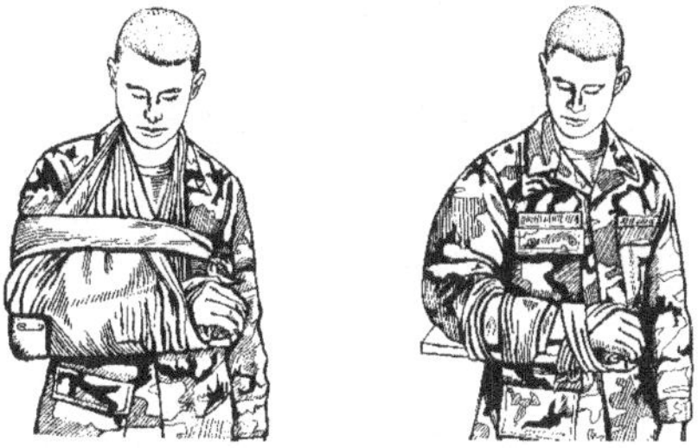

Figure 4-14. Board splint applied to fractured forearm.

Figure 4-15. Fractured forearm or wrist splinted with sticks and supported with tail of shirt and strips of material.

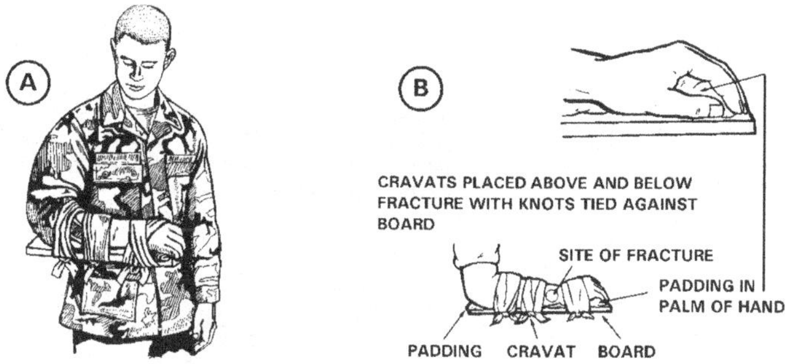

Figure 4-16. Board splint applied to fractured wrist and hand (Illustrated A—B).

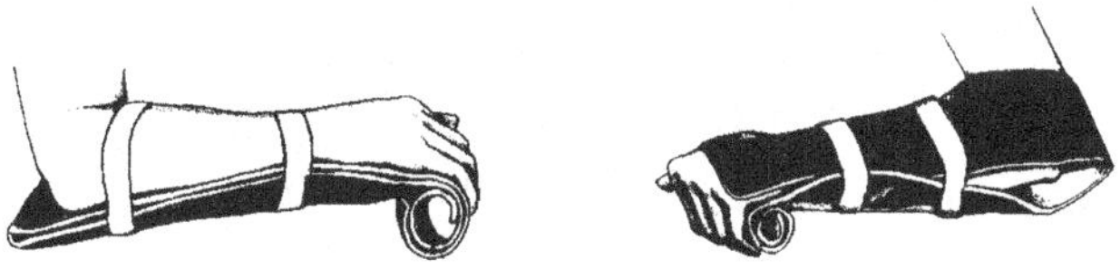

Figure 4-17. SAM® splint applied to fractured wrist or forearm.

4-8. Lower Extremity Fractures

Figures 4-18 through 4-24 show how to apply splints to immobilize fractures of the lower extremities. *Although padding is not visible in some of the illustrations, it is always preferable to apply padding along the injured part for the length of the splint and especially where it touches any bony parts of the body.*

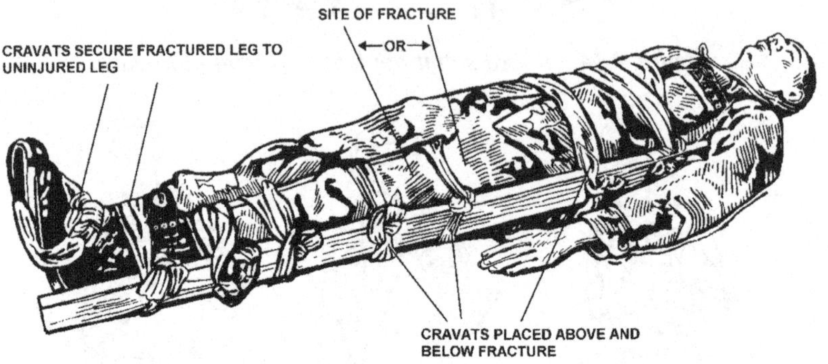

Figure 4-18. Board splints applied to fractured hip or thigh.

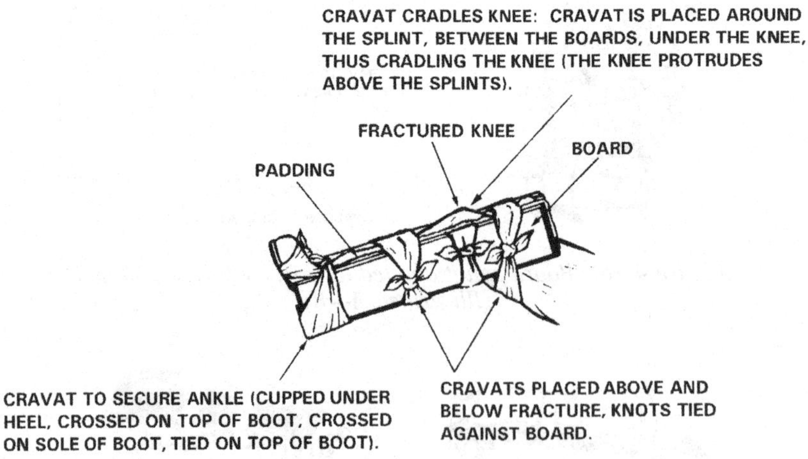

Figure 4-19. Board splint applied to fractured or dislocated knee.

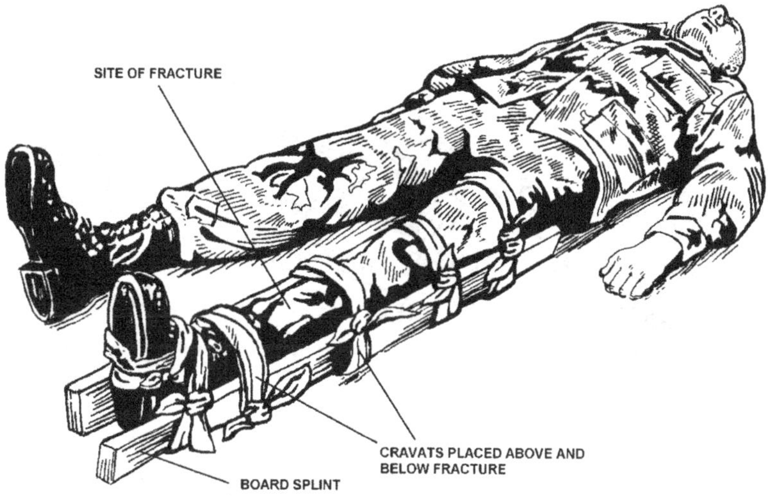

Figure 4-20. Board splints applied to fractured lower leg or ankle.

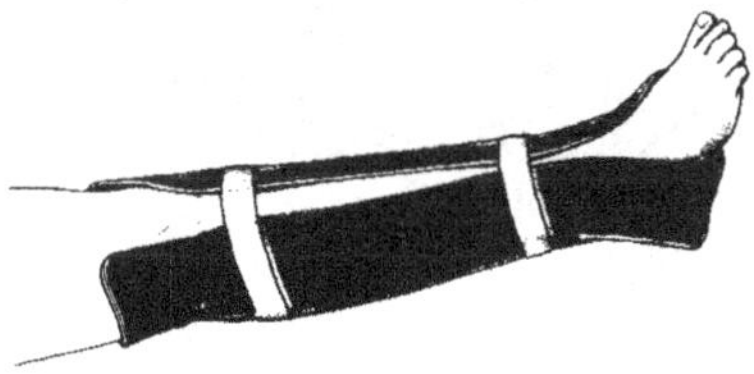

Figure 4-21. SAM® splint applied to fractured lower leg or ankle.

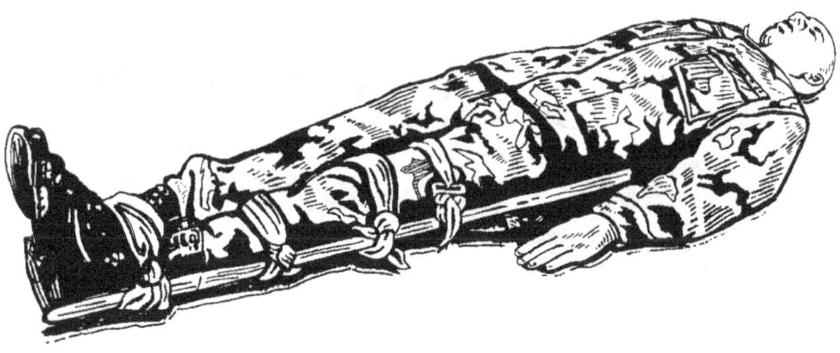

Figure 4-22. Improvised splints applied to fractured lower leg or ankle.

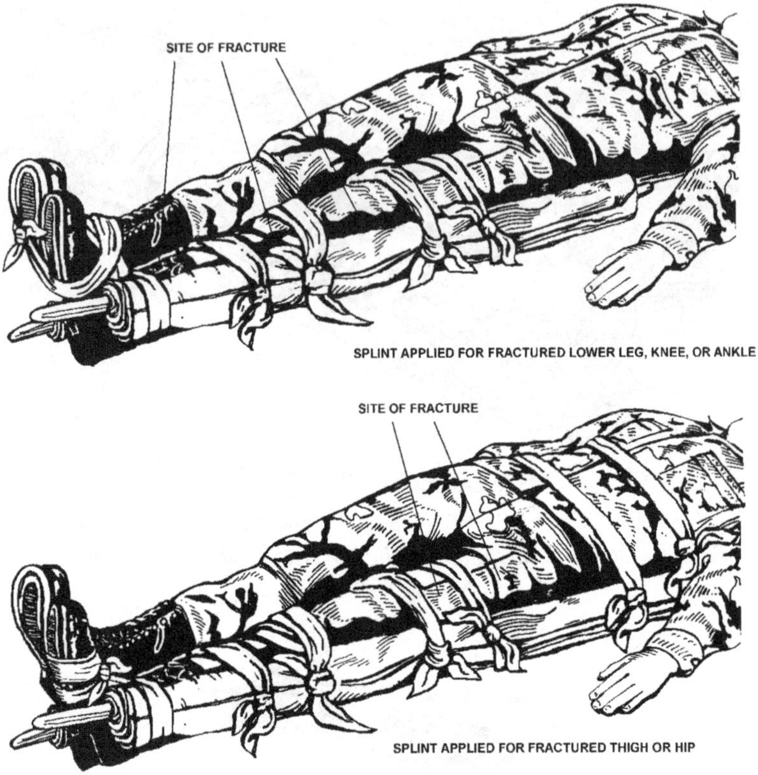

Figure 4-23. Poles rolled in a blanket and used as splints applied to fractured lower extremity.

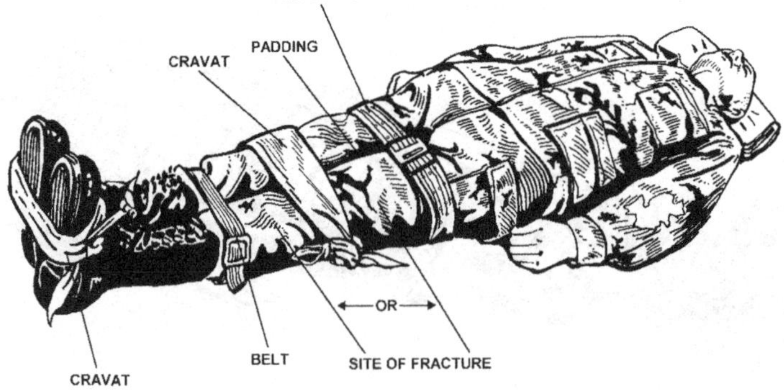

Figure 4-24. Uninjured leg used as splint for fractured leg (anatomical splint).

4-9. Jaw, Collarbone, and Shoulder Fractures

a. Apply a cravat to immobilize a fractured jaw as illustrated in Figure 4-25. Direct all bandaging support to the top of the casualty's head, not to the back of his neck. If incorrectly placed, the bandage will pull the casualty's jaw back and interfere with his breathing.

Figure 4-25. Fractured jaw immobilized.

> **WARNING**
>
> **Casualties with lower jaw (mandible) fractures cannot be laid flat on their backs because facial muscles will relax and may cause an airway obstruction.**

b. Apply two belts, a sling, and a cravat to immobilize a fractured collarbone, as illustrated in Figure 4-26.

Figure 4-26. Application of belts, sling, and cravat to immobilize a fractured collarbone.

c. Apply a sling and a cravat to immobilize a fractured or dislocated shoulder, using the technique illustrated in Figure 4-27.

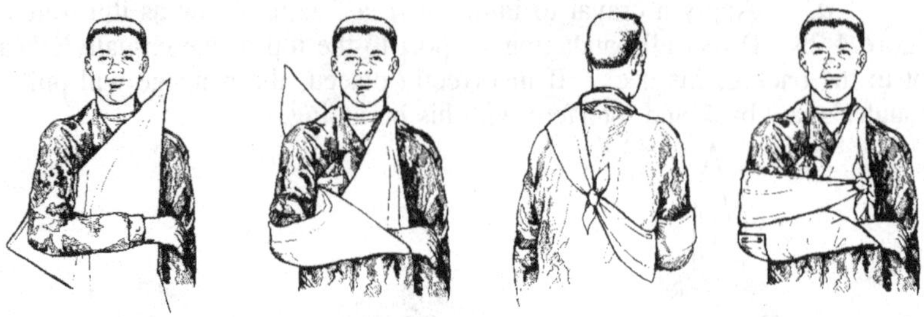

Figure 4-27. Application of sling and cravat to immobilize a fractured or dislocated shoulder.

4-10. Spinal Column Fractures

It is often impossible to be sure a casualty has a fractured spinal column. Be suspicious of any back injury, especially if the casualty has fallen or if his back has been sharply struck or bent. If a casualty has received such an injury and does not have feeling in his legs or cannot move them, you can be reasonably sure that he has a severe back injury, which should be managed as a fracture. Remember, that the possibility of a neck fracture or injury to the back should always be suspected, and it is often impossible to be sure if a casualty has a fractured spinal column. If the spine is fractured, bending it can cause the sharp bone fragments to bruise or cut the spinal cord and result in permanent paralysis or death (Figure 4-28A). The spinal column must maintain normal spinal position at the lower back (lumbar region) to help remove pressure from the spinal cord.

a. If the casualty is not to be transported until medical personnel arrive—

- Caution him not to move. Ask him if he is in pain or if he is unable to move any part of his body.

- Leave him in the position in which he is found. **DO NOT** move any part of his body, unless he is in imminent danger.

- If the casualty is lying face up, slip a blanket or other supporting material under the arch of his lower back to help support the spine in a normal position (Figure 4-28B). Take care not to place so much bulky padding as to cause potential damage by causing undo pressure on the spine. If he is lying face down, **DO NOT** put anything under any part of his body.

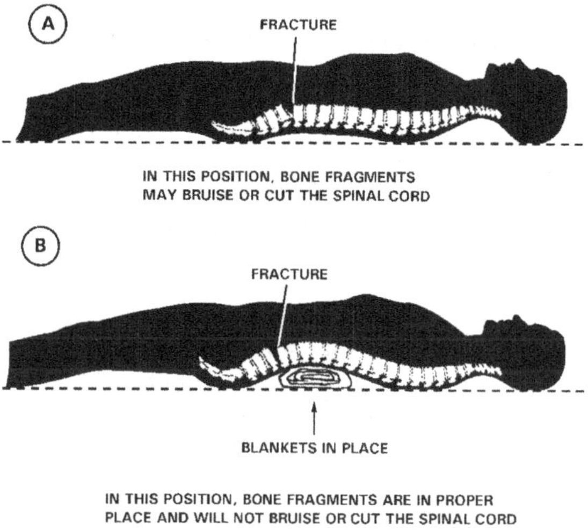

Figure 4-28. Spinal column must maintain a normal spine position.

b. If the casualty must be transported to a safe location before medical personnel arrive and if the casualty is in a—

- Face-up position, transport him by litter or use a firm substitute, such as a wide board or a door longer than his height. Loosely tie the casualty's wrists together over his waistline, using a cravat or a strip of cloth. Tie his feet together to prevent the accidental dropping or shifting of his legs. Lay a folded blanket across the litter where the arch of his back is to be placed. Using a four-man team (Figure 4-29), place the casualty on the litter without bending his spinal column or his neck.

Figure 4-29. Placing face-up casualty with fractured back onto litter.

• The number two man positions himself at the casualty's head. His responsibility is to provide manual in-line (neutral) stabilization of the head and neck. The number three, and four men position themselves on one side of the casualty; all kneel on one knee along the side of the casualty. The number one man positions himself to the opposite side of the casualty (or can be on the same side of number three and four). The numbers two, three, and four men gently place their hands under the casualty. The number one man on the opposite side places his hands under the injured part to assist.

• When all four men are in position to lift, the number two man commands, "**PREPARE TO LIFT**" and then, "**LIFT**." All men, in unison, gently lift the casualty about 8 inches. Once the casualty is lifted, the number one man recovers and slides the litter under the casualty, ensuring that the blanket is in proper position. The number one man then returns to his original lift position (Figure 4-29).

• When the number two man commands, "**LOWER CASUALTY**," all men, in unison, gently lower the casualty onto the litter.

• Facedown position, he must be transported in this same position. The four-man team lifts him onto a regular or improvised litter, keeping the spinal column in a normal spinal position. If a regular litter is used, first place a folded blanket on the litter at the point where the chest will be placed.

4-11. Neck Fractures

A fractured neck is extremely dangerous. Bone fragments may bruise or cut the spinal cord just as they might in a fractured back.

a. If the casualty is not to be transported until medical personnel arrive—

• Caution him not to move. Moving may cause permanent injury or death.

• Leave the casualty in the position in which he is found. If his neck and head (cervical spine) are in an abnormal position, immediately immobilize the neck and head.

• Keep his head still, if the casualty is lying face up, raise his shoulders slightly, and slip a roll of cloth that has the bulk of a bath towel under his neck (Figure 4-31). The roll should be thick enough to arch his neck only slightly, leaving the back of his head on the ground. **DO NOT** bend his neck or head forward. **DO NOT** raise or twist his head. Immobilize the casualty's head (Figure 4-32). Do this by padding heavy objects (such as rocks or his boots filled with dirt, sand, gravel, or rock) and placing them on each side of his head. If it is necessary to use boots, after filling them, tie the top tightly or stuff with pieces of cloth to secure the contents.)

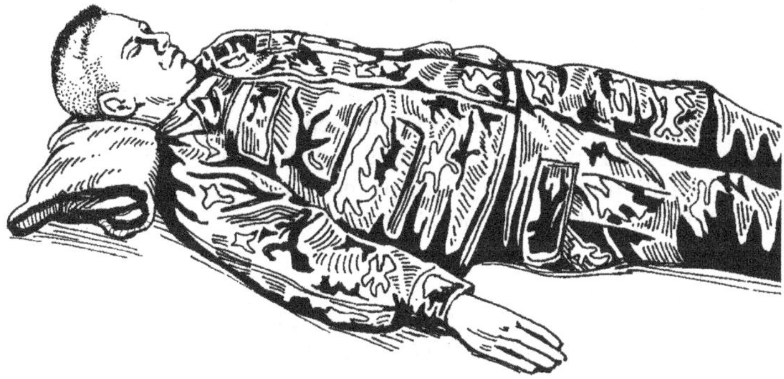

Figure 4-30. Casualty with roll of cloth (bulk) under neck.

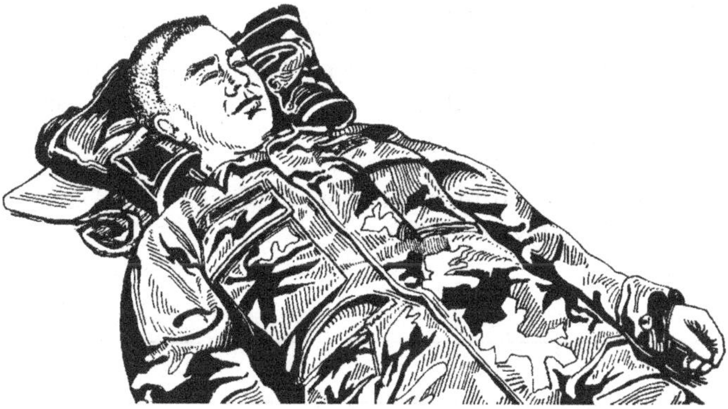

Figure 4-31. Immobilization of fractured neck.

- **DO NOT** move him if the casualty is lying face down. Immobilize the head and neck by padding heavy objects and placing them on each side of his head. DO NOT put a roll of cloth under the neck. **DO NOT** bend the neck or head, nor roll the casualty onto his back.

b. If the casualty must be prepared for transportation before medical personnel arrive—

- If the casualty has a fractured neck, at least two persons are needed because the casualty's head and trunk must be moved in unison. The two persons must work in close coordination (Figure 4-32) to avoid bending of the neck.

- A wide board is placed lengthwise beside the casualty. It should extend at least 4 inches beyond the casualty's head and feet (Figure 4-32A).

- If the casualty is lying face up, the number one man steadies the casualty's head and neck between his hands. At the same time, the number two man positions one foot and one knee against the board to prevent it from slipping. He then grasps the casualty underneath his shoulder and hip and gently slides him onto the board (Figure 4-32B).

- If the casualty is lying face down, the number one man steadies the casualty's head and neck between his hands, while the number two man gently rolls the casualty over onto the board (Figure 4-32C).

- The number one man continues to steady the casualty's head and neck. The number two man simultaneously raises the casualty's shoulders slightly, places padding under his neck, and immobilizes the casualty's head (Figures 4-32D—E).

- Any improvised supports are secured in position with a cravat or strip of cloth extended across the casualty's forehead and under the board (Figure 4-32D).

- The board is lifted onto a litter or blanket in order to transport the casualty (Figure 4-32E).

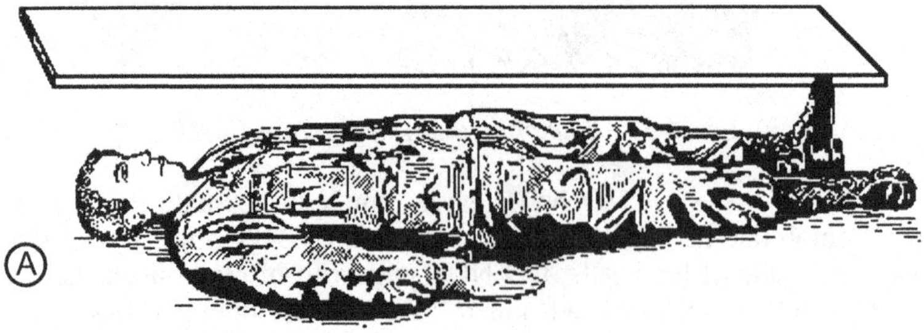

Figure 4-32. Preparing casualty with fractured neck for transportation (Illustrated A—E).

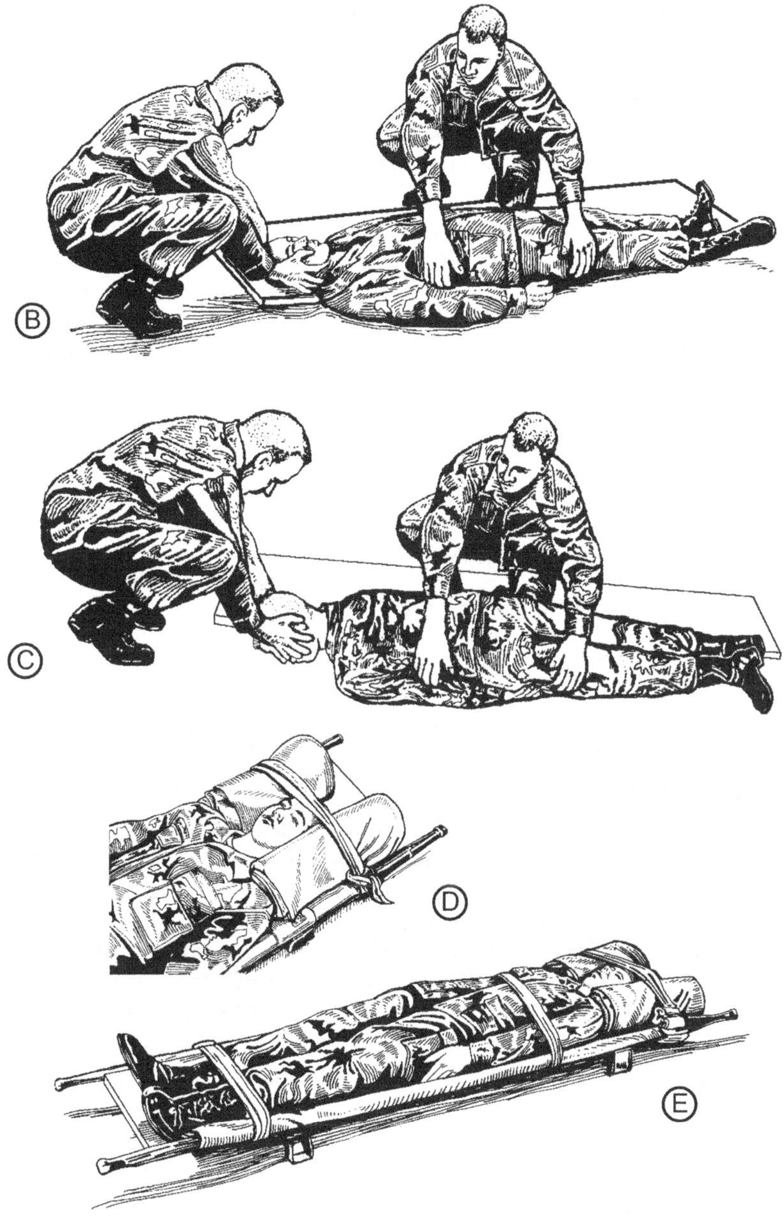

Figure 4-32. Preparing casualty with fractured neck for transportation (Illustrated A—E) (Continued).

CHAPTER 5

FIRST AID FOR CLIMATIC INJURIES

5-1. General

a. It is desirable, but not always possible, for an individual's body to become adjusted (acclimated) to an environment.

(1) The service members physical condition determines the amount of time their bodies need to adjust to the environment. Even those individuals in good physical condition need time before working or training in extremes of hot or cold weather. Climate-related injuries are usually preventable; prevention is both an individual and leadership responsibility.

(2) Several factors contribute to health and well-being in any environment—

- Diet.
- Sleep and rest.
- Exercise.
- Suitable clothing.

(3) Diet should be suited to an individual's needs in a particular climate. A special diet started for any purpose (such as weight reduction) should be done with appropriate medical supervision.

WARNING

Service members should use extreme caution when starting fad diets or taking over-the-counter herbal supplements. Medical records revealed that deaths and severe injuries occurred in individuals using dietary/ herbal supplements without medical monitoring.

NOTE

Weight loss and the use of weight loss supplements should be supervised by a trained health care provider.

(4) Specialized clothing and equipment (such as cold weather gear) for a specific environment should be obtained and used properly.

b. For information on the prevention of heat and cold injuries, refer to FM 21-10/Marine Corps Reference Publication (MCRP) 4-11.1D.

5-2. Heat Injuries

a. Heat injuries are environmental injuries. They may result when a service member—

- Is exposed to extreme heat, such as from the sun or from high temperatures.
- Does not wear proper clothing.
- Is in MOPP gear.
- Is inside closed spaces, such as inside an armored vehicle.
- Wears body armor.

b. Heat injury can be divided into three categories: heat cramps, heat exhaustion, and heatstroke.

c. Each service member must be able to recognize and give first aid for heat injuries.

WARNING

The heat casualty should be continually monitored for development of conditions which may require the performance of necessary basic lifesaving measures.

CAUTION

Do not use salt solutions in first aid procedures for heat injuries.

(1) Check the casualty for signs and symptoms of *cramping*.

- *Signs and symptoms.* Cramping is caused by an imbalance of chemicals (called electrolytes) in the body as a result of excessive sweating. This condition causes the casualty to exhibit:

- Cramping in the extremities (arms and legs).
- Abdominal (stomach) cramps.
- Excessive sweating.

NOTE

Thirst may or may not occur. Cramping can occur without the service member being thirsty.

- *First aid measures.*
  - Move the casualty to a cool, shady area or improvise shade if none is available.
  - Loosen his clothing (if not in a chemical environment).

NOTE

In a chemical environment, transport the heat casualty to a noncontaminated area as soon as the mission permits.

  - Have him slowly drink at least one canteen full of water. (The body absorbs cool water faster than warm or cold water; therefore, cool water is preferred if it is available.)
  - Seek medical assistance should cramps continue.

(2) Check the casualty for signs and symptoms of *heat exhaustion*.

- *Signs and symptoms.* Heat exhaustion is caused by loss of body fluids (dehydration) through sweating without adequate fluid replacement. It can occur in an otherwise fit individual who is involved in physical exertion in any hot environment especially if the service member is not acclimatized to that environment. These signs and symptoms are—
  - Excessive sweating with pale, moist, cool skin.
  - Headache.
  - Weakness.

- Dizziness.
- Loss of appetite.
- Cramping.
- Nausea (with or without vomiting).
- Urge to defecate.
- Chills (gooseflesh).
- Rapid breathing.
- Tingling of hands and/or feet.
- Confusion.

- *First aid measures.*

  - Move the casualty to a cool, shady area or improvise shade if none is available.
  - Loosen or remove his clothing and boots (unless in a chemical environment); pour water on him and fan him.
  - Have him slowly drink at least one canteen of water.
  - Elevate his legs.
  - If possible, the casualty should not participate in strenuous activity for the remainder of the day.
  - Monitor the casualty until the symptoms are gone, or medical assistance arrives.

(3) Check the casualty for signs and symptoms of *heatstroke*.

WARNING

Heatstroke is a medical emergency which may result in death if care is delayed.

- *Signs and symptoms.* A service member suffering from heatstroke has been exposed to high temperatures (such as direct sunlight) or been dressed in protective overgarments, which causes the body temperature to rise. Heatstroke occurs more rapidly in service members who are engaged in work or other physical activity in a high heat environment. Heatstroke is caused by a failure of the body's cooling mechanism which includes a decrease in the body's ability to produce sweat. The casualty's skin is red (flushed), hot, and dry. He may experience weakness, dizziness, confusion, headaches, seizures, nausea, stomach pains or cramps, and his respiration and pulse may be rapid and weak. Unconsciousness and collapse may occur suddenly.

- *First aid measures.* Cool casualty immediately by—

  - Moving him to a cool, shady area or improvising shade if none is available.

  - Loosening or removing his clothing (except in a chemical environment).

  - Spraying or pouring water on him; fanning him to permit the coolant effect of evaporation.

  - Massaging his extremities and skin, which increases the blood flow to those body areas, thus aiding the cooling process.

  - Elevating his legs.

  - Having him slowly drink at least one canteen full of water if he is conscious.

NOTE

Start cooling casualty immediately. Continue cooling while awaiting transportation and during transport to an MTF.

- *Medical assistance.* Seek medical assistance because the casualty should be transported to an MTF as soon as possible. Do not interrupt the cooling process or lifesaving measures to seek help; if someone else is present send them for help. The casualty should be continually monitored for development of conditions that may require the performance of necessary basic lifesaving measures.

d. *Table.* See Table 5-1 for further information.

Table 5-1. Heat Injuries

INJURIES	SIGNS AND SYMPTOMS	FIRST AID[1]
HEAT CRAMPS	THE CASUALTY EXPERIENCES MUSCLE CRAMPS OF THE ARMS, LEGS, AND/OR STOMACH. THE CASUALTY MAY ALSO HAVE EXCESSIVE SWEATING.	1. MOVE THE CASUALTY TO A COOL SHADY AREA OR IMPROVISE SHADE AND LOOSEN CLOTHING.[2] 2. HAVE HIM SLOWLY DRINK AT LEAST ONE CANTEEN FULL OF COOL WATER SLOWLY. 3. MONITOR THE CASUALTY AND GIVE HIM MORE WATER AS TOLERATED.
HEAT EXHAUSTION	THE CASUALTY EXPERIENCES HEAVY SWEATING WITH PALE, MOIST, COOL SKIN; HEADACHE, WEAKNESS, DIZZINESS, AND/OR LOSS OF APPETITE, HEAT CRAMPS, NAUSEA (WITH OR WITHOUT VOMITING), URGE TO DEFECATE, CHILLS (GOOSE-FLESH), RAPID BREATHING, CONFUSION, AND TINGLING OF THE HANDS AND/OR FEET.	1. MOVE THE CASUALTY TO A COOL, SHADY AREA OR IMPROVISE SHADE AND LOOSEN OR REMOVE HIS CLOTHING.[2] 2. POUR WATER ON HIM AND FAN HIM TO PERMIT THE COOLANT EFFECT OF EVAPORATION. 3. HAVE HIM SLOWLY DRINK AT LEAST ONE CANTEEN FULL OF COOL WATER. 4. ELEVATE THE CASUALTY'S LEGS. 5. SEEK MEDICAL ASSISTANCE IF SYMPTOMS CONTINUE; MONITOR UNTIL SYMPTOMS ARE GONE OR MEDICAL ASSISTANCE ARRIVES.
HEATSTROKE[3] (SUNSTROKE)	THE CASUALTY STOPS SWEATING (RED [FLUSHED] HOT, DRY SKIN). HE FIRST MAY EXPERIENCE HEADACHE, DIZZINESS, NAUSEA, FAST PULSE AND RESPIRATION, SEIZURES, AND MENTAL CONFUSION. HE MAY COLLAPSE	1. MOVE THE CASUALTY TO A COOL, SHADY AREA OR IMPROVISE SHADE AND LOOSEN OR REMOVE HIS CLOTHING, REMOVE THE OUTER GARMENTS AND PROTECTIVE

Table 5-1. Heat Injuries (Continued)

INJURIES	SIGNS AND SYMPTOMS	FIRST AID[1]
	AND SUDDENLY BECOME UNCONSCIOUS. **THIS IS A MEDICAL EMERGENCY.**	CLOTHING IF THE SITUATION PERMITS.[2] 2. START COOLING THE CASUALTY IMMEDIATELY. SPRAY OR POUR WATER ON HIM. FAN HIM. MASSAGE HIS EXTREMITIES AND SKIN. 3. ELEVATE HIS LEGS. 4. IF CONSCIOUS, HAVE HIM SLOWLY DRINK AT LEAST ONE CANTEEN FULL OF COOL WATER. 5. SEEK MEDICAL AID. CONTINUE COOLING WHILE AWAITING TRANSPORT AND CONTINUE FIRST AID WHILE EN ROUTE.

LEGEND:

1 THE FIRST AID PROCEDURE FOR HEAT RELATED INJURIES CAUSED BY WEARING INDIVIDUAL PROTECTIVE EQUIPMENT (IPE) IS TO MOVE THE CASUALTY TO A CLEAN AREA AND GIVE HIM WATER TO DRINK.
2 WHEN IN A CHEMICAL ENVIRONMENT, DO NOT LOOSEN OR REMOVE THE CASUALTY'S CLOTHING.
3 CAN BE FATAL IF NOT PROVIDED FIRST AID AND MEDICAL TREATMENT PROMPTLY.

5-3. Cold Injuries

Cold injuries are most likely to occur when conditions are moderately cold, but accompanied by wet or windy conditions. Cold injuries can usually be prevented. Well-disciplined and well-trained service members can be protected even in the most adverse circumstances. They and their leaders must know the hazards of exposure to the cold. They must know the importance of personal hygiene, exercise, care of the feet and hands, and the use of protective clothing.

a. Contributing Factors.

(1) Temperature, humidity, precipitation, and wind greatly increase likelihood of cold injuries, and the service members with wet clothing are at great risk of cold injuries. Riverine operations (river, swamp, and stream crossings) increase likelihood of cold injuries. Low temperatures and low relative humidity (dry cold) promote frostbite. Higher temperatures, together with moisture, promote immersion syndrome. Windchill accelerates the loss of body heat and may aggravate cold injuries.

(2) Relatively stationary activities such as being in an observation post or on guard duty increase the service member's vulnerability to cold injury. Also, a service member is more likely to receive a cold injury if he is—

- In contact with the ground (such as marching, performing guard duty, or engaging in other outside activities).
- Immobile for long periods (such as while riding in an unheated or open vehicle).
- Standing in water, such as in a foxhole.
- Out in the cold for days without being warmed.
- Deprived of an adequate diet and rest.
- Not able to take care of his personal hygiene.

(3) Physical fatigue contributes to apathy, which leads to inactivity, personal neglect, carelessness, and reduced heat production. In turn, these increase the risk of cold injury. Service members with prior cold injuries have a higher-than-normal risk of subsequent cold injury; not necessarily involving the body part previously injured.

(4) Depressed or unresponsive service members are also vulnerable because they are less active. These service members tend to be careless about precautionary measures, especially warming activities, when cold injury is a threat.

(5) Excessive use of alcohol or drugs leading to faulty judgment or unconsciousness in a cold environment increases the risk of becoming a cold injury casualty.

b. *Signs and Symptoms*. Once a service member becomes familiar with the factors that contribute to cold injury, he must learn to recognize cold injury signs and symptoms.

(1) Many service members suffer cold injury without realizing what is happening to them. They may be cold and generally uncomfortable. These service members often do not notice the injured part because it is already numb from the cold.

(2) Superficial cold injury usually can be detected by numbness or tingling sensations. These signs and symptoms often can be relieved simply by loosening boots or other clothing and by exercising to improve circulation. In more advanced cases involving deep cold injury, the service member often is not aware that there is a problem until the affected part feels like a stump or block of wood.

(3) Outward signs of cold injury include discoloration of the skin at the site of injury. In light-skinned persons, the skin first reddens and then becomes pale or waxy white. In dark-skinned persons, grayness in the skin is usually evident. An injured foot or hand feels cold to the touch. Swelling may be an indication of deep injury. Also note that blisters may occur after rewarming the affected parts. Service members should work in pairs (buddy teams) to check each other for signs of discoloration and other symptoms.

c. First Aid Measures. First aid for cold injuries depends on whether they are superficial or deep. Rewarming the affected part using body heat can adequately treat cases of superficial cold injury. (For example, this can be done by covering cheeks with hands, putting fingertips in armpits, or placing the casualty's feet under the clothing of a buddy [next to his belly].) The injured part should **NOT** be massaged, exposed to a fire or stove, rubbed with snow, slapped, chafed, or soaked in cold water. Walking on injured feet should be avoided. Deep cold injury (frostbite) is very serious and requires prompt first aid to avoid or to minimize the loss of parts or all of the fingers, toes, hands, or feet. The sequence for treating cold injuries depends on whether the condition is life-threatening. The first priority in managing cold injuries is to remove the casualty from the cold environment (such as building an improvised shelter). Other injuries the casualty may have are provided first aid simultaneously while waiting for transportation or evacuation. If the casualty is to be transported in a nonmedical vehicle, first aid measures should be continued en route to the MTF.

d. Conditions Caused by Cold. Conditions caused by cold include chilblain, immersion syndrome (immersion foot and trench foot), frostbite, snow blindness, dehydration, and hypothermia.

(1) *Chilblain.*

- *Signs and symptoms.* Chilblain is caused by repeated prolonged exposure of bare skin at temperatures from 60° Fahrenheit (F) to

32°F, or 20°F for acclimated, dry, unwashed skin. The area may be acutely swollen, red, tender, and hot with itchy skin. There may be no loss of skin tissue in untreated cases but continued exposure may lead to infected, ulcerated, or bleeding lesions.

- *First aid measures.* Within minutes, the area usually responds to locally applied body heat. Rewarm the affected part by applying firm steady pressure with your hands, or placing the affected part under your arms or against the stomach of a buddy. **DO NOT** rub or massage affected areas.

NOTE

Medical personnel should evaluate the injury, because signs and symptoms of tissue damage may be slow to appear.

(2) *Immersion syndrome (immersion foot and trench foot).* Immersion foot and trench foot are injuries that result from fairly long exposure of the feet to wet conditions at temperatures from approximately 32°F to 50°F. Inactive feet in damp or wet socks and boots, or tightly laced boots which impair circulation, are even more susceptible to injury. This injury can be very serious; it can lead to loss of toes or parts of the feet. If exposure of the feet has been prolonged and severe, the feet may swell so much that pressure closes the blood vessels and cuts off circulation. Should an immersion injury occur, dry the feet thoroughly and transport the casualty to an MTF by the fastest means possible.

- *Signs and symptoms.* At first, the parts of the affected foot are cold and painless, the pulse is weak, and numbness may be present. Second, the parts may feel hot, and burning and shooting pains may begin. In later stages, the skin is pale with a bluish cast and the pulse decreases. Other signs and symptoms that may follow are blistering, swelling, redness, heat, hemorrhaging (bleeding), and gangrene.

- *First aid measures.* First aid measures are required for all stages of immersion syndrome injury. Rewarm the injured part gradually by exposing it to warm air. Protect it from trauma and secondary infections. Dry, loose clothing or several layers of warm coverings are preferable to extreme heat. Under no circumstances should the injured part be exposed to an open fire. Elevate the injured part to relieve the swelling. Transport the casualty to an MTF as soon as possible. When the part is rewarmed, the casualty often feels a burning sensation and pain. Symptoms may persist for days or weeks even after rewarming.

NOTE

When providing first aid for immersion foot and trench foot—**DO NOT** massage the injured part. **DO NOT** moisten the skin. **DO NOT** apply heat or ice.

(3) *Frostbite*. Frostbite is the injury of tissue caused from exposure to cold, usually below 32°F depending on the windchill factor, duration of exposure, and adequacy of protection. Individuals with a history of cold injury are likely to suffer an additional cold injury. The body parts most easily frostbitten are the cheeks, nose, ears, chin, forehead, wrists, hands, and feet. Frostbite may involve only the skin (superficial), or it may extend to a depth below the skin (deep). Deep frostbite is very serious and requires prompt first aid to avoid or to minimize the loss of parts or all of the fingers, toes, hands, or feet.

- *Signs and symptoms.*
  - Loss of sensation (numb feeling) in any part of the body.
  - Sudden blanching (whitening) of the skin of the affected part, followed by a momentary tingling sensation.
  - Redness of skin in light-skinned service members; grayish coloring in dark-skinned service members.
  - Blisters.
  - Swelling or tender areas.
  - Loss of previous sensation of pain in affected area.
  - Pale, yellowish, waxy-looking skin.
  - Frozen tissue that feels solid (or wooden) to the touch.

CAUTION

Deep frostbite is a very serious injury and requires immediate first aid and subsequent medical treatment to avoid or minimize loss of body parts.

- *First aid measures.*

  - *Face, ears, and nose.* Cover the casualty's affected area with his and/or your bare hands until sensation and color return.

  - *Hands.* Open the casualty's field jacket and shirt. (In a chemical environment, do not loosen or remove the clothing and protective overgarments.) Place the affected hands under the casualty's armpits. Close the field jacket and shirt to prevent additional exposure.

  - *Feet.* Remove the casualty's boots and socks if he does not need to walk any further to receive additional treatment. (Thawing the casualty's feet and forcing him to walk on them will cause additional pain and injury.) Place the affected feet under clothing and against the body of another service member.

WARNING

DO NOT attempt to thaw the casualty's feet or other frozen areas if he will be required to walk or travel to an MTF for additional medical treatment. The possibility of additional injury from walking is less when the feet are frozen than when they are thawed. (However, if possible avoid walking.) Thawing in the field increases the possibilities of infection, gangrene, or other injury.

NOTE

Thawing may occur spontaneously during transportation to the MTF; this cannot be avoided since the body in general must be kept warm.

In all of the above areas, ensure that the casualty is kept warm and that he is covered (to avoid further injury). Seek medical treatment as soon as possible. Reassure the casualty, protect the affected area from further injury by covering it lightly with a blanket or any dry clothing, and seek shelter out of the wind. Remove or loosen constricting clothing (except in a contaminated environment) and increase insulation. Ensure the casualty exercises as much as possible, avoiding trauma to the injured part, and is prepared for pain when thawing occurs. Protect the frostbitten part from additional injury. **DO NOT**—

- Rub the injured part with snow or apply cold water soaks.

- Warm the part by massage or exposure to open fire because the frozen part may be burned due to the lack of feeling.

- Use ointments or other salves.

- Manipulate the part in any way to increase circulation.

- Use alcohol or tobacco because this reduces the body's resistance to cold.

NOTE

Remember, when freezing extends to a depth below the skin, it is a much more serious injury. Extra care is required to reduce or avoid the chances of losing all or part of the toes or feet. This also applies to the fingers and hands.

(4) *Snow blindness.* Snow blindness is the effect that glare from an ice field or snowfield has on the eyes. It is more likely to occur in hazy, cloudy weather than when the sun is shining. Glare from the sun will cause an individual to instinctively protect his eyes. However, in cloudy weather, he may be overconfident and expose his eyes longer than when the threat is more obvious. He may also neglect precautions such as the use of protective eyewear. Waiting until discomfort (pain) is felt before using protective eyewear is dangerous because a deep burn of the eyes may already have occurred.

- *Signs and symptoms.* Symptoms of snow blindness are a sensation of grit in the eyes with pain in and over the eyes, made worse by moving the eyeball. Other signs and symptoms are watering, redness, headache, and increased pain on exposure to light.

- *First aid measures.* First aid measures consist of blindfolding or covering the eyes with a dark cloth which stops painful eye movement. Complete rest is desirable. If further exposure to light is not preventable, the eyes should be protected with dark bandages or the darkest glasses available. Once unprotected exposure to sunlight stops, the condition usually heals in a few days without permanent damage. The casualty should be evacuated to the nearest MTF.

(5) *Dehydration.* Dehydration occurs when the body loses too much fluid. A certain amount of body fluid is lost through normal body processes. A normal daily intake of liquids replaces these losses. When individuals are engaged in any strenuous exercises or activities, fluid is lost

through sweating and this loss creates an imbalance of fluids in the body, and if not matched by rehydration it can contribute to dehydration. The danger of dehydration is as prevalent in cold regions as it is in hot regions. In hot weather, the individual is aware of his body losing fluids through sweat. In cold weather, however, it is extremely difficult to realize that this condition exists since sweating is not as apparent as in a hot environment. The danger of dehydration in cold weather operations is a serious problem. In cold climates, sweat evaporates so rapidly or is absorbed so thoroughly by layers of heavy clothing that it is rarely visible on the skin. Dehydration also occurs during cold weather operations because drinking is inconvenient. Dehydration will weaken or incapacitate a casualty for a few hours, or sometimes, several days. Because rest is an important part of the recovery process, casualties must take care that limited movement during their recuperative period does not enhance the risk of becoming a cold injury casualty.

- *Signs and symptoms.* The symptoms of cold weather dehydration are similar to those encountered in heat exhaustion. The mouth, tongue, and throat become parched and dry, and swallowing becomes difficult. The casualty may have nausea (with or without vomiting) along with extreme dizziness and fainting. The casualty may also feel generally tired and weak and may experience muscle cramps. Focusing the eyes may also become difficult.

- *First aid measures.* The casualty should be kept warm and his clothes should be loosened (if not in a chemical environment) to allow proper circulation. Shelter from wind and cold must be provided. Fluid replacement should begin immediately and the service member transported to an MTF as soon as possible.

(6) *Hypothermia (general cooling).* When exposed to prolonged cold weather a service member may become both mentally and physically numb, thus neglecting essential tasks or requiring more time and effort to achieve them. Under some conditions (particularly cold water immersion), even a service member in excellent physical condition may die in a matter of minutes. The destructive influence of cold on the body is called *hypothermia.* This means bodies lose heat faster than they can produce it. Hypothermia can occur from exposure to temperatures either above or below freezing, especially from immersion in cold water, wet-cold conditions, or from the effect of wind. Physical exhaustion and insufficient food intake may also increase the risk of hypothermia. General cooling of the entire body to a temperature below 95°F is caused by continued exposure to low or rapidly dropping temperatures, cold moisture, snow, or ice. Fatigue, poor physical condition, dehydration, faulty blood circulation, alcohol or other drug use, trauma, and immersion can cause hypothermia. Remember, cold

may affect the body systems slowly and almost without notice. Service members exposed to low temperatures for extended periods may suffer ill effects even if they are well protected by clothing.

- *Signs and symptoms*. As the body cools, there are several stages of progressive discomfort and impairment. A sign that is noticed immediately is shivering. Shivering is an attempt by the body to generate heat. The pulse is faint or very difficult to detect. People with temperatures around 90°F may be drowsy and mentally slow. Their ability to move may be hampered, stiff, and uncoordinated, but they may be able to function minimally. Their speech may be slurred. As the body temperature drops further, shock becomes evident as the person's eyes assume a glassy state, breathing becomes slow and shallow, and the pulse becomes weaker or absent. The person becomes very stiff and uncoordinated. Unconsciousness may follow quickly. As the body temperature drops even lower, the extremities freeze, and a deep (or core) body temperature (below 85°F) increases the risk of irregular heart action. This irregular heart action or heart standstill can result in sudden death.

- *First aid measures*. Except in cases of the most severe hypothermia (marked by coma or unconsciousness and a weak pulse), first aid measures for hypothermia are directed towards protecting the casualty from further loss of body heat. For the casualty who is conscious, first aid measures are directed at rewarming the body evenly and without delay. Provide heat by using a hot water bottle or field expedient or another service member's body heat.

CAUTION

DO NOT expose the casualty to an open fire, as he may become burned.

NOTE

When using a hot water bottle or field expedient (canteen filled with warm water), the bottle or canteen must be wrapped in cloth prior to placing it next to the casualty. This will reduce the chance of burning the casualty's skin.

Always call or send for help as soon as possible and protect the casualty immediately with dry clothing or a sleeping bag. Then, move him to a warm place. Evaluate other injuries and provide first aid as required. First aid measures can be performed while the casualty is waiting transportation or while he is en route. In the case of an accidental breakthrough into ice water, or other hypothermic accident, strip the casualty of wet clothing immediately and bundle him into a sleeping bag. Rescue breathing should be started at once if the casualty's breathing has stopped or is irregular or shallow. Warm liquids (**NOT HOT**) may be given gradually if the casualty is conscious. **DO NOT** force liquids on an unconscious or semiconscious casualty because he may choke. The casualty should be transported on a litter because the exertion of walking may aggravate circulation problems. Medical personnel should immediately treat any hypothermia casualty. Hypothermia is life threatening until normal body temperature has been restored. The first aid measures for a casualty with severe hypothermia are based upon the following principles: attempt to avoid further heat loss, handle the casualty gently, and transport the casualty as soon as possible to the nearest MTF. If at all possible, the casualty should be evacuated by medical personnel.

WARNING

Rewarming a severely hypothermic casualty is extremely dangerous in the field due to the possibility of such complications as rewarming, shock and disturbances in the rhythm of the heartbeat. These conditions require treatment by medical personnel.

NOTE

Resuscitation of casualties with hypothermic complications is difficult if not impossible to do outside of an MTF setting.

CAUTION

The casualty is unable to generate his own body heat. Therefore, merely placing him in a blanket or sleeping bag is not sufficient.

e. *Table.* See Table 5-2 for further information.

Table 5-2. Injuries Caused by Cold and Wet Conditions

INJURIES	SIGNS/SYMPTOMS	FIRST AID
CHILBLAIN	RED SWOLLEN, HOT, TENDER, ITCHING SKIN. CONTINUED EXPOSURE MAY LEAD TO INFECTED (ULCERATED OR BLEEDING) SKIN LESIONS.	1. AREA USUALLY RESPONDS TO LOCALLY APPLIED REWARMING (BODY HEAT). 2. DO NOT RUB OR MASSAGE AREA. 3. SEEK MEDICAL AID.
IMMERSION SYNDROME (IMMERSION FOOT/TRENCH FOOT)	AFFECTED PARTS ARE COLD, NUMB, AND PAINLESS. PARTS MAY THEN BE HOT, WITH BURNING AND SHOOTING PAINS. ADVANCED STAGE: SKIN PALE WITH BLUISH CAST; PULSE DECREASES; BLISTERING, SWELLING, HEAT, HEMORRHAGING, AND GANGRENE MAY FOLLOW.	1. GRADUAL REWARMING BY EXPOSURE TO WARM AIR. 2. DO NOT MASSAGE OR MOISTEN SKIN. 3. PROTECT AFFECTED PARTS FROM TRAUMA. 4. DRY FEET THOROUGHLY, AVOID WALKING. 5. SEEK MEDICAL AID.
FROSTBITE	LOSS OF SENSATION (NUMB FEELING) IN ANY PART OF THE BODY. SUDDEN BLANCHING (WHITENING) OF THE SKIN OF THE AFFECTED PART, FOLLOWED BY A MOMENTARY TINGLING SENSATION. REDNESS OF SKIN IN LIGHT-SKINNED SERVICE MEMBERS; GRAYISH COLORING IN DARK-SKINNED SERVICE MEMBERS. BLISTERS. SWELLING OR TENDER AREAS. LOSS OF PREVIOUS SENSATION OF PAIN IN THE AFFECTED AREA. PALE YELLOWISH, WAXY-LOOKING SKIN. FROZEN TISSUE THAT FEELS SOLID (WOODEN) TO THE TOUCH.	1. WARM THE AREA AT THE FIRST SIGN OF FROSTBITE, USING FIRM, STEADY PRESSURE OF THE HAND, UNDERARM, OR ABDOMEN. 2. FACE, EARS, NOSE: COVER AREA WITH HANDS (CASUALTY'S OWN OR BUDDY'S). 3. HANDS: OPEN FIELD JACKET AND PLACE CASUALTY'S HANDS AGAINST HIS BODY, THEN CLOSE THE JACKET TO PREVENT HEAT LOSS. 4. FEET: REMOVE THE CASUALTY'S BOOTS AND SOCKS AND PLACE HIS FEET AGAINST THE BODY OF ANOTHER SERVICE MEMBER.

Table 5-2. Injuries Caused by Cold and Wet Conditions (Continued)

INJURIES	SIGNS/SYMPTOMS	FIRST AID
		5. WARNING: DO NOT ATTEMPT TO THAW THE CASUALTY'S FEET OR OTHER FROZEN AREAS IF HE WILL BE REQUIRED TO WALK OR TRAVEL TO AN MTF FOR ADDITIONAL TREATMENT. THE POSSIBILITY OF INJURY FROM WALKING IS LESS WHEN THE FEET ARE FROZEN THAN WHEN THEY HAVE BEEN THAWED. (HOWEVER, IF POSSIBLE AVOID WALKING.) THAWING IN THE FIELD INCREASES THE POSSIBILITY OF INFECTION, GANGRENE, OR INJURY. 6. LOOSEN OR REMOVE CONSTRICTING CLOTHING AND REMOVE ANY JEWELRY. 7. INCREASE INSULATION (COVER WITH BLANKET OR OTHER DRY MATERIAL). ENSURE CASUALTY EXERCISES AS MUCH AS POSSIBLE, AVOIDING TRAUMA TO INJURED PART.
SNOW BLINDNESS	EYES MAY FEEL SCRATCHY. WATERING, REDNESS, HEADACHE, AND INCREASED PAIN WITH EXPOSURE TO LIGHT CAN OCCUR.	1. COVER THE EYES WITH A DARK CLOTH. 2. SEEK MEDICAL AID.
DEHYDRATION	SIMILAR TO HEAT EXHAUSTION (REFER TO TABLE 5-1).	1. KEEP WARM. 2. CASUALTY NEEDS FLUID REPLACEMENT, REST, AND PROMPT MEDICAL AID.
HYPOTHERMIA	CASUALTY IS COLD. SHIVERING. CORE TEMPERATURE IS LOW. CONSCIOUSNESS MAY BE	**MILD HYPOTHERMIA** 1. REWARM BODY

Table 5-2. Injuries Caused by Cold and Wet Conditions (Continued)

INJURIES	SIGNS/SYMPTOMS	FIRST AID
	ALTERED. UNCOORDINATED MOVEMENTS MAY OCCUR. SHOCK AND COMA MAY RESULT AS BODY TEMPERATURE DROPS.	EVENLY AND WITHOUT DELAY. (NEED TO PROVIDE HEAT SOURCE; CASUALTY'S BODY UNABLE TO GENERATE HEAT.) 2. KEEP DRY, PROTECT FROM THE ELEMENTS. 3. WARM (NOT HOT) LIQUIDS MAY BE GIVEN GRADUALLY (TO CONSCIOUS CASUALTIES ONLY). 4. BE PREPARED TO START BASIC LIFE SUPPORT MEASURES FOR THE CASUALTY. 5. SEEK MEDICAL TREATMENT IMMEDIATELY. **SEVERE HYPOTHERMIA** 1. STABILIZE THE TEMPERATURE. 2. ATTEMPT TO AVOID FURTHER HEAT LOSS. 3. HANDLE THE CASUALTY GENTLY. 4. EVACUATE TO THE NEAREST MTF AS SOON AS POSSIBLE. **5. WARNING: HYPOTHERMIA IS A MEDICAL EMERGENCY. PROMPT MEDICAL ATTENTION IS NECESSARY.**

CHAPTER 6

FIRST AID FOR BITES AND STINGS

6-1. General

Snakebites, insect bites, or stings can cause intense pain and/or swelling. If not treated promptly and correctly, they can cause serious illness or death. The severity of a snakebite depends upon: whether the snake is poisonous or nonpoisonous, the type of snake, the location of the bite, and the amount of venom injected. Bites from humans and other animals, such as dogs, cats, bats, raccoons, and rats, can cause severe bruises and infection and tears or lacerations of tissue. Awareness of the potential sources of injuries can reduce or prevent them from occurring. Knowledge and prompt application of first-aid measures can lessen the severity of injuries from bites and stings and keep the service member from becoming a serious casualty.

6-2. Types of Snakes

a. Nonpoisonous Snakes. There are approximately 130 different varieties of nonpoisonous snakes in the United States. They have oval-shaped heads and round eyes. Unlike poisonous snakes, discussed below, nonpoisonous snakes do not have fangs with which to inject venom. Figure 6-1 depicts the characteristics of a nonpoisonous snake.

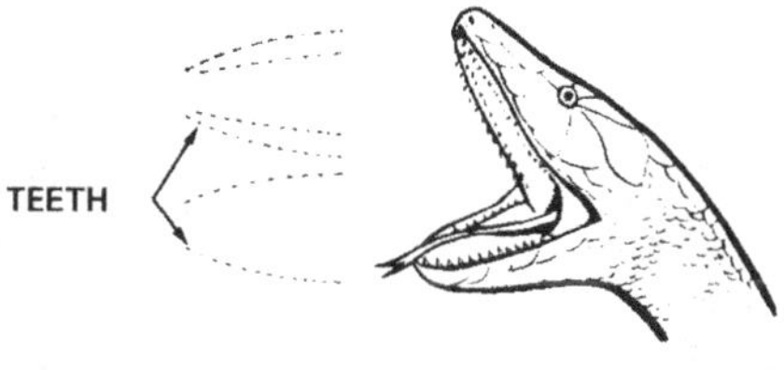

Figure 6-1. Characteristics of nonpoisonous snake.

b. Poisonous Snakes. Poisonous snakes are found throughout the world, primarily in tropical to moderate climates. Within the United States, there are four kinds: rattlesnakes, copperheads, water moccasins (cottonmouth), and coral snakes. Poisonous snakes in other parts of the world include sea snakes, the fer-de-lance, the bushmaster, and the tropical rattlesnake in tropical Central America; the Malayan pit viper in the tropical Far East; the cobra in Africa and Asia; the mamba (or black mamba) in central and southern Africa; and the krait in India and Southeast Asia. Refer to Figure 6-2 for characteristics of a poisonous pit viper.

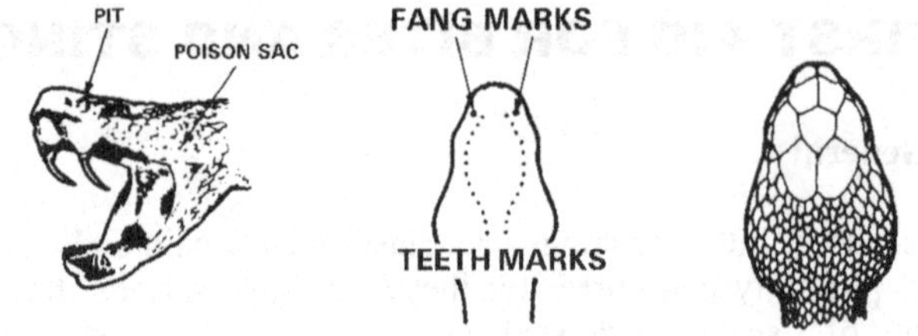

Figure 6-2. Characteristics of poisonous pit viper.

c. Pit Vipers (Poisonous). Figure 6-3 depicts a variety of poisonous snakes.

Figure 6-3. Poisonous snakes.

(1) Rattlesnakes, bushmasters, copperheads, fer-de-lance, Malayan pit vipers, and water moccasins (cottonmouth) are called pit vipers because of the small, deep pits between the nostrils and eyes on each side of the head (Figure 6-2). In addition to their long, hollow fangs, these snakes have other identifying features: thick bodies, slit-like pupils of the eyes, and flat, almost triangular-shaped heads. Color markings and other identifying characteristics, such as rattles or a noticeable white interior of the mouth (cottonmouth), also help distinguish these poisonous snakes. Further

identification is provided by examining the bite pattern of the wound for signs of fang entry. Occasionally there will be only one fang mark, as in the case of a bite on a finger or toe where there is no room for both fangs, or when the snake has broken off a fang.

(2) The casualty's condition provides the best information about the seriousness of the situation, or how much time has passed since the bite occurred. Pit viper bites are characterized by severe burning pain. Discoloration and swelling around the fang marks usually begins within 5 to 10 minutes after the bite. If only minimal swelling occurs within 30 minutes, the bite will almost certainly have been from a nonpoisonous snake or possibly from a poisonous snake which did not inject venom. The venom destroys blood cells, causing a general discoloration of the skin. Blisters and numbness in the affected area follow this reaction. Other signs, which can occur, are weakness, rapid pulse, nausea, shortness of breath, vomiting, and shock.

d. Corals, Cobras, Kraits, and Mambas. Corals (Figure 6-4), cobras (Figure 6-5), kraits, and mambas all belong to the same group even though they are found in different parts of the world. All four inject their venom through short, grooved fangs, leaving a characteristic bite pattern.

Figure 6-4. Coral snake.

(1) The small coral snake, found in the Southeastern US, is brightly colored with bands of red, yellow (or almost white), and black completely encircling the body. Other nonpoisonous snakes have the same coloring, but on the coral snake found in the US, the red ring always touches the yellow ring. To know the difference between a harmless snake and the coral snake found in the United States, remember the following:

"Red on yellow will kill a fellow,
Red on black, venom will lack."

Figure 6-5. Cobra snake.

(2) The venom of corals, cobras, kraits, and mambas produces symptoms different from those of pit vipers. Because there is only minimal pain and swelling, many people believe that the bite is not serious. Delayed reactions in the nervous system normally occur between 1 to 7 hours after the bite. Symptoms include blurred vision, drooping eyelids, slurred speech, drowsiness, and increased salivation and sweating. Nausea, vomiting, shock, respiratory difficulty, paralysis, convulsions, and coma will usually develop if the bite is not treated promptly.

e. Sea Snakes. Sea snakes (Figure 6-6) are found in the warm water areas of the Pacific and Indian oceans, along the coasts, and at the mouths of some larger rivers. Their venom is **VERY** poisonous, but their fangs are only 1/4 inch long. The first aid outlined for land snakes also applies to sea snakes.

Figure 6-6. Sea snake.

6-3. Snakebites

a. Poisonous snakes **DO NOT** always inject venom when they bite or strike a person. However, all snakes may carry tetanus (lockjaw); anyone bitten by a snake, whether poisonous or nonpoisonous, should immediately seek medical attention.

- Poison is injected from the venom sacs through grooved or hollow fangs. Depending on the species, these fangs are either long or short. Pit vipers have long hollow fangs. These fangs are folded against the roof of the mouth and extend when the snake strikes. This allows them to strike quickly and then withdraw. Cobras, coral snakes, kraits, mambas, and sea snakes have short, grooved fangs. These snakes are less effective in their attempts to bite, since they must chew after striking to inject enough venom (poison) to be effective. Figure 6-7 depicts the characteristics of a poisonous snakebite.

- In the event you are bitten, attempt to identify and/or kill the snake. Take it to medical personnel for inspection/identification. This provides valuable information to medical personnel who deal with snakebites. **TREAT ALL SNAKEBITES AS POISONOUS**.

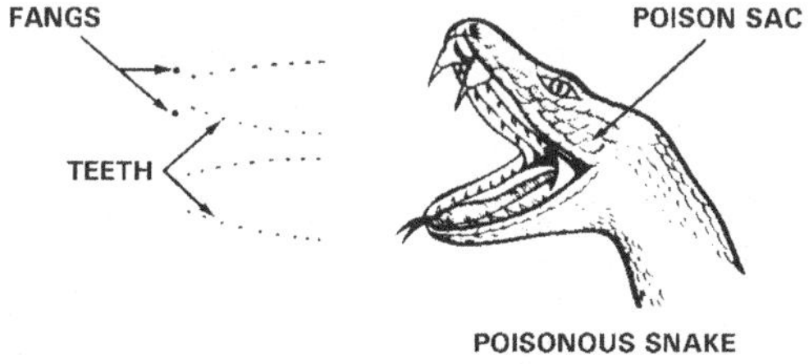

Figure 6-7. Characteristics of poisonous snakebite.

b. The venoms of different snakes cause different effects. Pit viper venom (hemotoxin [blood toxin]) destroys tissue and blood cells. Cobras, adders, and coral snakes inject powerful venom (neurotoxin [nerve toxin]) which affect the central nervous system, causing respiratory paralysis. Water moccasins and sea snakes have venom that is both hemotoxic and neurotoxic.

c. The identification of poisonous snakes is very important since medical treatment will be different for each type of venom. *Unless it can be positively identified, the snake should be killed and saved.* When this is not possible or when doing so is a serious threat to others, identification may sometimes be difficult since many venomous snakes resemble harmless varieties. When dealing with snakebite problems in foreign countries, seek advice, professional or otherwise, which may help identify species in the particular area of operations.

d. Get the casualty to an MTF as soon as possible and with minimum movement. Until evacuation or treatment is possible, have the casualty lie quietly and not move any more than necessary. If the casualty has been bitten on an extremity, **DO NOT** elevate the limb; keep the extremity level with the body. Keep the casualty comfortable and reassure him. If the casualty is alone when bitten, he should go to the medical facility himself rather than wait for someone to find him. Unless the snake has been positively identified, attempt to kill it and send it with the casualty. Be sure that retrieving the snake does not endanger anyone or delay transporting the casualty.

(1) If the bite is on an arm or leg, place a constricting band (narrow cravat [swathe], or narrow gauze bandage) one to two fingerbreadths above and below the bite (Figure 6-8). If the bite is on the hand or foot, place a single band above the wrist or ankle. The band should be tight enough to stop the flow of blood near the skin, but not tight enough to interfere with circulation. In other words, it should not have a tourniquet-like affect. If no swelling is seen, place the bands about 1 inch from either side of the bite. If swelling is present, put the bands on the unswollen part at the edge of the swelling. If the swelling extends beyond the band, move the band to the new edge of the swelling. (If possible, leave the old band on, place a new one at the new edge of the swelling, and then remove and save the old one in case the process has to be repeated.)

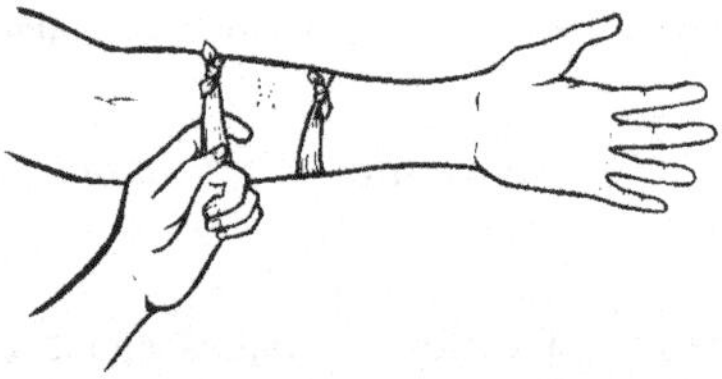

Figure 6-8. Constricting band.

CAUTION

DO NOT attempt to cut open the bite nor suck out the venom. If the venom should seep through any damaged or lacerated tissues in your mouth, you could immediately lose consciousness or even die.

(2) If the bite is located on an arm or leg, immobilize it at a level below the heart. **DO NOT** elevate an arm or leg even with or above the level of the heart.

> **CAUTION**
>
> When a splint is used to immobilize the arm or leg, take **EXTREME** care to ensure the splinting is done properly and does not bind. Watch it closely and adjust it if any changes in swelling occur.

(3) When possible, clean the area of the bite with soap and water. **DO NOT** use ointments of any kind.

(4) **NEVER** give the casualty food, alcohol, stimulants (coffee or tea), drugs, or tobacco.

(5) Remove rings, watches, or other jewelry from the affected limb.

6-4. Human or Animal Bites

Human or other land animal bites may cause lacerations or bruises. In addition to damaging tissue, bites always present the possibility of infection.

a. Human Bites. Human bites that break the skin may become seriously infected since the mouth is heavily contaminated with bacteria. Medical personnel **MUST** treat all human bites.

b. Animal Bites. Land animal bites can result in both infection and disease. Tetanus, rabies, and various types of fevers can follow an untreated animal bite. Because of these possible complications, the animal causing the bite should, if possible, be captured or killed (without damaging its head) so that it can be tested for disease.

c. First Aid.

(1) Cleanse the wound thoroughly with soap.

(2) Flush it well with water.

(3) Cover it with a sterile dressing.

(4) Immobilize the injured arm or leg, if appropriate.

(5) Transport the casualty immediately to an MTF.

NOTE

If unable to capture or kill the animal, provide medical personnel with any information that will help identify it.

6-5. Marine (Sea) Animals

With the exception of sharks and barracuda, most marine animals will not deliberately attack. The most frequent injuries from marine animals are wounds by biting, stinging, or puncturing. Wounds inflicted by marine animals can be very painful, but are rarely fatal.

a. Sharks, Barracuda, and Alligators. Wounds from these marine animals can involve major trauma as a result of bites and lacerations. Bites from large marine animals are potentially the most life threatening of all injuries from marine animals. Major wounds from these animals can be treated by controlling the bleeding, preventing shock, giving basic life support, splinting the injury, and by securing prompt medical aid.

b. Turtles, Moray Eels, and Corals. These animals normally inflict minor wounds. Treat by cleansing the wound(s) thoroughly and by splinting if necessary.

c. Jellyfish, Portuguese Man-of-War, Anemones, and Others. This group of marine animals inflict injury by means of stinging cells in their tentacles. Contact with the tentacles produces burning pain with a rash and small hemorrhages on the skin. Shock, muscular cramping, nausea, vomiting, and respiratory distress may also occur. Gently remove the clinging tentacles with a towel and wash or treat the area. Use diluted ammonia or alcohol, meat tenderizer, and talcum powder. If symptoms become severe or persist, seek medical assistance.

d. Spiny Fish, Urchins, Stingrays, and Cone Shells. These animals inject their venom by puncturing the skin with their spines. General signs and symptoms include swelling, nausea, vomiting, generalized cramps, diarrhea, muscular paralysis, and shock. Deaths are rare. Treatment consists of soaking the wounds in hot water (when available) for 30 to 60 minutes. This inactivates the heat sensitive toxin. In addition, further first aid measures (controlling bleeding, applying a dressing, and so forth) should be carried out as necessary.

CAUTION

Be careful not to scald the casualty with water that is too hot because the pain of the wound will mask the normal reaction to heat.

6-6. Insect (Arthropod) Bites and Stings

An insect bite or sting can cause great pain, allergic reaction, inflammation, and infection. If not treated correctly, some bites/stings may cause serious illness or even death. When an allergic reaction is not involved, first aid is a simple process. In any case, medical personnel should examine the casualty at the earliest possible time. It is important to properly identify the spider, bee, or creature that caused the bite/sting, especially in cases of allergic reaction.

a. *Types of Insects*. The insects found throughout the world that can produce a bite or sting are too numerous to mention in detail. Commonly encountered stinging or biting insects include brown recluse spiders (Figure 6-9), black widow spiders (Figure 6-10), tarantulas (Figure 6-11), scorpions (Figure 6-12), urticating caterpillars, bees, wasps, centipedes, conenose beetles (kissing bugs), ants, and wheel bugs. Upon being reassigned, especially to overseas areas, take the time to become acquainted with the types of insects to avoid.

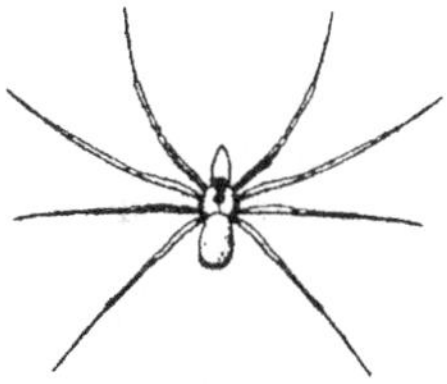

Figure 6-9. Brown recluse spider.

Figure 6-10. Black widow spider.

Figure 6-11. Tarantula.

Figure 6-12. Scorpion.

b. Signs and Symptoms. Discussed in paragraphs (1) and (2) below are the most common effects of insect bites/stings. They can occur alone or in combination with the others.

(1) *Less serious.* Commonly seen signs/symptoms are pain, irritation, swelling, heat, redness, and itching. Hives or wheals (raised areas of the skin that itch) may occur. These are the least severe of the allergic reactions that commonly occur from insect bites/stings. They are usually dangerous only if they affect the air passages (mouth, throat, nose, and so forth), which could interfere with breathing. The bites/stings of bees, wasps, ants, mosquitoes, fleas, and ticks are usually not serious and normally produce mild and localized symptoms. A tarantula's bite is usually no worse than that of a bee sting. Scorpions are rare and their stings (except for a specific species found only in the Southwest desert) are painful but usually not dangerous.

(2) *Serious.* Emergency allergic or hypersensitive reactions sometimes result from the stings of bees, wasps, and ants. Many people are allergic to the venom of these particular insects. Bites or stings from these insects may produce more serious reactions, to include generalized itching and hives, weakness, anxiety, headache, breathing difficulties, nausea, vomiting, and diarrhea. Very serious allergic reactions (called *anaphylactic shock*) can lead to complete collapse, shock, and even death. Spider bites (particularly from the black widow and brown recluse spiders) can also be

serious. Venom from the black widow spider affects the nervous system. This venom can cause muscle cramps, a rigid, nontender abdomen, breathing difficulties, sweating, nausea, and vomiting. The brown recluse spider generally produces local rather than system-wide problems; however, local tissue damage around the bite can be severe and can lead to an ulcer and even gangrene.

c. *First Aid.* There are certain principles that apply regardless of what caused the bite/sting. Some of these are—

- If there is a stinger present (for example, from a bee), remove the stinger by scraping the skin's surface with a fingernail or knife. **DO NOT** squeeze the sac attached to the stinger because it may inject more venom.

- Wash the area of the bite/sting with soap and water (alcohol or an antiseptic may also be used) to help reduce the chances of an infection and remove traces of venom.

- Remove jewelry from bitten extremities because swelling may occur.

- In most cases of insect bites the reaction will be mild and localized; use ice or cold compresses (if available) on the site of the bite/ sting. This will help reduce swelling, ease the pain, and slow the absorption of venom. Meat tenderizer (to neutralize the venom) or calamine lotion (to reduce itching) may be applied locally. If necessary, seek medical assistance.

- In more serious reactions (severe and rapid swelling, allergic symptoms, and so forth) treat the bite/sting like you would treat a snakebite; that is, apply constricting bands above and below the site.

- Be prepared to perform basic life-support measures, such as rescue breathing.

- Reassure the casualty and keep him calm.

- In serious reactions, attempt to capture the insect for positive identification; however, be careful not to become a casualty yourself.

- If the reaction to the bite/sting appears serious, seek medical assistance.

WARNING

Insect bites/stings may cause *anaphylactic shock* (a shock caused by a severe allergic reaction). This is a life-threatening event and a TRUE MEDICAL EMERGENCY. Be prepared to perform the basic life-support measures and to immediately transport the casualty to an MTF.

NOTE

Be aware that some allergic or hypersensitive individuals may carry identification or emergency insect bite treatment kits. If the casualty is having an allergic reaction and has such a kit, administer the medication in the kit according to the instructions which accompany the kit.

d. Supplemental Information. For additional information concerning biting insects, see FM 21-10.

6-7. First Aid for Bites and Stings

See the table below for information on bites and stings.

Table 6-1. First Aid Measures for Bites and Stings

TYPES	FIRST AID MEASURES
SNAKEBITE	1. MOVE CASUALTY AWAY FROM THE SNAKE. 2. REMOVE JEWELRY FROM THE AFFECTED AREA, IF APPLICABLE. 3. REASSURE CASUALTY AND KEEP HIM QUIET. 4. APPLY CONSTRICTING BAND, 1-2 FINGERBREADTHS FROM THE BITE. YOU SHOULD BE ABLE TO INSERT A FINGER BETWEEN THE BAND AND THE SKIN. a. *ARM OR LEG BITE.* PLACE ONE BAND ABOVE AND ONE BAND BELOW THE BITE SITE. b. *HAND OR FOOT BITE.* PLACE ONE BAND ABOVE THE WRIST OR ANKLE.

Table 6-1. First Aid Measures for Bites and Stings

TYPES	FIRST AID MEASURES
	5. IMMOBILIZE THE AFFECTED PART IN A POSITION BELOW THE LEVEL OF THE HEART. 6. KILL THE SNAKE (IF POSSIBLE, WITHOUT DAMAGING ITS HEAD OR ENDANGERING YOURSELF) AND SEND IT WITH THE CASUALTY. 7. SEEK MEDICAL ASSISTANCE IMMEDIATELY.
BROWN RECLUSE SPIDER OR BLACK WIDOW SPIDER BITE	1. KEEP CASUALTY QUIET. 2. REMOVE ALL JEWELRY FROM AFFECTED PART, IF APPLICABLE. 3. WASH THE AREA. 4. APPLY ICE OR FREEZE PACK, IF AVAILABLE. 5. SEEK MEDICAL ASSISTANCE.
TARANTULA BITE OR SCORPION STING OR ANT BITE	1. WASH THE AREA. 2. REMOVE ALL JEWELRY FROM AFFECTED PART, IF APPLICABLE. 3. APPLY ICE OR FREEZE PACK, IF AVAILABLE. 4. APPLY BAKING SODA, CALAMINE LOTION, OR MEAT TENDERIZER (IF AVAILABLE) TO BITE SITE TO RELIEVE PAIN AND ITCHING. 5. IF THE SITE OF THE BITE IS ON THE FACE, NECK (POSSIBLE AIRWAY PROBLEMS), OR GENITAL AREA, OR IF LOCAL REACTION SEEMS SEVERE, OR IF THE STING IS BY THE DANGEROUS TYPE OF SCORPION FOUND IN THE SOUTHWEST UNITED STATES DESERT, KEEP THE CASUALTY AS QUIET AS POSSIBLE. SEEK MEDICAL ASSISTANCE.
BEE STING	1. IF THE STINGER IS PRESENT, REMOVE BY SCRAPING WITH A KNIFE OR FINGERNAIL. DO NOT SQUEEZE VENOM SAC ON STINGER; MORE VENOM MAY BE INJECTED. 2. REMOVE ALL JEWELRY FROM AFFECTED PART, IF APPLICABLE.

Table 6-1. First Aid Measures for Bites and Stings

TYPES	FIRST AID MEASURES
	3. WASH THE AREA. 4. APPLY ICE OR FREEZE PACK, IF AVAILABLE. 5. IF ALLERGIC SIGNS OR SYMPTOMS APPEAR, BE PREPARED TO PERFORM BASIC LIFE SUPPORT MEASURES. SEEK IMMEDIATE MEDICAL ASSISTANCE.

CHAPTER 7

FIRST AID IN A NUCLEAR, BIOLOGICAL, AND CHEMICAL ENVIRONMENT

7-1. General

American forces have not been exposed to NBC weapons/agents on the battlefield since World War I. In future conflicts and wars we can expect the use of such agents. Nuclear, biological, and chemical weapons will rapidly degrade unit effectiveness by forcing troops to wear protective clothing and by creating confusion and fear. Through training in protective procedures and first aid, units can maintain their effectiveness on the integrated battlefield.

7-2. First Aid Materials

You may be issued the following materials to protect, decontaminate, and use as first aid for NBC exposure. You must know how to use the items; some items are described in *a* through *d* below. It is equally important that you know when to use them.

a. Nerve Agent Pyridostigmine Pretreatment (NAPP). You may be issued a blister pack of pretreatment tablets when your commander directs. The NAPP is a pretreatment; it is not an antidote. It improves the effectiveness of the nerve agent antidote. When ordered to take the pretreatment you must take one tablet every 8 hours, mission permitting. This must be taken prior to exposure to nerve agents, since it may take several hours to develop adequate blood levels.

NOTE

Commanders must follow investigational new drug protocols for use of the NAPP.

b. M291 Skin Decontaminating Kit. The M291 Skin Decontaminating Kit (Figure 7-1) contains six packets of XE-555 decontaminant resin.

WARNING

For external use only. May be slightly irritating to the eyes. Keep decontaminating powder out of eyes. Use water to wash toxic agent out of eyes.

c. Nerve Agent Antidote Kit, MARK I. Each service member is issued three MARK Is for use in first aid for nerve agent poisoning (Figure 7-2 and paragraph 7-6).

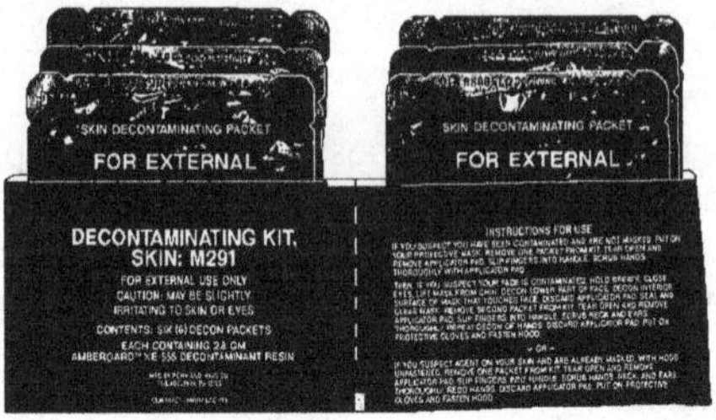

Figure 7-1. M291 Skin Decontamination Kit.

d. Antidote Treatment, Nerve Agent, Autoinjector. A new nerve agent antidote injection device, Antidote Treatment, Nerve Agent, Autoinjector (ATNAA) is currently under development that will replace the MARK I. The ATNAA is a multichambered device with the atropine and pralidoxime chloride in separate chambers. Both antidotes will be administered through a single needle.

7-3. Classification of Chemical and Biological Agents

a. Chemical agents are classified according to the primary physiological effects they produce, such as blistering, choking, vomiting, and incapacitating agents.

b. Biological warfare agents are classified according to the effect they have on man. The effects include their ability to incapacitate and cause death. Most biological warfare agents are delivered as aerosols that effect the respiratory tract; some can be delivered by releasing infected insects, by contaminating food and water, and by injection (injecting material in individuals by terrorist, not mass exposure). These agents are found in living organisms such as fungi, bacteria, and viruses.

WARNING

Swallowing water or food contaminated with nerve, blister, and other chemical agents and with some biological agents can be fatal. NEVER consume water or food that is suspected of being contaminated until it has been tested and found safe for consumption by medical personnel.

7-4. Conditions for Masking Without Order or Alarm

a. Once an attack with a chemical or biological agent is detected or suspected, or information is available that such an agent is about to be used, you must **STOP BREATHING** and mask immediately. **DO NOT WAIT** to receive an order or alarm under the following circumstances:

- Your position is hit by artillery missiles, rockets that produce vapors, smoke, and mists, and aerial sprays.
- Smoke or vapor cloud from an unknown source is present or approaching.
- A suspicious odor, liquid, or solid is present.
- A chemical or biological warfare agent attack is occurring.
- You are entering an area known or suspected of being contaminated.
- When casualties are being received from an area where chemical or biological agents have reportedly been used.
- You have one or more of the following symptoms:
  - An unexplained runny nose.
  - A sudden unexplained headache.
  - A feeling of choking or tightness in the chest or throat.
  - Dimness of vision.
  - Irritation of the eyes.
  - Difficulty in or increased rate of breathing without obvious reasons.
  - Sudden feeling of depression.
  - Dread, anxiety, or restlessness.
  - Dizziness or light-headedness.

- Slurred speech.
- Unexplained laughter or unusual behavior is noted in others.
- Numerous unexplained ill personnel.
- Service members suddenly collapsing without evident cause.
- Animals or birds exhibiting unusual behavior or suddenly dying.

b. For further information on protection and masking procedures, refer to FM 3-4, FM 4-02.7, FM 8-284, and FM 8-285.

7-5. First Aid for a Chemical Attack

Your field protective mask gives protection against biological and chemical warfare agents as well as radiological fallout. With practice you can mask in 9 seconds or less, or put on your mask with hood within 15 seconds.

a. Stop breathing. Don your mask, seal it properly, and clear and check it; then resume breathing. Give the alarm, and continue the mission. Keep your mask on until the "all clear" signal has been given.

NOTE

Keep your mask on until the area is no longer hazardous and you are told to unmask.

b. If symptoms of nerve agent poisoning (paragraph 7-7) appear, immediately give yourself one MARK I or ATNAA.

CAUTION

Do not inject a nerve agent antidote until you are sure you need it.

c. If your eyes and face become contaminated, you must immediately try to get under cover. You need shelter to prevent further contamination while performing decontamination procedures on your face. If no overhead cover is available, put your poncho over your head before beginning the decontamination process. Then you put on the remaining protective clothing. If vomiting occurs, the mask should be lifted momentarily and drained—with your eyes closed and while holding your breath—then replaced, cleared, and sealed.

d. If nerve agents are used, mission permitting, watch for persons needing nerve agent antidotes and immediately follow procedures outlined in paragraph 7-8*b* or *c*.

e. Decontaminate your skin immediately and clothing and equipment as soon as the mission permits.

7-6. Background Information on Nerve Agents

a. Nerve agents are among the deadliest of chemical agents. Nerve agents enter the body by inhalation, by ingestion, and through the skin. Depending on the route of entry and the amount, nerve agents can produce injury or death within minutes. Nerve agents can achieve their effects with small amounts. Nerve agents are absorbed rapidly, and the effects are felt immediately upon entry into the body. You will be issued three MARK Is or three ATNAAs and one Convulsant Antidote for Nerve Agent (CANA). Each MARK I consists of one atropine autoinjector and one pralidoxime chloride (2 PAM Cl) autoinjector (Figure 7-2A). Each ATNAA consist of a multichambered autoinjector with the atropine and pralidoxime chloride in separate chambers (Figure 7-2C). The CANA is a single autoinjector with flanges (Figure 7-2B). Procedures for use of both the MARK I and ATNAA are described below. You will use either the MARK I or the ATNAA in self-aid and buddy aid as issued.

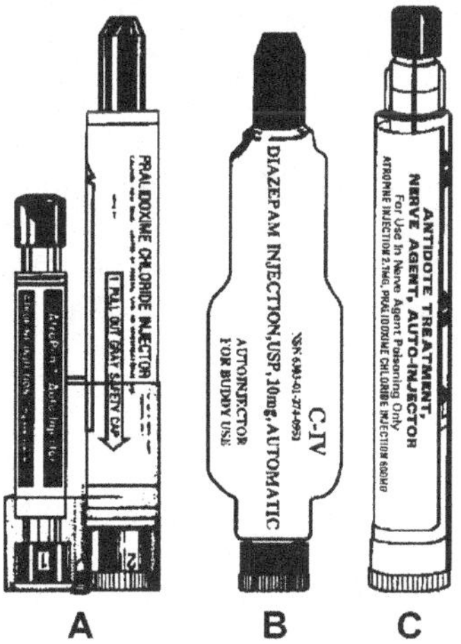

Figure 7-2. Nerve Agent Antidote Kit, MARK I, CANA, and ATNAA.

b. When you have the signs and symptoms of nerve agent poisoning, you should immediately put on the protective mask and then inject yourself with one set of the MARK I or ATNAA. Do not administer the CANA. You should inject yourself in the outer (lateral) thigh muscle (Figure 7-3) or if you are thin, in the upper outer (lateral) part of the buttocks (Figure 7-4).

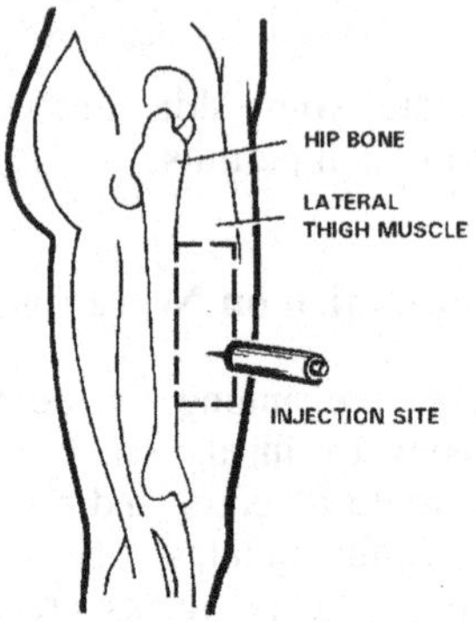

Figure 7-3. Thigh injection site.

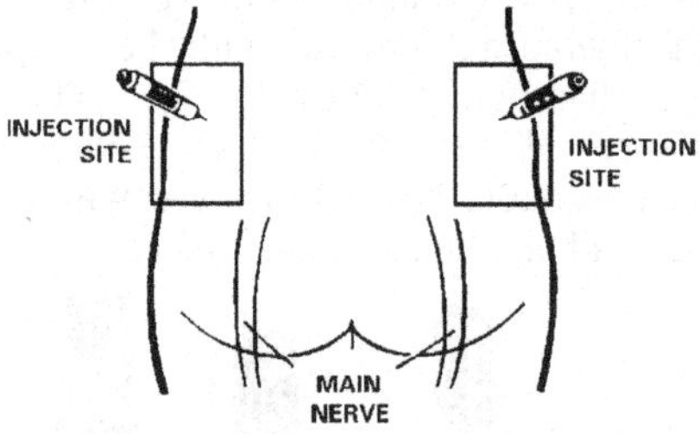

Figure 7-4. Buttocks injection site.

c. Also, you may come upon an unconscious chemical agent casualty who will be unable to care for himself and who will require first aid. You should be able to successfully—

(1) Mask him if he is unmasked.

(2) Inject him, if necessary, with all of **HIS** autoinjectors.

(3) Decontaminate his skin.

(4) Seek medical assistance.

7-7. Signs and Symptoms of Nerve Agent Poisoning

The symptoms of nerve agent poisoning are grouped as **MILD**—those that you recognize and for which you can perform self-aid, and **SEVERE**—those which require buddy aid.

a. MILD Signs and Symptoms.

- Unexplained runny nose.
- Unexplained sudden headache.
- Sudden drooling.
- Difficulty seeing (dimness of vision and miosis).
- Tightness in the chest or difficulty in breathing.
- Localized sweating and muscular twitching in the area of contaminated skin.
- Stomach cramps.
- Nausea.
- Tachycardia followed by bradycardia. (*Tachycardia* is an abnormally rapid heartbeat with a heart rate of over 100 beats per minute. *Bradycardia* is a slow heart rate of less than 60 beats per minute.)

b. SEVERE Signs and Symptoms.

- Strange or confused behavior.
- Wheezing, dyspnea (difficulty in breathing), and coughing.
- Severely pinpointed pupils.
- Red eyes with tearing.
- Vomiting.
- Severe muscular twitching and general weakness.
- Involuntary urination and defecation.

- Convulsions.
- Unconsciousness.
- Respiratory failure.
- Bradycardia.

7-8. First Aid for Nerve Agent Poisoning

First aid for nerve agent poisoning consists of administering the MARK I or ATNAA and CANA.

a. Injection Site. The injection site for administering the antidotes is normally in the outer thigh muscle. The thigh injection site is the area about a hand's width above the knee to a hand's width below the hip joint (Figure 7-3). It is important that the injection be given into a large muscle area. If the individual is thinly built, then the injections should be administered into the upper outer quarter (quadrant) of the buttock (Figure 7-4). Injecting in the buttocks of a thinly built individual avoids injury to the thighbone.

b. Self-Administer MARK I. If you experience any or all of the nerve agent **MILD** symptoms (paragraph 7-7*a*), you must **IMMEDIATELY** put on your protective mask and self-administer one MARK I (Figure 7-2A). Follow the procedure given in Table 7-1. The MARK I is carried in your protective mask carrier, pocket of the MOPP overgarment, or other location as specified in your unit tactical standing operating procedure (TSOP). (In cold weather, the MARK I should be stored in an inside pocket of your clothing to protect the antidote from freezing. A frozen MARK I cannot be immediately used to provide you with antidote, when needed. (However, the MARK I can still be used after complete thawing.)

Table 7-1. Self Aid for Nerve Agent Poisoning

MARK I*	ATNAA*
STEP 1. OBTAIN ONE MARK I.**	STEP 1. OBTAIN ONE ATNAA.**
STEP 2. CHECK INJECTION SITE.	STEP 2. CHECK INJECTION SITE.
STEP 3. HOLD MARK I AT EYE LEVEL WITH NONDOMINANT HAND WITH THE LARGE INJECTOR ON TOP (FIGURE 7-5A).	STEP 3. HOLD ATNAA WITH DOMINANT HAND (FIGURE 7-12A).

Table 7-1. Self Aid for Nerve Agent Poisoning (Continued)

MARK I*	ATNAA*
STEP 4. GRASP SMALL INJECTOR (ATROPINE) (FIGURE 7-5B) AND REMOVE FROM CLIP (FIGURE 7-5C).	STEP 4. GRASP SAFETY CAP WITH NONDOMINANT HAND AND REMOVE FROM INJECTOR (FIGURE 7-12B).
STEP 5. CLEAR HARD OBJECTS FROM INJECTION SITE.	STEP 5. CLEAR HARD OBJECTS FROM INJECTION SITE.
STEP 6. INJECT ATROPINE AT INJECTION SITE APPLYING EVEN PRESSURE TO THE INJECTOR (FIGURE 7-6 OR 7-7). HOLD IN PLACE FOR 10 SECONDS.	STEP 6. INJECT ATNAA AT INJECTION SITE APPLYING EVEN PRESSURE TO THE INJECTOR (FIGURE 7-14 OR 7-15). HOLD IN PLACE FOR 10 SECONDS.
STEP 7. HOLD USED INJECTOR WITH NONDOMINANT HAND.	STEP 7. BEND NEEDLE OF USED INJECTOR BY PRESSING ON A HARD SURFACE TO FORM A HOOK.
STEP 8. GRASP THE LARGE (2 PAM CI) INJECTOR (FIGURE 7-8B) AND PULL IT FROM CLIP (FIGURE 7-8C). DROP CLIP TO GROUND.	STEP 8. ATTACH USED INJECTOR TO BLOUSE POCKET FLAP OF BDO/JSLIST (FIGURE 7-16).
STEP 9. INJECT 2 PAM CI AT INJECTION SITE APPLYING EVEN PRESSURE TO THE INJECTOR (FIGURE 7-6 OR 7-7). HOLD IN PLACE FOR 10 SECONDS.	STEP 9. MASSAGE INJECTION SITE, MISSION PERMITTING.
STEP 10. BEND THE NEEDLES OF ALL USED INJECTORS BY PRESSING ON A HARD SURFACE TO FORM A HOOK.	
STEP 11. ATTACH ALL USED INJECTORS TO BLOUSE POCKET FLAP OF BDO/JSLIST (FIGURE 7-9).	
STEP 12. MASSAGE INJECTION SITE, MISSION PERMITTING.	

* USE STEPS LISTED FOR TYPE OF ANTIDOTE DEVICE ISSUED.

** ONLY ADMINISTER ONE MARK I OR ATNAA AS SELF-AID. DO NOT SELF-ADMINISTER CANA.

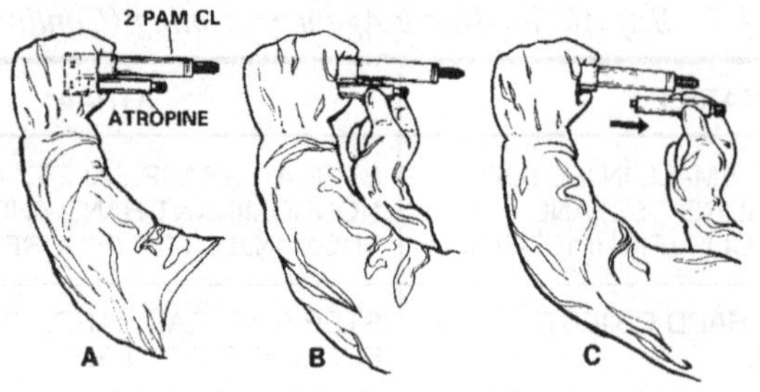

Figure 7-5. Removing the atropine autoinjector from the MARK I clip.

CAUTION

DO NOT cover or hold the needle end with your hand, thumb, or fingers—you might accidentally inject yourself. An accidental injection into the hand **WILL NOT** deliver an effective dose of the antidote, especially if the needle goes through the hand.

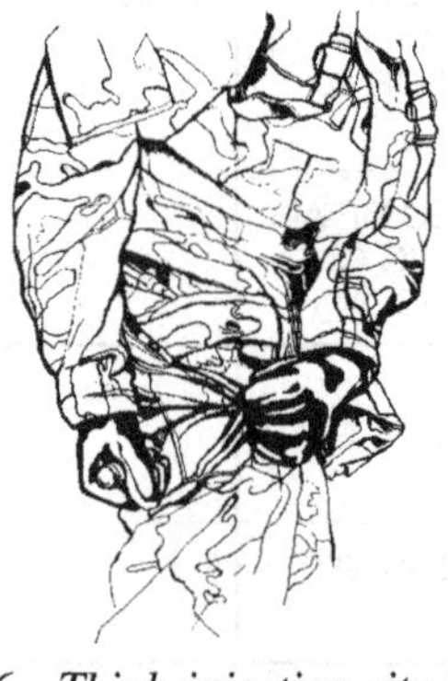

Figure 7-6. Thigh injection site for self-aid.

NOTE

If you are thinly built, inject yourself into the upper outer quadrant of the buttock (Figure 7-7). There is a nerve that crosses the buttocks; hitting this nerve can cause paralysis. Therefore, you must only inject into the *upper outer quadrant* of the buttock.

Figure 7-7. Buttocks injection site for self-aid.

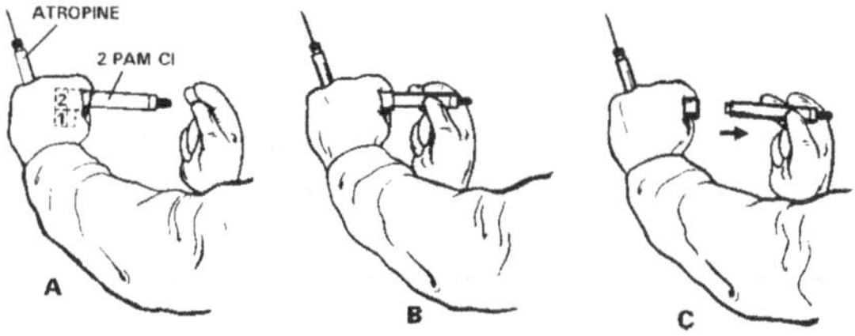

Figure 7-8. Removing the 2 PAM Cl autoinjector from the MARK I clip.

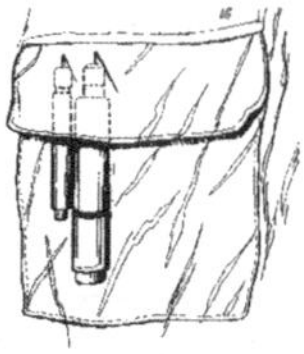

Figure 7-9. One set of used MARK I autoinjectors attached to pocket flap.

NOTES

1. **DO NOT** give yourself another set of injections. If you are able to walk without assistance, know who you are, and where you are, you **WILL NOT** need the second set of injections. (If not needed, giving yourself a second set of MARK I injections or ATNAA may create a nerve agent antidote overdose, which could cause incapacitation [inability to perform mission or defend yourself].)

2. If you continue to have symptoms of nerve agent poisoning, seek someone else (a buddy) to check your symptoms and administer the additional sets of injections, if required.

c. Buddy Evaluation and Buddy Aid. Service members may seek assistance after self-aid (self-administering one MARK I or ATNAA) or may become incapacitated after self-aid. A buddy must evaluate the individual to determine if additional antidotes are required to counter the effects of the nerve agent. Also, service members may experience **SEVERE** symptoms of nerve agent poisoning (paragraph 7-7*b*); they will not be able to treat themselves. In either case, other service members must perform buddy aid as quickly as possible. Before initiating buddy aid, determine if one set of MARK I autoinjectors has already been used so that no more than three sets of the antidote are administered. Buddy aid also includes administering the CANA with the third MARK I or ATNAA to prevent convulsions. Follow the procedures indicated in Table 7-2.

WARNING

Squat, DO NOT kneel, when masking the casualty or administering the nerve agent antidote to the casualty. Kneeling may force the chemical agent into or through your protective clothing.

CAUTION

DO NOT use your own MARK I, ATNAA, or CANA on a casualty. If you use your own, you may not have any antidote if needed for self-aid.

WARNING

DO NOT inject into areas close to the hip, knee, or thigh-bone.

Table 7-2. Buddy Aid/Combat Lifesaver Aid for Nerve Agent Casualty.

MARK I*	ATNAA*	CANA**
STEP 1. MASK THE CASUALTY AND POSITION HIM ON HIS SIDE (SWIMMER'S POSITION).	STEP 1. MASK THE CASUALTY AND POSITION HIM ON HIS SIDE (SWIMMER'S POSITION).	STEP 1. OBTAIN BUDDY'S CANA.
STEP 2. POSITION YOURSELF NEAR THE CASUALTY'S THIGH.	STEP 2. POSITION YOURSELF NEAR THE CASUALTY'S THIGH.	STEP 2. CHECK INJECTION SITE.

Table 7-2. Buddy Aid/Combat Lifesaver Aid for Nerve Agent Casualty (Continued).

MARK I*	ATNAA*	CANA**
STEP 3. OBTAIN BUDDY'S THREE OR REMAINING MARK Is.	STEP 3. OBTAIN BUDDY'S THREE OR REMAINING ATNAAs.	STEP 3. HOLD CANA IN A CLOSED FIST WITH DOMINANT HAND (FIGURE 7-12A).
STEP 4. CHECK INJECTION SITE.	STEP 4. CHECK INJECTION SITE.	STEP 4. GRASP SAFETY CAP WITH NONDOMINANT HAND AND REMOVE FROM INJECTOR (FIGURE 7-12B).
STEP 5. HOLD MARK I WITH NONDOMINANT HAND (FIGURE 7-5A).	STEP 5. HOLD ATNAA IN A CLOSED FIST WITH DOMINANT HAND (FIGURE 7-12A).	STEP 5. CLEAR HARD OBJECTS FROM INJECTION SITE.
STEP 6. GRASP SMALL INJECTOR (ATROPINE) AND REMOVE FROM CLIP (FIGURE 7-5B).	STEP 6. GRASP SAFETY CAP WITH NONDOMINANT HAND AND REMOVE FROM INJECTOR (FIGURE 7-12B).	STEP 6. INJECT CANA AT INJECTION SITE BY APPLYING EVEN PRESSURE TO THE INJECTOR, NOT A JABBING MOTION (FIGURE 7-14 OR 7-15). HOLD IN PLACE FOR 10 SECONDS.
STEP 7. CLEAR HARD OBJECTS FROM INJECTION SITE.	STEP 7. CLEAR HARD OBJECTS FROM INJECTION SITE.	STEP 7. BEND NEEDLE OF INJECTOR BY PRESSING ON A HARD SURFACE TO FORM A HOOK.
STEP 8. INJECT ATROPINE AT INJECTION SITE BY APPLYING EVEN PRESSURE TO THE INJECTOR, NOT A JABBING MOTION (FIGURE 7-10 OR 7-11). HOLD IN PLACE FOR 10 SECONDS.	STEP 8. INJECT ATNAA AT INJECTION SITE BY APPLYING EVEN PRESSURE TO THE INJECTOR, NOT A JABBING MOTION (FIGURE 7-14 OR 7-15). HOLD IN PLACE FOR 10 SECONDS.	STEP 8. ATTACH USED INJECTOR TO BLOUSE POCKET FLAP OF BDO/JSLIST (FIGURE 7-16).
STEP 9. HOLD USED INJECTOR BETWEEN LITTLE FINGER AND RING FINGER OF NONDOMINANT HAND (FIGURE 7-5A).	STEP 9. BEND NEEDLE OF INJECTOR BY PRESSING ON A HARD SURFACE TO FORM A HOOK.	STEP 9. MASSAGE INJECTION SITE, MISSION PERMITTING.
STEP 10. PULL LARGE INJECTOR (2 PAM CI) FROM CLIP (FIGURE 7-5C). DROP CLIP TO GROUND.	STEP 10. ATTACH ALL USED INJECTORS TO BLOUSE POCKET FLAP OF BDO/JSLIST (FIGURE 7-16).	

Table 7-2. Buddy Aid/Combat Lifesaver Aid for Nerve Agent Casualty (Continued).

MARK I*	ATNAA*	CANA**
STEP 11. INJECT 2 PAM CI AT INJECTION SITE BY APPLYING EVEN PRESSURE TO THE INJECTOR, NOT A JAB-BING MOTION (FIGURE 7-10 OR 7-11). HOLD IN PLACE FOR 10 SECONDS.	STEP 11. MASSAGE INJECTION SITE, MISSION PERMITTING.	
STEP 12. REPEAT STEPS ABOVE FOR REMAINING MARK Is.		
STEP 13. BEND THE NEEDLES OF ALL USED INJECTORS BY PRESSING ON A HARD SURFACE TO FORM A HOOK.		
STEP 14. ATTACH ALL USED INJECTORS TO BLOUSE POCKET FLAP OF BDO/JSLIST (FIGURE 7-13).		
STEP 15. MASSAGE INJECTION SITE, MISSION PERMITTING.		

* USE STEPS LISTED FOR TYPE OF ANTIDOTE DEVICE ISSUED.
** CANA IS USED IN BUDDY AID/CLS AID ONLY. DO NOT USE IN SELF-AID.

NOTE

If the casualty is thinly built, inject the antidote into the buttock. Only inject the antidote into the upper outer portion of the casualty's buttock (Figure 7-11). This avoids hitting the nerve that crosses the buttocks (Figure 7-4). Hitting this nerve can cause paralysis.

Figure 7-10. Injecting the casualty's thigh (Mark I or CANA).

Figure 7-11. Injecting the casualty's buttocks (Mark I or CANA).

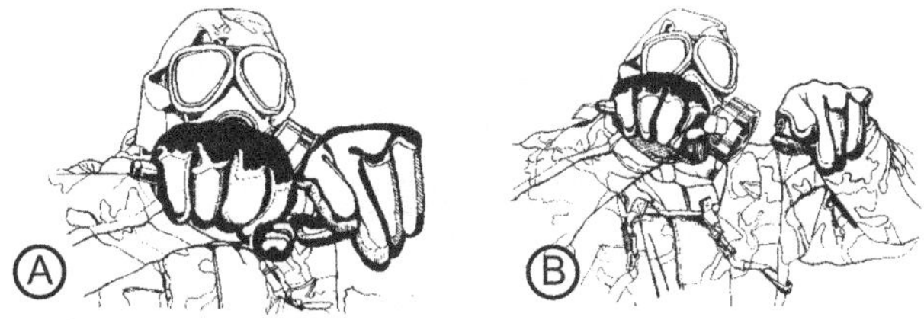

Figure 7-12. Preparing CANA or ATNAA for injection.

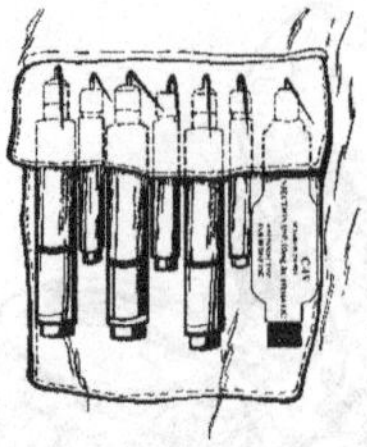

Figure 7-13. Three sets of used MARK I autoinjectors and one CANA autoinjector attached to pocket flap.

d. Self-Administer Antidote Treatment Nerve Agent Autoinjector. If you experience any or all of the nerve agent **MILD** symptoms (paragraph 7-7*b*), you must **IMMEDIATELY** self-administer one ATNAA following the procedure given Table 7-1.

NOTE

If you are thinly-built, inject yourself into the upper outer quarter (quadrant) of the buttock (Figure 7-15). There is a nerve that crosses the buttocks; hitting this nerve can cause paralysis. Therefore, you must only inject into the upper outer quarter (quadrant) of the buttocks.

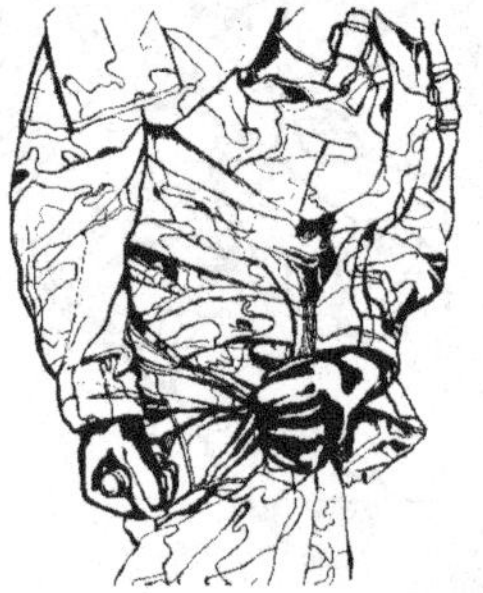

Figure 7-14. Self-administration of ATNAA (thigh).

Figure 7-15. Self-administration of ATNAA (buttock).

NOTE

If you continue to have symptoms of nerve agent poisoning, seek someone else (a buddy) to check your symptoms and administer your remaining sets of injections, if required.

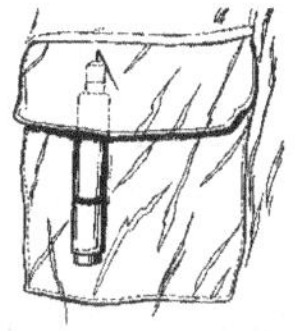

Figure 7-16. Used ATNAA attached to clothing.

e. Buddy Assistance. Service members may seek assistance after self-aid (self-administering one ATNAA) or may become incapacitated after self-aid. A buddy must evaluate the individual to determine if additional antidotes are required to counter the effects of the nerve agent. Also, service members may experience **SEVERE** symptoms of nerve agent poisoning (paragraph 7-7*b*); they will not be able to treat themselves. In either case, other service members must perform buddy aid as quickly as possible. Before initiating buddy aid, determine if one ATNAA has already been used so that no more than three ATNAA are administered. Buddy aid also includes administering the CANA with the third ATNAA to prevent convulsions. Follow the procedures indicated in Table 7-2.

WARNING

Squat, DO NOT kneel, when masking the casualty or administering the nerve agent antidotes to the casualty. Kneeling may force any chemical agent on your overgarment into or through your protective clothing.

Figure 7-17. Buddy injecting casualty's outer thigh (ATNAA or CANA).

NOTE

If the casualty is thinly built, inject the antidote into the buttocks (Figure 7-18). Only inject the antidote into the upper outer portion of the casualty's buttocks. This avoids hitting the nerve that crosses the buttocks (Figure 7-4). Hitting this nerve can cause paralysis.

WARNING

DO NOT inject into areas close to the hip, knee, or thighbone.

Figure 7-18. Buddy injecting casualty's buttocks (*ATNAA or CANA*).

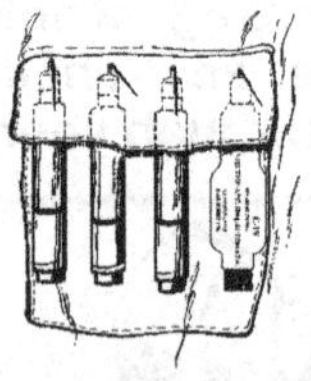

Figure 7-19. Three used ATNAAs and one CANA autoinjector attached to clothing.

f. *Combat Lifesaver.*

(1) The combat lifesaver must check to verify if the individual has received three sets of MARK I or ATNAAs. If not, the combat lifesaver performs first aid as described for buddy aid above. If the

individual has received the initial three sets of MARK I, then the combat lifesaver may administer additional atropine injections at approximately 15 minute intervals until atropinization is achieved (that is a heart rate above 90 beats per minute, reduced bronchial secretions, and reduced salivations). Administer additional atropine at intervals of 30 minutes to 4 hours to maintain atropinization or until the casualty is placed under the care of medical personnel. Check the heart rate by lifting the casualty's mask hood and feeling for a pulse at the carotid artery. Request medical assistance as soon as the tactical situation permits.

(2) The CLS should administer additional CANA to casualties suffering convulsions. Administer a second, and if needed, a third CANA at 5 to 10 minute intervals for a maximum of three injections (30 milligrams diazepam). Follow the steps and procedures described in buddy aid for administering the CANA. **DO NOT** give more than two additional injections for a total of three (one self-aid plus two by the CLS).

7-9. Blister Agents

Blister agents (vesicants) include mustard (H and HD), nitrogen mustards (HN), lewisite (L), and other arsenicals, mixtures of mustards and arsenicals, and phosgene oxime (CX). Blister agents may act on the eyes, mucous membranes, lungs, and skin. They burn and blister the skin or any other body parts they contact. Even relatively low doses may cause serious injury. Blister agents damage the respiratory tract (nose, sinuses, and windpipe) when inhaled and cause vomiting and diarrhea when absorbed. Lewisite and CX cause immediate pain on contact. However, mustard agents are deceptive as there is little or no pain at the time of exposure. Thus, in some cases, signs of injury may not appear for several hours after exposure.

a. Protective Measures. Your protective mask with hood and protective overgarment provide protection against blister agents. If it is known or suspected that blister agents are being used, **STOP BREATHING**, put on your mask and your protective overgarment.

CAUTION

Large drops of liquid vesicants on the protective overgarment ensemble may penetrate it if allowed to stand for an extended period. Remove large drops as soon as possible.

b. Signs and Symptoms of Blister Agent Poisoning.

(1) Immediate and intense pain upon contact with L, LH (lewisite and mustard) mixture, and CX. No initial pain upon contact with mustard.

(2) Inflammation and blisters (burns) resulting in tissue destruction. The severity of a chemical burn is directly related to the concentration of the agent and the duration of contact with the skin. The longer the agent is in contact with the tissue, the more serious the injury will be.

(3) Vomiting and diarrhea. Exposure to high concentrations of vesicants may cause vomiting or diarrhea.

(4) Death. The blister agent vapors absorbed during ordinary field exposure will probably not cause enough internal body (systemic) damage to result in death. However, death may occur from prolonged exposure to high concentrations of vapor or from extensive liquid contamination over wide areas of the skin, particularly *when decontamination is neglected or delayed.*

c. First Aid Measures.

(1) Use your M291 Skin Decontaminating Kit to decontaminate your skin and use water to flush contaminated eyes. Decontamination of vesicants must be done immediately (within 1 minute is best).

(2) If blisters form, cover them loosely with a field dressing and secure the dressing.

CAUTION

Blisters are actually burns. **DO NOT** attempt to decontaminate the skin where blisters have formed, as the agent has already been absorbed.

(3) If you receive blisters over a wide area of the body, you are considered seriously burned. Seek medical assistance immediately.

(4) If vomiting occurs, the mask should be lifted momentarily and drained—while the eyes are closed and the breath is held—and replaced, cleared, and sealed.

(5) Remember, if vomiting or diarrhea occurs after having been exposed to blister agents, seek medical assistance immediately.

7-10. Choking Agents (Lung-Damaging Agents)

Chemical agents that attack lung tissue, primarily causing fluid buildup (pulmonary edema), are classified as choking agents (lung-damaging agents). This group includes phosgene (CG), diphosgene (DP), chlorine (Cl), and chloropicrin (PS). Of these four agents, CG is the most dangerous and is more likely to be employed by the enemy in future conflict.

a. Protective Measures. Your protective mask gives adequate protection against choking agents.

b. Signs and Symptoms. During and immediately after exposure to choking agents (depending on agent concentration and length of exposure), you may experience some or all of the following signs and symptoms:

- Tears (lacrimation).
- Coughing.
- Choking.
- Tightness of chest.
- Nausea and vomiting.
- Headaches.

c. Self-Aid.

(1) The protective mask should be put on immediately when any of the conditions described in *b* above exist. Another indication of a CG attack is an odor like newly mown hay; however, **DO NOT** rely upon odor as indication of a chemical attack.

(2) If some CG is inhaled, normal combat duties should be continued unless there is difficulty in breathing, nausea, vomiting, or more than the usual shortness of breath during exertion. If any of the above symptoms occur and the mission permits, remain at quiet rest until medical evacuation is accomplished.

d. Death. With ordinary field exposure to choking agents, death will probably not occur. *However, prolonged exposure to high concentrations of the vapor and neglect or delay in masking can be fatal.*

7-11. Cyanogen (Blood) Agents

Cyanogen agents interfere with proper oxygen utilization in the body. Hydrogen cyanide (AC) and cyanogen chloride (CK) are the primary agents in this group.

a. *Protective Measures.* Your protective mask with a fresh filter gives adequate protection against field concentrations of cyanogen agent vapor. The protective overgarments, as well as the mask, are needed when exposed to liquid AC.

b. *Signs and Symptoms.* During and immediately after exposure to cyanogen agents (depending on agent concentration and length of exposure), you may experience some or all of the following signs and symptoms:

- Tearing (lacrimation).
- Eye, nose, and throat irritation.
- Sudden stimulation of breathing (unable to hold breath).
- Nausea.
- Coughing.
- Tightness of chest.
- Headache.
- Light-headedness (dizziness).
- Unconsciousness.

c. *First Aid.*

(1) *Hydrogen cyanide.* During any chemical attack, if you get a sudden stimulation of breath or detect an odor like bitter almonds, **PUT ON YOUR MASK IMMEDIATELY**. Speed is absolutely essential since this agent acts so rapidly that within a few seconds its effects will make it impossible for service members to put on their mask by themselves. Stop breathing until the mask is on, if at all possible. This may be very difficult since the agent strongly stimulates respiration.

(2) *Cyanogen chloride.* Put your mask on immediately if you experience any irritation of the eyes, nose, or throat. Service members

who are unable to mask should be masked by the nearest service member (buddy).

d. *Medical Assistance.* If you suspect that you have been exposed to blood agents, seek medical assistance immediately.

7-12. Incapacitating Agents

An incapacitating agent is a chemical agent which produces temporary, disabling conditions which persist for hours to days after exposure. Unlike riot control agents, which usually are momentary or fleeting in action, incapacitating agents have a persistent effect. It is likely that smoke-producing munitions or aerosols will disseminate such agents, thus making breathing their means of entry into the body. The protective mask is, therefore, essential.

a. There are no specific first aid measures to relieve the symptoms of incapacitating agents. Supportive first aid and physical restraint may be indicated. If the casualty is stuporous or comatose, be sure that respiration is unobstructed; then turn him on his side in case vomiting should occur. Complete cleansing of the skin with soap and water should be done as soon as possible; or, the M291 Skin Decontaminating Kit can be used if washing is impossible. Remove weapons and other potentially harmful items from service members who are suspected of having these symptoms. Harmful items include cigarettes, matches, medications, and small items that might be swallowed accidentally. Delirious (confused) persons have been known to attempt to eat items bearing only a superficial resemblance to food.

b. Incapacitating agents (anticholinergic drugs BZ type) may produce alarming dryness and coating of the lips and tongue; however, there is usually no danger of immediate dehydration. Fluids should be given sparingly, if at all, because of the danger of vomiting and because of the likelihood of temporary urinary retention due to paralysis of bladder muscles.

c. If the body temperature is elevated and mucous membranes are dry, immediate and vigorous cooling (as for heatstroke) is indicated. Methods that can be used to cool the skin are spraying with cool water and air circulation (fanning); applying alcohol soaked cloths and air circulation; and providing maximum exposure to air in a shaded area, along with maximum air circulation. Such cases are usually a result of anticholinergic poisoning. Rapid evacuation should be accomplished since medical treatment with the appropriate medication may be lifesaving.

CAUTION

DO NOT use **ice** for cooling the skin.

d. Reassurance and a firm, but friendly, attitude by individuals providing first aid will be beneficial if the casualty appears to comprehend what is being said. Conversation is a waste of time if the service member is incoherent or cannot understand what is being said. In such cases, the less said, the better it is—these casualties will benefit more from prompt and vigorous restraint and evacuation to an MTF.

7-13. Incendiaries

Incendiaries can be grouped as WP, thickened gasoline, metal, and oil and metal. You must learn to protect yourself against these incendiaries.

a. White phosphorus is used primarily as a smoke producer but can be used for its incendiary effect to ignite field expedients and combustible materials. The burns from WP are usually multiple, deep, and variable in size. When particles of WP get on the skin or clothing, they continue to burn until deprived of air. They also have a tendency to stick to a surface and must be brushed off or picked out.

(1) If burning particles of WP strike and stick to your clothing, quickly take off the contaminated clothing before the WP burns through to the skin.

(2) If burning WP strikes your skin, smother the flame with water, a wet cloth, or mud.

NOTE

Since WP is soluble in oil, **DO NOT** use grease, oily ointments, or eye ointments to smother the flame.

(3) Keep the WP particles covered with a wet material to exclude air until you can remove them or have them removed from your skin.

(4) Remove the WP particles from the skin by brushing them with a wet cloth and by picking them out with a knife, bayonet, stick, or other available object.

(5) Seek medical assistance when the mission permits.

b. Thickened fuel mixtures (napalm) have a tendency to cling to clothing and body surfaces, thereby producing prolonged exposure and severe burns. The first aid for these burns is the same as for other heat burns. The heat and irritating gases given off by these combustible mixtures may cause lung damage, which must be treated by medical personnel.

c. Metal incendiaries pose special problems. Thermite particles on the skin should be immediately cooled with water and then removed. The first aid for these burns is the same as for other heat burns. Particles of magnesium on the skin burn quickly and deeply. Like other metal incendiaries, they must be removed. Ordinarily, medical personnel should do the complete removal of these particles as soon as possible. Immediate medical treatment is required.

d. Oil and metal incendiaries have much the same effect on contact with the skin and clothing as those discussed (*b* and *c* above). First aid measures for burns are discussed in Chapter 3.

7-14. Biological Agents and First Aid

a. Biological attacks can result in combat ineffectiveness by introducing disease-causing organisms into a troop population.

b. Once a disease is identified, first aid or medical treatment is initiated, depending on the seriousness of the disease. First aid measures are concerned with observable symptoms of the disease such as diarrhea or vomiting.

7-15. Toxins

Toxins are alleged to have been used in past conflicts. Witnesses and victims have described the agent as toxic rain (or yellow rain) because it was reported to have been released from aircraft as a yellow powder or liquid that covered ground, structures, vegetation, and people.

a. *Signs and Symptoms.* The occurrence of the symptoms from toxins may appear in a period of a few minutes to several hours depending on the particular toxin, the service member's susceptibility, and the amount of toxin inhaled, ingested, or deposited on the skin. Symptoms from toxins usually involve the central nervous system but are often preceded by less prominent symptoms, such as nausea, vomiting, diarrhea, cramps, or stomach irritation and burning sensation. Typical neurological symptoms often develop rapidly in severe cases; for example, visual disturbances, inability to swallow, speech difficulty, lack of muscle coordination, and sensory abnormalities (numbness of mouth, throat, or extremities). Yellow rain (mycotoxins) also may have hemorrhagic symptoms, which could include any or all of the following:

- Dizziness.
- Severe itching or tingling of the skin.
- Formation of multiple, small, hard blisters.
- Coughing up blood.
- Shock (which could result in death).

b. *Self-Aid.* Upon recognition of an attack employing toxins, you must immediately take the following actions:

(1) Stop breathing, put on your protective mask with hood, and then resume breathing. Next, put on your protective clothing.

(2) Should severe itching of the face become unbearable, quickly—

- Loosen the cap on your canteen.
- Take and hold a deep breath and lift your mask.
- While holding your breath, close your eyes and flush your face with generous amounts of water.

CAUTION

DO NOT rub or scratch your eyes. Try not to let the water run onto your clothing or protective overgarment.

- Put your protective mask back on, seat it properly, clear it, and check it for a seal; then resume breathing.
- Decontaminate your skin by bathing with soap and water as soon as the mission permits.

- Change clothing and decontaminate your protective mask using soap and water. Replace the filters if directed.

(3) If vomiting occurs, the mask should be lifted momentarily and drained—while the eyes are closed and the breath is held—and replaced, cleared, and sealed.

c. *Medical Assistance.* If you suspect that you have been exposed to toxins, you should seek medical assistance immediately.

7-16. Nuclear Detonation

a. Three types of injuries may result from a nuclear detonation. These are thermal, blast, and radiation injuries. Many times the casualty will have a combination of these types of injuries. First aid for thermal and blast injuries is provided based on observable injuries, such as burns, hemorrhage, or fractures.

b. The signs and symptoms of radiation illness in the initial phase include the rapid onset of nausea, vomiting, and malaise (tiredness). The only first aid procedure for radiological casualties is decontamination.

CHAPTER 8

FIRST AID FOR PSYCHOLOGICAL REACTIONS

8-1. General

Psychological first aid is as natural and reasonable as physical first aid and is just as familiar. When you were hurt as a child, the understanding attitude of your parents did as much as the psychological effect of a bandage. Later, your disappointment or grief was eased by supportive words from a friend. Certainly, taking a walk and talking things out with a friend are familiar ways of dealing with an emotional crisis. The same natural feelings that make us want to help a person who is injured make us want to give a helping hand to a buddy who is upset. *Psychological first aid* really means nothing more complicated than assisting people with emotional distress whether it results from physical injury, disease, or excessive stress. Emotional distress is not always as visible as a wound or a broken bone. However, overexcitement, severe fear, excessive worry, deep depression, misdirected irritability, and anger are signs that stress has reached the point of interfering with effective coping. The more noticeable the symptoms become, the more urgent the need for you to be of help and the more important it is for you to know *how* to help.

8-2. Importance of Psychological First Aid

You must know how to give psychological first aid to be able to help yourself, your buddies, and your unit in order to keep performing the mission. Psychological first aid measures are simple and easy to understand. Your decision of what to do depends upon your ability to observe the service member and understand his needs. Making the best use of resources requires ingenuity on your part. A stress reaction resulting in poor judgment can cause injury or even death to yourself or others on the battlefield. It can be even more dangerous if other persons are affected by the judgment of an emotionally upset service member. If it is detected early enough, the affected service member stands a good chance of remaining in his unit as an effective member. If it is not detected early and if the service member becomes more emotionally upset, he may become a threat to himself and to others.

8-3. Situations Requiring Psychological First Aid

- Psychological first aid (buddy aid) is most needed at the first sign that a service member cannot perform the mission because of emotional distress. Stress is inevitable in combat, in hostage and terrorist situations, and in civilian disasters such as floods, hurricanes, or industrial accidents. Most emotional reactions to such situations are temporary, and the service member can still carry on with encouragement. Painful or disruptive symptoms may last for minutes, hours, or days. However, if the stress symptoms are seriously disabling, they may be psychologically contagious and endanger not only the emotionally upset service member but also the entire unit.

- Sometimes people continue to function well during a disastrous event, but suffer from emotional scars which impair their job performance or quality of life at a later time. Painful memories and dreams may recur for months and years and still be considered a normal reaction. However, if the memories are so painful that the person must avoid all situations which arouse them, becomes socially withdrawn, or shows symptoms of anxiety, depression, or substance abuse, he needs treatment. Experience with police, firemen, emergency medical technicians, and others who deal with disasters has proved that the routine application of psychological first aid to all the participants, including those who have functioned well, greatly reduces the likelihood of future serious post-traumatic stress disorders (PTSDs).

8-4. Interrelationship of Psychological and Physical First Aid

Psychological first aid should go hand in hand with physical first aid. The discovery of a physical injury or cause for an inability to function does not rule out the possibility of a psychological injury (or vice versa). The person suffering from pain, shock, fear of serious injury, or fear of death does not respond well to joking, indifference, or fearful-tearful attention. Fear and anxiety may take as high a toll of the service member's strength as does the loss of blood.

8-5. Goals of Psychological First Aid

The goals of psychological first aid are to—

- Be supportive; assist the service member in dealing with his stress reaction.
- Prevent, and if necessary control, behavior harmful to himself and to others.
- Return the service member to duty as soon as possible after dealing with the stress reaction.

8-6. Respect for Others' Feelings

a. Accept the service member you are trying to help without censorship or ridicule. Respect his right to his own feelings. Even though your feelings, beliefs, and behavior are different, DO NOT blame or make light of him for the way he feels or acts. Your purpose is to help him in this tough situation, not to be his critic. A person DOES NOT WANT to be upset and worried. When he seeks help, he needs and expects consideration of his fears, not abrupt dismissal or ridicule.

b. Realize that people are the products of a wide variety of factors. All people DO NOT react the same way to the same situations. Each individual has complex needs and motivations, both conscious and unconscious, that are uniquely his own. Often the one thing that finally causes the person to become overloaded by a stressful situation is not the stressor itself, but some other problem.

8-7. Emotional and Physical Disability

a. Accept emotional disability as being just as real as physical disability. If a service member's ankle is seriously sprained in a fall, no one expects him to run right away. A service member's emotions may be temporarily strained by the overwhelming stress of battle or other traumatic incident. DO NOT demand that he pull himself together immediately and carry on without a break. Some individuals can pull themselves together immediately, but others cannot. The service member whose emotional stability has been disrupted has a disability just as real as the service member who has sprained his ankle. There is an unfortunate tendency in many people to regard as real only what they can see, such as a wound or bleeding. Some people tend to assume that damage involving a person's mind and emotions is just imagined, that he is not really sick or injured, and that he could overcome his trouble by using his will power.

b. The terms *it's all in your head, snap out of it,* and *get control of yourself* are often used by people who believe they are being helpful. Actually, these terms are expressions of hostility because they show lack of understanding. They only emphasize weakness and inadequacy. Such terms are of no use in psychological first aid.

c. Every physically injured person has some emotional reaction to the fact that he is injured.

(1) It is normal for an injured person to feel upset. The more severe the injury, the more insecure and fearful he becomes, especially

if the injury is to a body part which is highly valued. For example, an injury to the eyes or the genitals, even though relatively minor, is likely to be extremely upsetting. An injury to some other part of the body may be especially disturbing to an individual for his own particular reason. For example, an injury of the hand may be a terrifying blow to a surgeon or an injury to the eye of a pilot.

(2) An injured service member always feels less secure, more anxious, and more afraid not only because of what has happened to him but because of what he imagines may happen as a result of his injury. This fear and insecurity may cause him to be irritable, uncooperative, or unreasonable. As you help him, always keep in mind that such behavior has little or nothing to do with you personally. He needs your patience, reassurance, encouragement, and support.

8-8. Combat and Other Operational Stress Reactions

Stress reaction is a temporary emotional disorder or inability to function, experienced by a previously normal service member as a reaction to the overwhelming or cumulative stress of combat. Stress reaction gets better with reassurance, rest, physical replenishment, and activities that restore confidence. All service members are likely to feel stress reaction under conditions of intense and/or prolonged stress. They may even become stress reaction casualties, unable to perform their mission for hours or days. Other combat and operational stress reactions (COSRs) may result in negative behavior, but are not termed *stress reaction,* as they need more intensive treatment. These negative COSRs may result in misconduct stress behaviors such as drug and alcohol abuse, criminal acts, looting, desertion, and self-inflicted wounds. These harmful COSRs can often be prevented by good psychological first aid. Service members who commit misconduct stress behaviors may require disciplinary action rather than medical treatment.

8-9. Reactions to Stress

Most service members react to stressful incidents after the situation has passed. All service members feel some fear. This fear may be greater than they have experienced at any other time, or they may be more aware of their fear. In such a situation, they should not be surprised if they feel shaky or become sweaty, nauseated, or confused. These reactions are normal and are not a cause for concern. However, some reactions, either short- or long-term, will cause problems if left unchecked. See paragraph 8-13 for more information.

a. Emotional Reactions.

(1) The most obvious combat stress reaction (CSR) is inefficient performance. This can be demonstrated by—

- Slow thinking (or reaction time).
- Difficulty recognizing priorities and seeing what needs to be done.
- Difficulty getting started.
- Indecisiveness and having trouble focusing attention.
- Tendency to do familiar tasks and be preoccupied with familiar details. (This can reach the point where the person is very passive, such as just sitting or wandering about not knowing what to do.)

(2) A less common reaction may be uncontrolled emotional outbursts; this can be demonstrated by crying, screaming, or laughing. Some service members will react in the opposite way. They will be very withdrawn and silent and try to isolate themselves from everyone. These service members should be encouraged to remain with their assigned unit. Uncontrolled reactions may appear by themselves or in any combination (the person may be crying uncontrollably one minute and then laughing the next). In this state, the person is restless and cannot keep still. He may run about, apparently without purpose. Inside, he feels a great rage or fear and his physical acts may show this. In his anger he may indiscriminately strike out at others.

b. Loss of Adaptability.

(1) In a desperate attempt to get away from the danger, which has overwhelmed him, a service member may panic and become confused. His mental ability may be so impaired he cannot think clearly or even follow simple commands. His judgment may be faulty and he may not be aware of his actions, such as standing up in his fighting position during an attack.

(2) In other cases, overwhelming stress may produce symptoms that are often associated with head injuries. For example, the service member may appear dazed or be found wandering around aimlessly. He may appear confused and disoriented and may seem to have a complete or partial loss of memory. In such cases, especially when no eyewitnesses can provide evidence that the service member has NOT suffered a head injury, it is necessary for him to be rapidly medically evacuated. **DO NOT** allow the

service member to expose himself to further personal danger until the cause of the problem has been determined.

c. *Sleep Disturbance and Repetition of Dreams.* A person who has been overwhelmed by stress often has difficulty sleeping. The service member may experience nightmares related to the stressors. Remember that nightmares, in themselves, are not considered abnormal when they occur soon after a period of intensive stress. As time passes, the nightmares usually become less frequent and less intense. In extreme cases, a service member, even when awake, may think repeatedly of the incident, feel as though it is happening again, and act out parts of his stress over and over again. For some persons, this repetitious reexperiencing of the stressful event may be necessary for eventual recovery; therefore, it should not be discouraged or viewed as abnormal. For the person reexperiencing the event, such reaction may be disruptive. The service member needs to be encouraged to *ventilate* about the incident. Ventilation is a technique where the service member is given the opportunity to talk extensively, often repetitiously about the experience.

8-10. Severe Stress or Stress Reaction

You do not need specialized training to recognize severe stress or stress reaction that will cause problems for the service member, the unit, or the mission. Reactions that are less severe, however, are more difficult to detect. To determine whether a person needs help, you must observe him to see whether he is doing something meaningful, performing his duties, taking care of himself, behaving in an unusual fashion, or acting out of character.

8-11. Application of Psychological First Aid

The emotionally disturbed service member has built a barrier against fear. He does this for his own protection, although he is probably not aware that he is doing it. If he finds that he does not have to be afraid and that there are normal, understandable things about him, he will feel safer in dropping this barrier. Persistent efforts to make him realize that you want to understand him will be reassuring, especially if you remain calm. Nothing can cause an emotionally disturbed person to become even more fearful than feeling that others are afraid of him. Try to remain calm. Familiar things, such as a cup of coffee, the use of his name, attention to a minor wound, being given a simple job to do, or the sight of familiar people and activities, will add to his ability to overcome his fear. He may not respond well if you get excited, angry, or abrupt.

a. Ventilation. After the service member becomes calmer, he is likely to have dreams about the stressful event. He also may think about it when he is awake or even repeat his personal reaction to the event. One benefit of this natural pattern is that it helps him master the stress by going over it just as one masters the initial fear of parachuting from an aircraft by doing it over and over again. Eventually, it is difficult to remember how frightening the event was initially. In giving first aid to the emotionally disturbed service member, you should let him follow this natural pattern. Encourage him to talk. Be a good listener. Let him tell, in his own words, what actually happened. If home front problems or worries have contributed to the stress, it will help him to talk about them. Your patient listening will prove to him that you are interested in him, and by describing his personal problem, he can work at mastering his fear. If he becomes overwhelmed in the telling, suggest a cup of coffee or a break. Whatever you do, assure him that you will listen again as soon as he is ready. Do try to help put the service member's perception of what happened back into realistic perspective; but DO NOT argue about it.

b. Activity.

(1) A person who is emotionally disturbed as the result of a combat action is a casualty of anxiety and fear. He is disabled because he has become temporarily overwhelmed by his anxiety. A good way to control fear is through activity. Almost all service members, for example, experience a considerable sense of anxiety and fear while they are poised, awaiting the opening of a big offensive; but this is normally relieved, and they actually feel better once they begin to move into action. They take pride in effective performance and pleasure in knowing that they are good service members, perhaps being completely unaware that overcoming their initial fear was their first major accomplishment.

(2) Useful activity is very beneficial to the emotionally disturbed service member who is not physically incapacitated. After you help a service member get over his initial fear, help him to regain some self-confidence. Make him realize his job is continuing by finding him something useful to do. Encourage him to be active. Get him to help load trucks, clean up debris, or dig fighting positions. If possible, get him back to his usual duty. Seek out his strong points and help him apply them. Avoid having him just sit around. You may have to provide direction by telling him what to do and where to do it. The instructions should be clear and simple and should be repeated. A person who has panicked is likely to argue. Respect his feelings, but point out more immediate, obtainable, and demanding needs. Channel his excessive energy and, above all, DO NOT argue. If you cannot get him interested in doing more profitable work, it may be necessary to enlist aid in controlling his overactivity before it spreads to the group and

results in more panic. Prevent the spread of such infectious feelings by restraining and segregating if necessary.

(3) Involvement in activity helps a service member in three ways; he—

- Forgets himself.
- Has an outlet for his excessive tensions.
- Proves to himself he can do something useful.

c. Rest. There are times, particularly in combat, when physical exhaustion is a principal cause for emotional reactions. A unit sleep plan should be established and implemented. When possible, service members should be given a safe and relatively comfortable area in which to sleep. Examples would be an area away from heavy traffic, noise, and congestion or a place that is clean and dry and protected from environmental conditions. The more uninterrupted sleep a service member gets the better he will be able to function in the tactical environment.

d. Hygiene. Field hygiene is an important ingredient in a service member's morale. A service member who is dirty and unkempt will not function as well as a service member who has had the opportunity to bathe and put on clean, dry clothing. During combat, unit leaders should stress the importance of personal hygiene. Good personal hygiene not only improves morale, it also is a preventive measure against disease and nonbattle injury (DNBI).

e. Group Activity. You have probably already noticed that a person works, faces danger, and handles serious problems better if he is a member of a closely-knit group. Each service member in the team supports the other team members. Esprit de corps is built because the service members have the same interests, goals, and mission, and as a result they are more productive; furthermore, they are less worried because everyone is involved. It is this spirit that takes a strategic hill in battle. It is so powerful that it is one of the most effective tools you have in your *psychological first aid bag*. Getting the service member back into the team or squad activities will reestablish his sense of belonging and security and will go far toward making him a useful member of the unit.

8-12. Reactions and Limitations

Up to this point the discussion has been primarily about the feelings of the emotionally distressed service member. What about your feelings toward him? Whatever the situation, you will have emotional reactions (conscious or unconscious) toward this service member. Your reactions can either help or hinder your ability to help him. When you are tired or worried, you may very easily become impatient with him if he is unusually slow or exaggerates. You may even feel resentful toward him. At times when many physically wounded lie about you, it will be especially natural for you to resent disabilities that you cannot see. Physical wounds can be seen and easily accepted. Emotional reactions are more difficult to accept as injuries. On the other hand, will you tend to be overly sympathetic? Excessive sympathy for an incapacitated person can be as harmful as negative feelings in your relationship with him. He needs strong help, but not your sorrow. To overwhelm him with pity will make him feel even more inadequate. You must expect your buddy to recover, to be able to return to duty, and to become a useful service member again. This expectation should be displayed in your behavior and attitude as well as in what you say. If he can see your calmness, confidence, and competence, he will be reassured and will feel a sense of greater security.

8-13. Stress Reactions

See Tables 8-1, 8-2, and 8-3 for more information.

Table 8-1. Mild Stress Reaction

PHYSICAL SIGNS*	EMOTIONAL SIGNS*
1. TREMBLING, TEARFUL 2. JUMPINESS, NERVOUSNESS 3. COLD SWEAT, DRY MOUTH 4. POUNDING HEART, DIZZINESS 5. INSOMNIA, NIGHTMARES 6. NAUSEA, VOMITING, DIARRHEA 7. FATIGUE 8. THOUSAND-YARD STARE 9. DIFFICULTY THINKING, SPEAKING, AND COMMUNICATING	1. ANXIETY, INDECISIVENESS 2. IRRITABLE, COMPLAINING 3. FORGETFUL, UNABLE TO CONCENTRATE 4. EASILY STARTLED BY NOISE, MOVEMENT 5. GRIEF, TEARFUL 6. ANGER, BEGINNING TO LOSE CONFIDENCE IN SELF AND UNIT
SELF- AND BUDDY AID	
1. CONTINUE MISSION PERFORMANCE, FOCUS ON IMMEDIATE MISSION. 2. EXPECT SERVICE MEMBER TO PERFORM ASSIGNED DUTIES. 3. REMAIN CALM AT ALL TIMES; BE DIRECTIVE AND IN CONTROL. 4. LET SERVICE MEMBER KNOW HIS REACTION IS NORMAL, AND THAT THERE IS NOTHING SERIOUSLY WRONG WITH HIM. 5. KEEP SERVICE MEMBER INFORMED OF THE SITUATION, OBJECTIVES, EXPECTATIONS, AND SUPPORT. CONTROL RUMORS. 6. BUILD SERVICE MEMBER'S CONFIDENCE, TALK ABOUT SUCCEEDING. 7. KEEP SERVICE MEMBER PRODUCTIVE (WHEN NOT RESTING) THROUGH RECREATIONAL ACTIVITIES, EQUIPMENT MAINTENANCE.	

8. ENSURE SERVICE MEMBER MAINTAINS GOOD PERSONAL HYGIENE.
9. ENSURE SERVICE MEMBER EATS, DRINKS, AND SLEEPS AS SOON AS POSSIBLE.
10. LET SERVICE MEMBER TALK ABOUT HIS FEELINGS. DO NOT "PUT DOWN" HIS FEELINGS OF GRIEF OR WORRY. GIVE PRACTICAL ADVICE AND PUT EMOTIONS INTO PERSPECTIVE.

* MOST OR ALL OF THESE SIGNS ARE PRESENT IN MILD STRESS REACTION. THEY CAN BE PRESENT IN ANY NORMAL SERVICE MEMBER IN COMBAT YET HE CAN STILL DO HIS JOB.

Table 8-2. More Serious Stress Reaction

PHYSICAL SIGNS*	EMOTIONAL SIGNS*
1. CONSTANTLY MOVES AROUND	1. RAPID AND/OR INAPPROPRIATE TALKING
2. FLINCHING OR DUCKING AT SUDDEN SOUNDS	2. ARGUMENTATIVE, RECKLESS MOVEMENTS/ACTIONS
3. SHAKING, TREMBLING (WHOLE BODY OR ARMS)	3. INATTENTIVE TO PERSONAL HYGIENE
4. CANNOT USE PART OF BODY, NO PHYSICAL REASON (HAND, ARM, LEGS)	4. INDIFFERENT TO DANGER
5. CANNOT SEE, HEAR, OR FEEL (PARTIAL OR COMPLETE LOSS)	5. MEMORY LOSS
6. PHYSICAL EXHAUSTION, CRYING	6. SEVERE STUTTERING, MUMBLING, OR CANNOT SPEAK AT ALL
7. FREEZING UNDER FIRE, OR TOTAL IMMOBILITY	7. INSOMNIA, NIGHTMARES
8. VACANT STARES, STAGGERS, SWAYS WHEN STANDS	8. SEEING OR HEARING THINGS THAT DO NOT EXIST
9. PANIC RUNNING UNDER FIRE	9. RAPID EMOTIONAL SHIFTS
	10. SOCIAL WITHDRAWAL
	11. APATHETIC
	12. HYSTERICAL OUTBURSTS
	13. FRANTIC OR STRANGE BEHAVIOR

TREATMENT PROCEDURES**

1. IF A SERVICE MEMBER'S BEHAVIOR ENDANGERS THE MISSION, SELF, OR OTHERS, DO WHATEVER IS NECESSARY TO CONTROL HIM.
2. IF THE SERVICE MEMBER IS UPSET, CALMLY TALK HIM INTO COOPERATING.
3. IF CONCERNED ABOUT THE SERVICE MEMBER'S RELIABILITY:
   - UNLOAD HIS WEAPON.
   - TAKE WEAPON IF SERIOUSLY CONCERNED.
   - PHYSICALLY RESTRAIN HIM ONLY WHEN NECESSARY FOR SAFETY OR TRANSPORTATION.
4. REASSURE EVERYONE THAT THE SIGNS ARE PROBABLY JUST STRESS REACTION AND WILL QUICKLY IMPROVE.
5. IF STRESS REACTION SIGNS CONTINUE:
   - GET THE SERVICE MEMBER TO A SAFER PLACE.
   - DO NOT LEAVE THE SERVICE MEMBER ALONE, KEEP SOMEONE HE KNOWS WITH HIM.
   - NOTIFY SENIOR NONCOMMISSIONED OFFICER (NCO) OR OFFICER.
   - HAVE THE SERVICE MEMBER EXAMINED BY MEDICAL PERSONNEL.

Table 8-2. More Serious Stress Reaction (Continued)

TREATMENT PROCEDURES**

6. GIVE THE SERVICE MEMBER EASY TASKS TO DO WHEN NOT SLEEPING, EATING, OR RESTING.
7. ASSURE THE SERVICE MEMBER HE WILL RETURN TO FULL DUTY IN 24 HOURS; AND, RETURN HIM TO NORMAL DUTIES AS SOON AS HE IS READY.

* THESE SIGNS ARE PRESENT IN ADDITION TO THE SIGNS OF MILD STRESS REACTION.

** DO THESE PROCEDURES IN ADDITION TO THE SELF- AND BUDDY AID CARE.

Table 8-3. Preventive Measures to Combat Stress Reaction

1. WELCOME NEW MEMBERS INTO YOUR TEAM, GET TO KNOW THEM QUICKLY. IF YOU ARE NEW, BE ACTIVE IN MAKING FRIENDS.
2. BE PHYSICALLY FIT (STRENGTH, ENDURANCE, AND AGILITY).
3. KNOW AND PRACTICE LIFESAVING SELF- AND BUDDY AID.
4. PRACTICE RAPID RELAXATION TECHNIQUES (FM 22-51).
5. HELP EACH OTHER OUT WHEN THINGS ARE TOUGH AT HOME OR IN THE UNIT.
6. KEEP INFORMED; ASK YOUR LEADER QUESTIONS, IGNORE RUMORS.
7. WORK TOGETHER TO GIVE EVERYONE FOOD, WATER, SHELTER, HYGIENE, AND SANITATION.
8. SLEEP WHEN MISSION AND SAFETY PERMIT; LET EVERYONE GET TIME TO SLEEP.
   - SLEEP ONLY IN SAFE PLACES AND BY STANDING OPERATING PROCEDURE (SOP).
   - IF POSSIBLE, SLEEP 6 TO 9 HOURS PER DAY.
   - TRY TO GET AT LEAST 4 HOURS SLEEP PER DAY.
   - GET GOOD SLEEP BEFORE GOING ON SUSTAINED OPERATIONS.
   - CATNAP WHEN YOU CAN, BUT ALLOW TIME TO WAKE UP FULLY.
   - CATCH UP ON SLEEP AFTER GOING WITHOUT.

APPENDIX A

FIRST AID CASE AND KITS, DRESSINGS, AND BANDAGES

A-1. First Aid Case with Field Dressings and Bandages

Every service member is issued a first aid case (Figure A-1A) with a field first aid dressing encased in a plastic wrapper (Figure A-1B). He carries it at all times for his use. The field first aid dressing is a standard sterile (germ-free) compress or pad with bandages attached (Figure A-1C). This dressing is used to cover the wound, to protect against further contamination, and to stop bleeding (pressure dressing). When a service member administers first aid to another person, he must remember to use the wounded person's dressing; he may need his own later. The service member must check his first aid case regularly and replace any used or missing dressing. The field first aid dressing may normally be obtained from his unit supply.

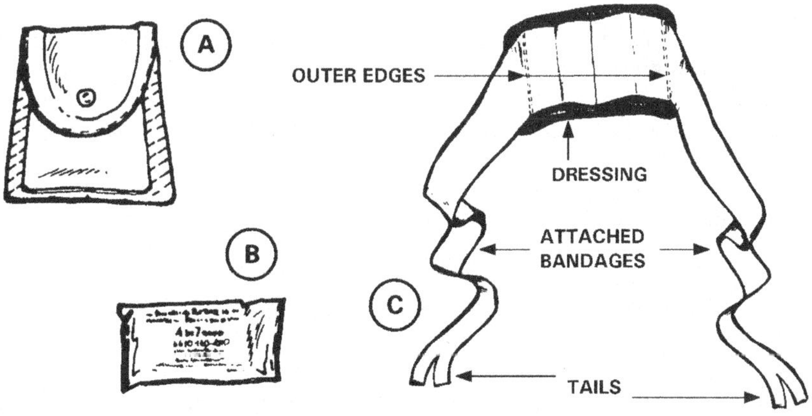

Figure A-1. Field first aid case and dressing (Illustrated A—C).

A-2. General Purpose First Aid Kits

General-purpose first aid kits are listed in the common table of allowances (CTA) 8-100. The operators, crew, and passengers carry these kits on Army vehicles, aircraft, and boats for use. Individuals designated by unit TSOP to be responsible for these kits are required to check them regularly and replace all items used. The general-purpose kit and its contents can be obtained through the unit supply system.

NOTE

Periodically check the dressings (for holes or tears in the packaging) and the medicines (for expiration date) that are in the first aid kits. If necessary, replace defective or outdated items.

A-3. Dressings

Dressings are sterile pads or compresses used to cover wounds. They usually are made of gauze or cotton wrapped in gauze (Figure A-1C). In addition to the standard field first aid dressing, other dressings such as sterile gauze compresses and small sterile compresses on adhesive strips may be available under CTA 8-100.

A-4. Standard Bandages

a. Standard bandages are made of gauze or muslin and are used over a sterile dressing to secure the dressing in place, to close off its edge from dirt and germs, and to create pressure on the wound and control bleeding. A bandage can also support an injured part or secure a splint.

b. Tailed bandages may be attached to the dressing as indicated on the field first aid dressing (Figure A-1C).

A-5. Triangular and Cravat (Swathe) Bandages

a. Triangular and cravat (or swathe) bandages (Figure A-2) are fashioned from a triangular piece of muslin (37 by 37 by 52 inches) provided in the general-purpose first aid kit. If it is folded into a strip, it is called a cravat. Two safety pins are packaged with each bandage. These bandages are valuable in an emergency since they are easily applied.

b. To improvise a triangular bandage, cut a square of available material, slightly larger than 3 feet by 3 feet, and *fold it diagonally*. If two bandages are needed, cut the material along the diagonal fold.

c. A cravat can be improvised from such common items as T-shirts, other shirts, bed linens, trouser legs, scarfs, or any other item made of pliable and durable material that can be folded, torn, or cut to the desired size.

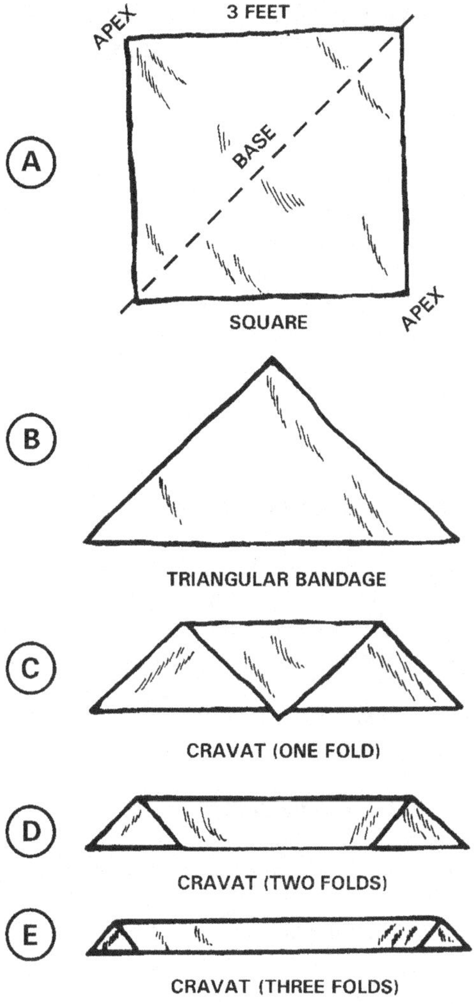

Figure A-2. Triangular and cravat bandages (Illustrated A—E).

APPENDIX B

RESCUE AND TRANSPORTATION PROCEDURES

B-1. General

A basic principle of first aid is to evaluate the casualty's injuries and administer first aid before moving him. However, adverse situations or conditions may jeopardize the lives of both the rescuer and the casualty if this is done. It may be necessary *first to rescue* the casualty before first aid can be effectively or safely given. The life and/or the well-being of the casualty will depend as much upon the manner in which he is *rescued and transported,* as it will upon the first aid and medical treatment he receives. Rescue actions must be done quickly and safely. Careless or rough handling of the casualty during rescue operations can aggravate his injuries.

B-2. Principles of Rescue Operations

a. When faced with the necessity of rescuing a casualty who is threatened by hostile action, fire, water, or any other immediate hazard, DO NOT take action without first determining the extent of the hazard and your ability to handle the situation. DO NOT become a casualty.

b. The rescuer must evaluate the situation and analyze the factors involved. This evaluation involves three major steps:

- Identify the task.
- Evaluate circumstances of the rescue.
- Plan the action.

B-3. Considerations

a. First determine if a rescue attempt is actually needed. It is a waste of time, equipment, and personnel to rescue someone not in need of rescuing. It is also a waste to look for someone who is not lost or needlessly risk the lives of the rescuer(s). In planning a rescue, attempt to obtain the following information:

- Who, what, where, when, why, and how the situation happened?

- How many casualties are involved and the nature of their injuries?

- What is the tactical situation?

- What are the terrain features and the location of the casualties?

- Will there be adequate assistance available to aid in the rescue/evacuation?

- Can first aid and/or medical treatment be provided at the scene; will the casualties require movement to a safer location?

- What specialized equipment will be required for the rescue operation?

- Is the rescue area contaminated? Will decontamination equipment and materiel be required for casualties, rescue personnel, and rescue equipment?

- How much time is available?

b. The time element can play a significant role in how the rescue is attempted. If the casualties are in imminent danger of losing their lives (such as near a burning vehicle or in a burning building) the time available will be relatively short and will sometimes cause a rescuer to compromise planning stages and/or the first aid which can be given. However, if the casualty is in a relatively secure area and his physical condition is strong, more deliberate planning can take place. A realistic estimate of time available must be made as quickly as possible to determine action time remaining. The key elements are the casualty's physical and mental condition, the tactical situation, and the environment.

B-4. Plan of Action

a. The casualty's ability to endure is of primary importance in estimating the time available. Age, physical condition, and extent of wounds and/or injuries will differ from casualty to casualty. Therefore, to determine the time available, you will have to consider—

- Endurance time of the casualty.

- Extent of injuries.

- Type of situation.
- Personnel and/or equipment availability.
- Weather.
- Terrain (natural and man-made).
- Environment (contaminated or uncontaminated).

b. In respect to terrain, you must consider altitude and visibility. In some cases, the casualty may be of assistance because he knows more about the particular terrain or situation than you do. Maximum use of secure/reliable trails or roads is essential.

c. When taking weather into account, ensure that blankets and/or rain gear are available. Even a mild rain can complicate a normally simple rescue. In high altitudes and/or extreme cold and gusting winds, the time available is critically shortened. Be prepared to provide shelter and warmth for the casualty as well as the rescuers.

B-5. Proper Handling of Casualties

a. You may have saved the casualty's life through the application of appropriate first aid measures. However, his life can be lost through rough handling or careless transportation procedures. Before you attempt to move the casualty—

- Evaluate the type and extent of his injuries.
- Ensure that dressings over wounds are adequately reinforced.
- Ensure that fractured bones are properly immobilized and supported to prevent them from cutting through muscle, blood vessels, and skin.

b. Based upon your evaluation of the type and extent of the casualty's injury and your knowledge of the various manual carries, you must select the best possible method of manual transportation. If the casualty is conscious, tell him how he is to be transported. This will help allay his fear of movement and gain his cooperation and confidence.

c. Buddy aid for chemical agent casualties includes those actions required to prevent an incapacitated casualty from receiving additional injury from the effects of chemical hazards. If a casualty is physically unable to decontaminate himself or administer the proper chemical agent antidote, the casualty's buddy assists him and assumes responsibility for his care. Buddy-aid includes—

- Administering the proper chemical agent antidote.
- Decontaminating the incapacitated casualty's exposed skin.
- Ensuring that his protective ensemble remains correctly emplaced.
- Maintaining respiration.
- Controlling bleeding.
- Providing other standard first aid measures
- Transporting the casualty out of the contaminated area.

B-6. Positioning the Casualty

The first step in any manual carry is to position the casualty to be lifted. If he is conscious, he should be told how he is to be positioned and transported. This helps lessen his fear of movement and to gain his cooperation. It may be necessary to roll the casualty onto his abdomen, or his back, depending upon the position in which he is lying and the particular carry to be used.

a. To roll a casualty onto his abdomen, kneel at the casualty's uninjured side.

(1) Place his arms above his head; cross his ankle which is farther from you over the one that is closer to you.

(2) Place your hands on the shoulder which is farther from you; place your other hand in the area of his hip or thigh (Figure B-1).

(3) Roll him gently toward you onto his abdomen (Figure B-2).

b. To roll a casualty onto his back, follow the same procedure described in *a* above, except gently roll the casualty onto his back, rather than onto his abdomen.

Figure B-1. Positioning the casualty.

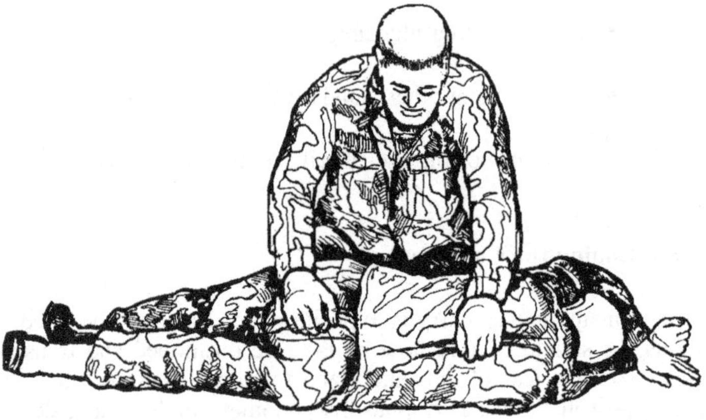

Figure B-2. Rolling casualty onto his abdomen.

B-7. Medical Evacuation and Transportation of Casualties

a. Medical evacuation of the sick and wounded (with en route medical care) is the responsibility of medical personnel who have been provided special training and equipment. Therefore, unless a good reason for you to transport a casualty arises, wait for some means of medical evacuation to be provided. When the situation is urgent and you are unable to obtain medical assistance or know that no medical evacuation assets are available, you will have to transport the casualty. For this reason, you must know how to transport him without increasing the seriousness of his condition.

b. Transporting a casualty by litter (FM 8-10-6) is safer and more comfortable for him than by manual means; it is also easier for you.

Manual transportation, however, may be the only feasible method because of the terrain or the combat situation; or it may be necessary to save a life. In these situations, the casualty should be transferred to a litter as soon as one can be made available or improvised.

B-8. Manual Carries

Casualties carried by manual means must be carefully and correctly handled, otherwise their injuries may become more serious or possibly fatal. Situation permitting, transport of a casualty should be organized and unhurried. Each movement should be performed as deliberately and gently as possible. Casualties should not be moved before the type and extent of injuries are evaluated and the required first aid is administered. The exception to this occurs when the situation dictates immediate movement for safety purposes (for example, it may be necessary to remove a casualty from a burning vehicle); that is, the situation dictates that the urgency of casualty movement outweighs the need to administer first aid. Manual carries are tiring for the bearers and involve the risk of increasing the severity of the casualty's injury. In some instances, however, they are essential to save the casualty's life. Although manual carries are accomplished by one or two bearers, the two-man carries are used whenever possible. They provide more comfort to the casualty, are less likely to aggravate his injuries, and are also less tiring for the bearers. The distance a casualty can be carried depends on many factors, such as—

- Nature of the casualty's injuries.
- Strength and endurance of the bearer(s).
- Weight of the casualty.
- Obstacles encountered during transport (natural or manmade).
- Type of terrain.

a. One-man Carries. These carries should be used when only one bearer is available to transport the casualty.

(1) The *fireman's carry* (Figure B-3) is one of the easiest ways for one individual to carry another. After an unconscious or disabled casualty has been properly positioned, he is raised from the ground, then supported and placed in the carrying position.

(*a*) After rolling the casualty onto his abdomen, straddle him. Extend your hands under his chest and lock them together.

(*b*) Lift the casualty to his knees as you move backward.

(*c*) Continue to move backward, thus straightening the casualty's legs and locking his knees.

(*d*) Walk forward, bringing the casualty to a standing position; tilt him slightly backward to prevent his knees from buckling.

(*e*) As you maintain constant support of the casualty with one arm, free your other arm, quickly grasp his wrist, and raise his arm high. Instantly pass your head under his raised arm, releasing it as you pass under it.

(*f*) Move swiftly to face the casualty and secure your arms around his waist. Immediately place your foot between his feet and spread them apart (approximately 6 to 8 inches).

(*g*) Grasp the casualty's wrist and raise his arm high over your head.

(*h*) Bend down and pull the casualty's arm over and down on your shoulder, bringing his body across your shoulders. At the same time, pass your arm between his legs.

(*i*) Grasp the casualty's wrist with one hand, and place your other hand on your knee for support.

(*j*) Rise with the casualty positioned correctly. Your other hand is free for use.

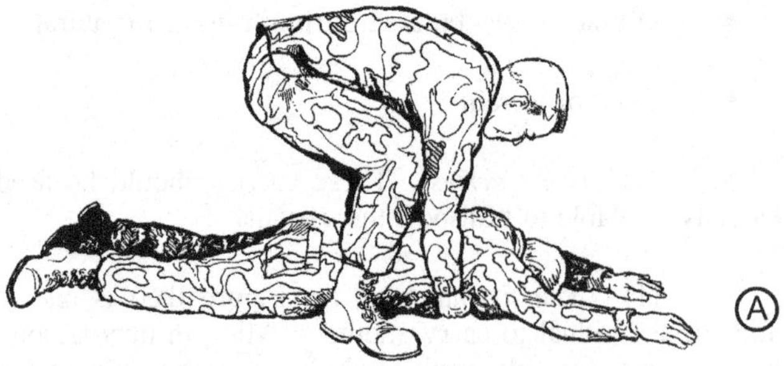

Figure B-3. Fireman's carry (Illustrated A—J).

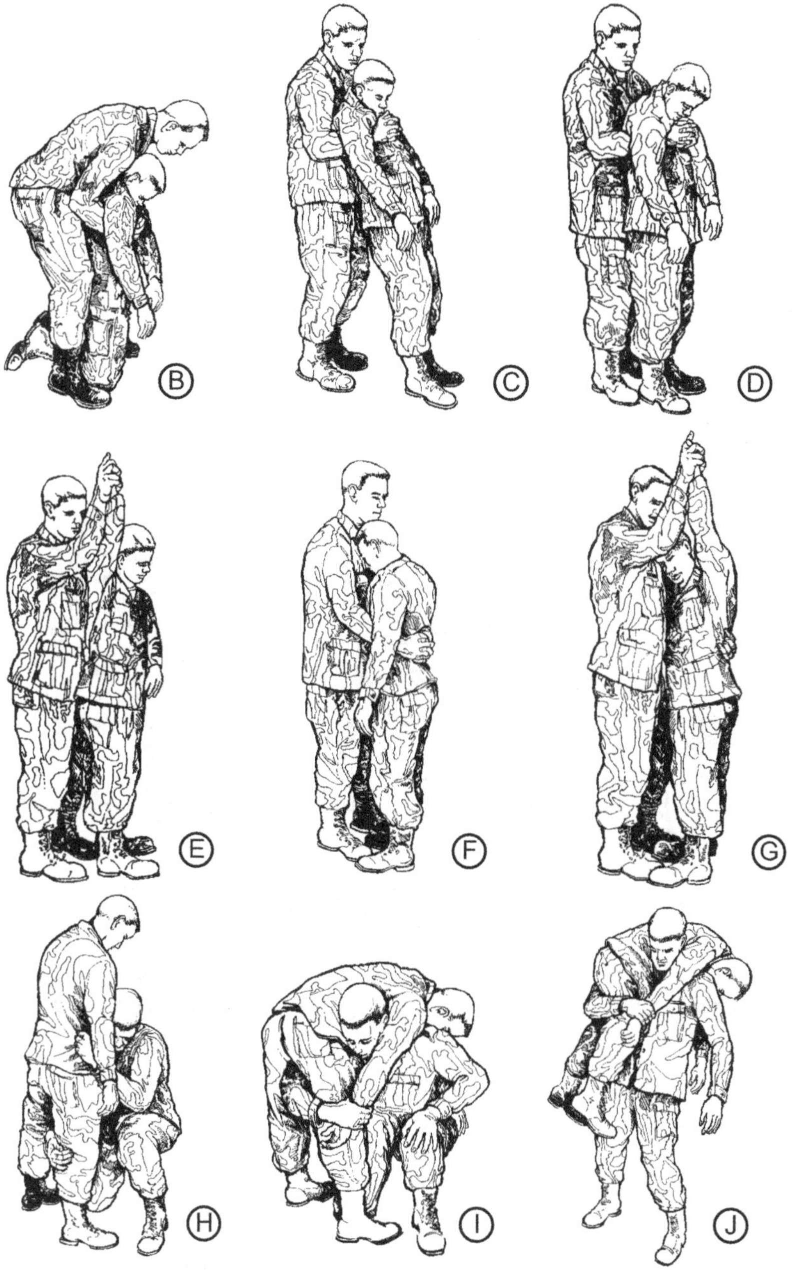

Figure B-3. Fireman's carry (Illustrated A—J) (Continued).

(2) The alternate method of the *fireman's carry* for raising a casualty from the ground is illustrated in Figure B-4; however, it should be used only when the bearer believes it to be safer for the casualty because of the location of his wounds. When the alternate method is used, care must be taken to prevent the casualty's head from snapping back and causing a neck injury. The steps for raising a casualty from the ground for the fireman's carry are also used in other one-man carries.

(*a*) Kneel on one knee at the casualty's head and face his feet. Extend your hands under his armpits, down his sides, and across his back.

(*b*) As you rise, lift the casualty to his knees. Then secure a lower hold and raise him to a standing position with his knees locked.

Figure B-4. Fireman's carry (alternate method) for lifting a casualty to a standing position (Illustrated A—B).

(3) In the *supporting carry* (Figure B-5), the casualty must be able to walk or at least hop on one leg, using the bearer as a crutch. This carry can be used to assist him as far as he is able to walk or hop.

(*a*) Raise the casualty from the ground to a standing position by using the fireman's carry.

(*b*) Grasp the casualty's wrist and draw his arm around your neck.

(*c*) Place your arm around his waist. The casualty is now able to walk or hop using you as a support.

Figure B-5. Supporting carry.

(4) The *arms carry* (Figure B-6) is useful in carrying a casualty for a short distance (up to 50 meters) and for placing him on a litter.

(*a*) Raise or lift the casualty from the ground to a standing position, as in the fireman's carry.

(*b*) Place one arm under the casualty's knees and your other arm around his back.

(*c*) Lift the casualty.

(*d*) Carry the casualty high to lessen fatigue.

Figure B-6. Arms carry.

(5) Only a conscious casualty can be transported by the *saddleback carry* (Figure B-7), because he must be able to hold onto the bearer's neck. To use this technique—

(*a*) Raise the casualty to an upright position, as in the fireman's carry.

(*b*) Support the casualty by placing an arm around his waist. Move to the casualty's side. Have the casualty put his arm around your neck and move in front of him with your back to support him.

(*c*) Have the casualty encircle his arms around your neck

(*d*) Stoop, raise him on your back and clasp your hands together beneath his thighs, if possible.

Figure B-7. Saddleback carry.

(6) In the *pack-strap carry* (Figure B-8), the casualty's weight rests high on the your back. This makes it easier for you to carry the casualty a moderate distance (50 to 300 meters). To eliminate the possibility of injury to the casualty's arms, you must hold his arms in a palms-down position.

(*a*) Lift the casualty from the ground to a standing position, as in the fireman's carry.

(*b*) Support the casualty with your arms around him and grasp his wrist closer to you.

(*c*) Place his arm over your head and across your shoulders.

(*d*) Move in front of him while still supporting his weight against your back.

(*e*) Grasp his other wrist and place this arm over your shoulder.

(*f*) Bend forward and raise or hoist the casualty as high on your back as possible so that his weight is resting on your back.

NOTE

Once the casualty is positioned on the bearer's back, the bearer remains as erect as possible to prevent straining or injuring his back.

Figure B-8. Pack-strap carry.

(7) The *pistol-belt carry* (Figure B-9) is the best one-man carry for a long distance (over 300 meters). The casualty is securely supported upon your shoulders by a belt. Both your hands and the casualty's (if conscious) are free for carrying a weapon or equipment, or climbing obstacles. With your hands free and the casualty secured in place, you are also able to creep through shrubs and under low-hanging branches.

(*a*) Link two pistol belts (or three, if necessary) together to form a sling. Place the sling under the casualty's thighs and lower back so that a loop extends from each side.

NOTE

If pistol belts are not available for use, other items such as a rifle sling, two cravat bandages, two litter straps, or any other suitable material, which will not cut or bind the casualty may be used.

(*b*) Lie face up between the casualty's outstretched legs. Thrust your arms through the loops and grasp his hands and trouser leg on his injured side.

(*c*) Roll toward the casualty's uninjured side onto your abdomen, bringing him onto your back. Adjust the sling, if necessary.

(*d*) Rise to a kneeling position. The belt will hold the casualty in place.

(*e*) Place one hand on your knee for support and rise to an upright position. (The casualty is supported on your shoulders.)

(*f*) Carry the casualty with your hands free for use in rifle firing, climbing, or surmounting obstacles.

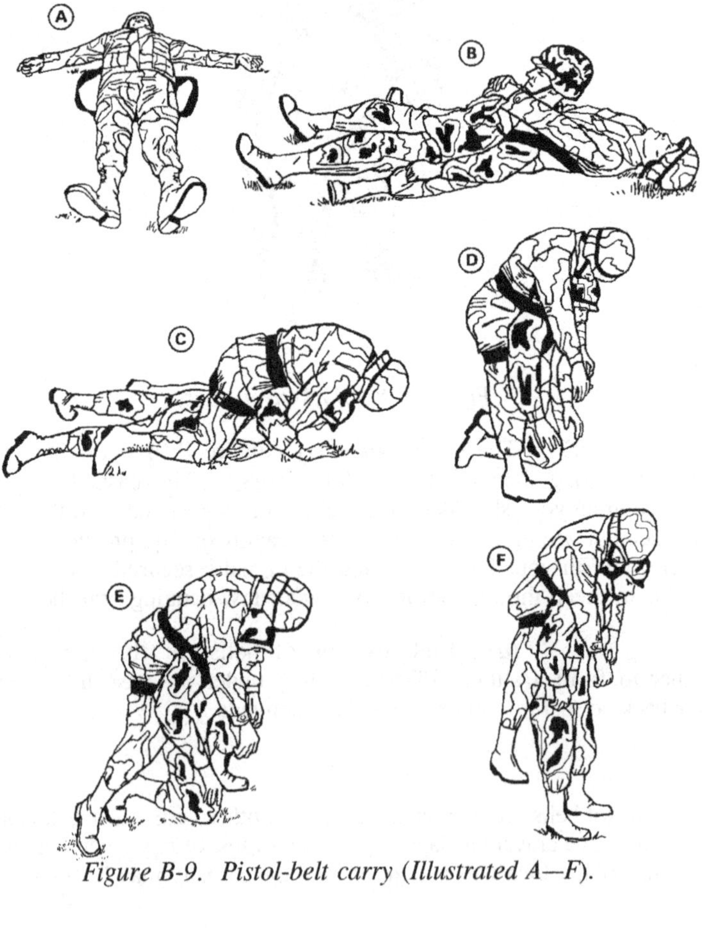

Figure B-9. Pistol-belt carry (Illustrated A—F).

(8) The *pistol-belt drag* (Figure B-10), as well as other drags, is generally used for short distances (up to 50 meters). This drag is useful in combat, since both the bearer and the casualty can remain closer to the ground than in any other drags.

(*a*) Extend two pistol belts or similar objects to their full length and join them together to make a continuous loop.

(*b*) Roll the casualty onto his back, as in the fireman's carry.

(*c*) Pass the loop over the casualty's head, and position it across his chest and under his armpits. Then cross the remaining portion of the loop, thus forming a figure eight. Keep tension on the belts so they do not come unhooked.

(*d*) Lie on your side facing the casualty.

(*e*) Slip the loop over your head and turn onto your abdomen. This enables you to drag the casualty as you crawl.

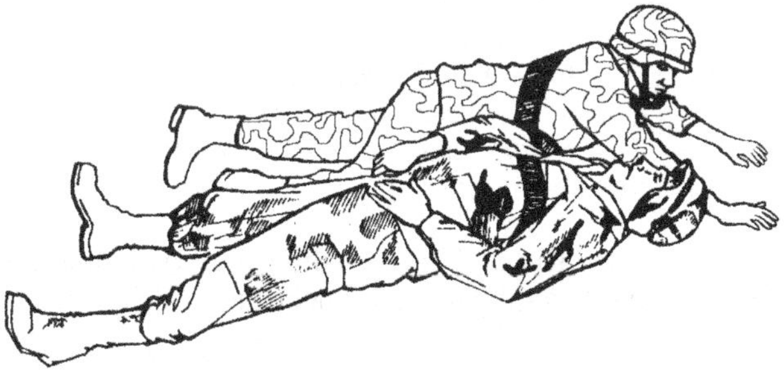

Figure B-10. Pistol-belt drag.

(9) The *neck drag* (Figure B-11) is useful in combat because the bearer can transport the casualty as he creeps behind a low wall or shrubbery, under a vehicle, or through a culvert. If the casualty is unconscious, his head must be protected from the ground. The neck drag cannot be used if the casualty has a broken arm.

NOTE

If the casualty is conscious, he may clasp his hands together around your neck.

(*a*) Tie the casualty's hands together at the wrists.

(*b*) Straddle the casualty in a kneeling face-to-face position.

(*c*) Loop the casualty's tied hands over and around your neck.

(*d*) Crawl forward dragging the casualty with you.

NOTE

If the casualty is unconscious, protect his head from the ground.

Figure B-11. Neck drag.

(10) The *cradle drop drag* (Figure B-12) is effective in moving a casualty up or down steps.

(*a*) Kneel at the casualty's head (with him lying on his back). Slide your hands, with palms up, under the casualty's shoulders and get a firm hold under his armpits.

(*b*) Rise (partially), supporting the casualty's head on one of your forearms. (You may bring your elbows together and let the casualty's head rest on both of your forearms.)

(*c*) Rise and drag the casualty backward. (The casualty is in a semisitting position.)

(*d*) Back down the steps, supporting the casualty's head and body and letting his hips and legs drop from step to step.

NOTE

If the casualty needs to be moved up the steps, you should back up the steps, using the same procedure.

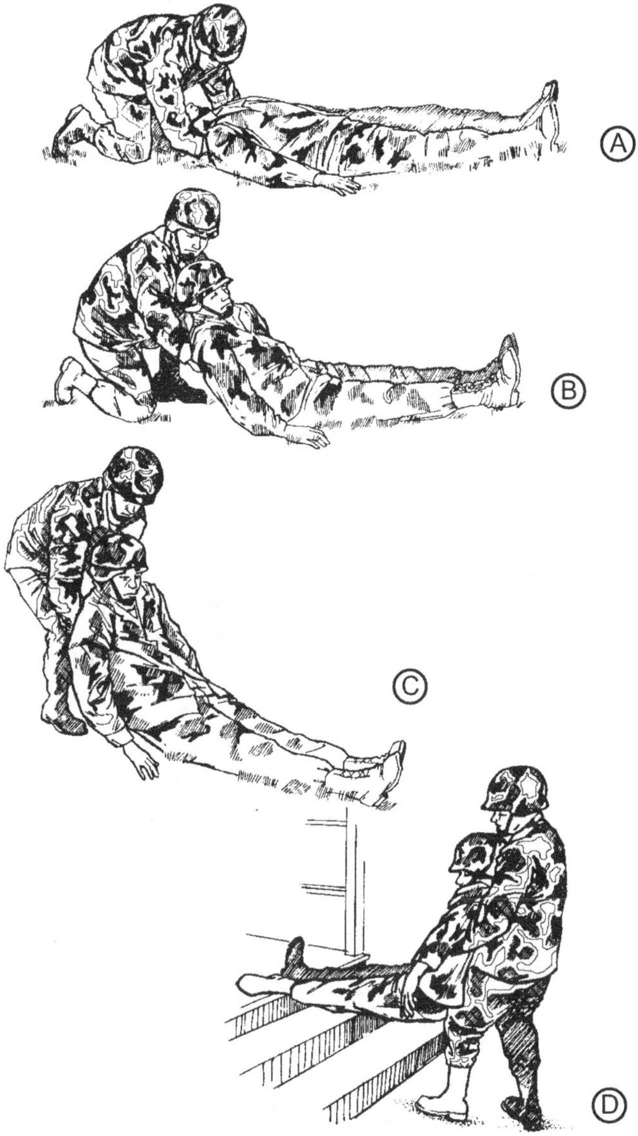

Figure B-12. Cradle-drop drag (Illustrated A—D).

(11) The *LBE carry using the bearer's LBE* can be used with a conscious casualty (Figure B-13).

(*a*) Loosen all suspenders on your LBE.

(*b*) Have the casualty place one leg into the loop formed by your suspenders and pistol belt.

(*c*) Squat in front of the standing casualty. Have him place his other leg into the loop, also.

(*d*) Have the casualty place his arms over your shoulders, lean forward onto your back, and lock his hands together.

(*e*) Stand up and lean forward into a comfortable position.

(*f*) Continue the mission.

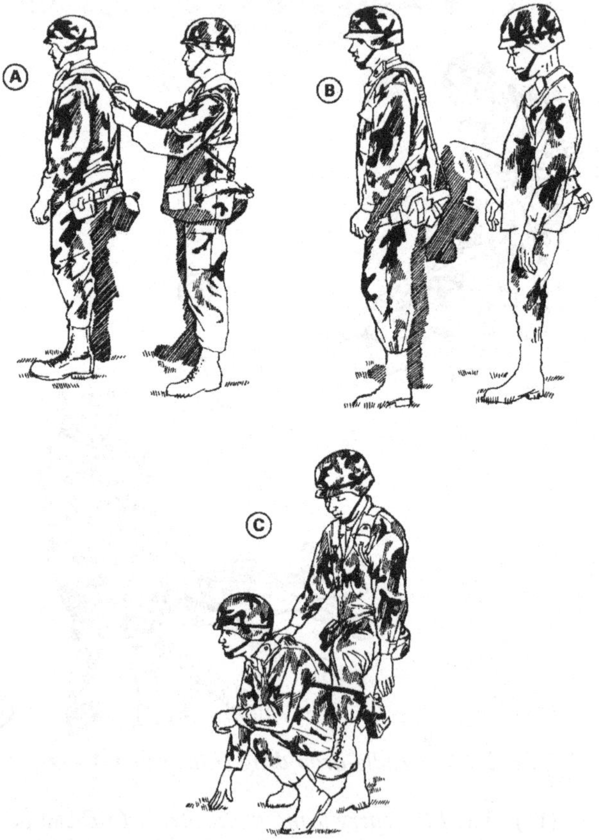

Figure B-13. Load bearing equipment carry using bearer's LBE (conscious casualty) (Illustrated A—F).

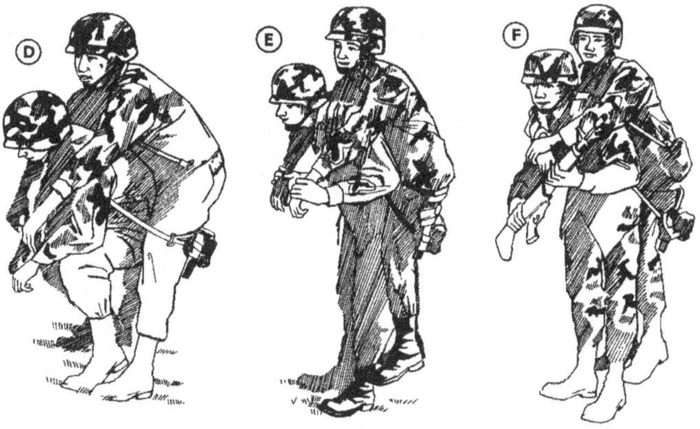

Figure B-13. Load bearing equipment carry using bearer's LBE (conscious casualty) (Illustrated A—F) (Continued).

(12) The *LBE carry using the bearer's LBE* can be used with an unconscious casualty or one who cannot stand (Figure B-14).

(*a*) Position the casualty on the flat of his back.

(*b*) Remove your LBE and loosen all suspender straps.

(*c*) Lift the casualty's leg and place it through the loop formed by your suspenders and pistol belt. Then place the other leg through the same loop. The LBE is moved up until the pistol belt is behind the casualty's thighs.

(*d*) Lay between the casualty's legs; work your arms through the LBE suspenders.

(*e*) Grasp the casualty's hand (on the injured side), and roll the casualty (on his uninjured side) onto your back.

(*f*) Rise to one knee and then push into a standing position.

(*g*) Bring the casualty's arms over your shoulders. Grasp his hands and secure them if the casualty is unconscious. If the casualty is conscious, have him lock his hands in front if he is able to do so.

(*h*) Lean forward into a comfortable position and continue the mission.

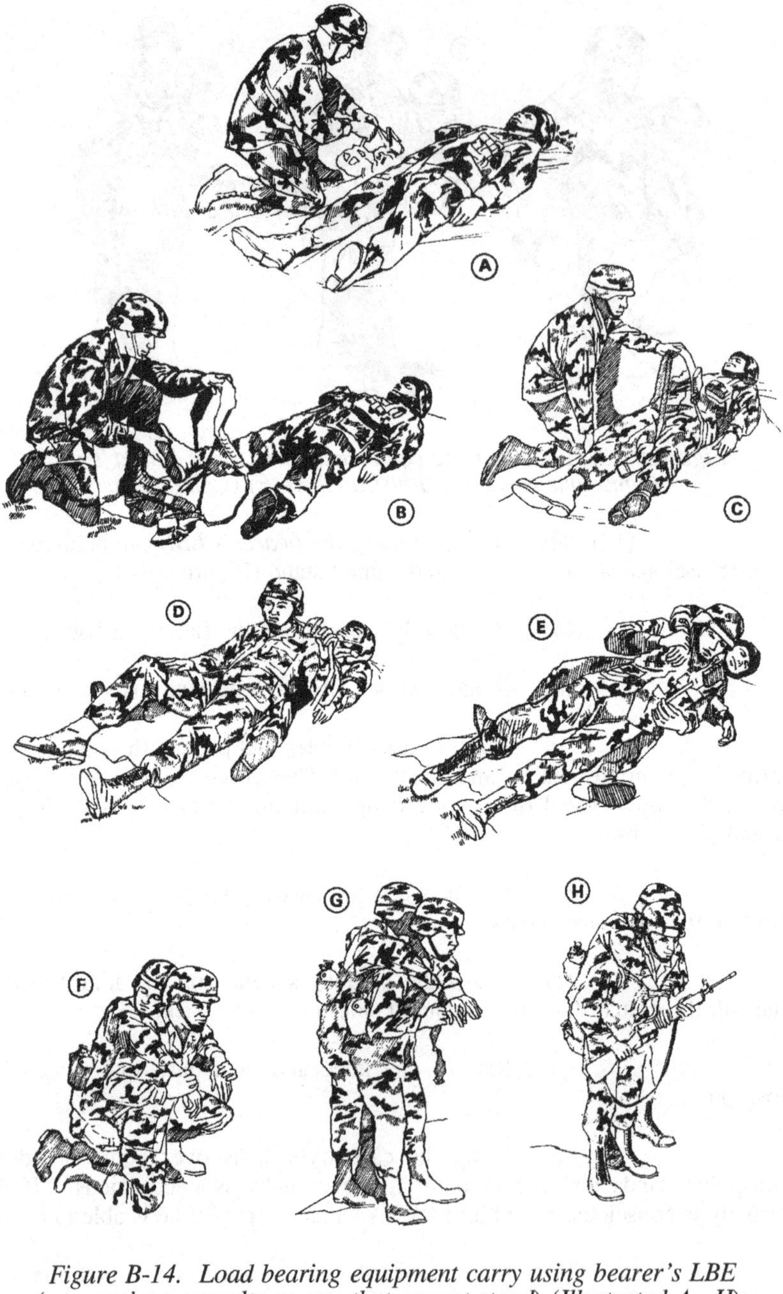

Figure B-14. Load bearing equipment carry using bearer's LBE (unconscious casualty or one that cannot stand) (Illustrated A—H).

(13) The *LBE carry using the casualty's LBE* (Figure B-15) can be used with a conscious or unconscious casualty.

(*a*) Position the casualty on his back with his LBE on.

(*b*) Loosen the casualty's two front suspenders.

(*c*) Position yourself between the casualty's legs, and slip your arms into the casualty's two front suspenders (up to your shoulders).

(*d*) Work his arms out of his LBE suspenders.

(*e*) Grasp the casualty's hand (on the injured side), and roll him (on his uninjured side) onto your back.

(*f*) Rise to one knee, then into a standing position.

(*g*) Grasp the casualty's hands and secure them, if the casualty is unconscious. Have the casualty lock his hands in front of you, if he is conscious.

(*h*) Lean forward into a comfortable position and continue the mission.

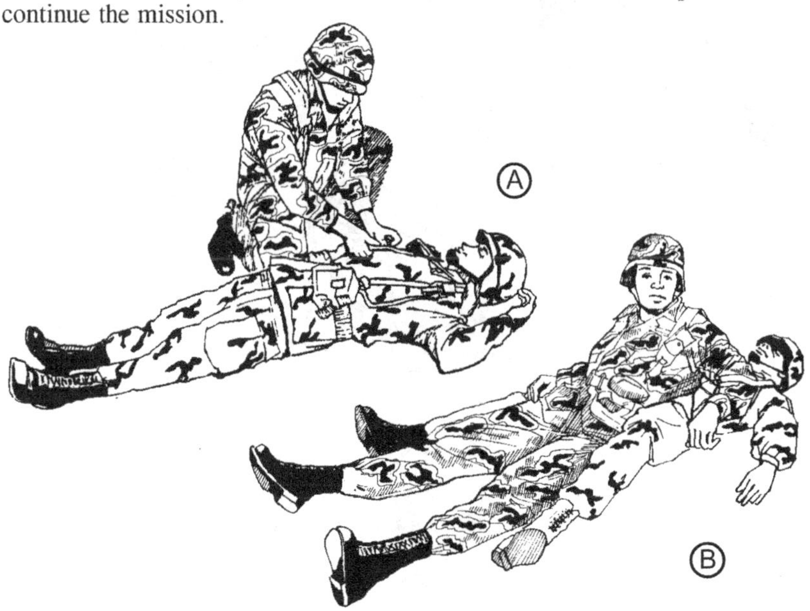

Figure B-15. Load bearing equipment carry using casualty's LBE (Illustrated A—G).

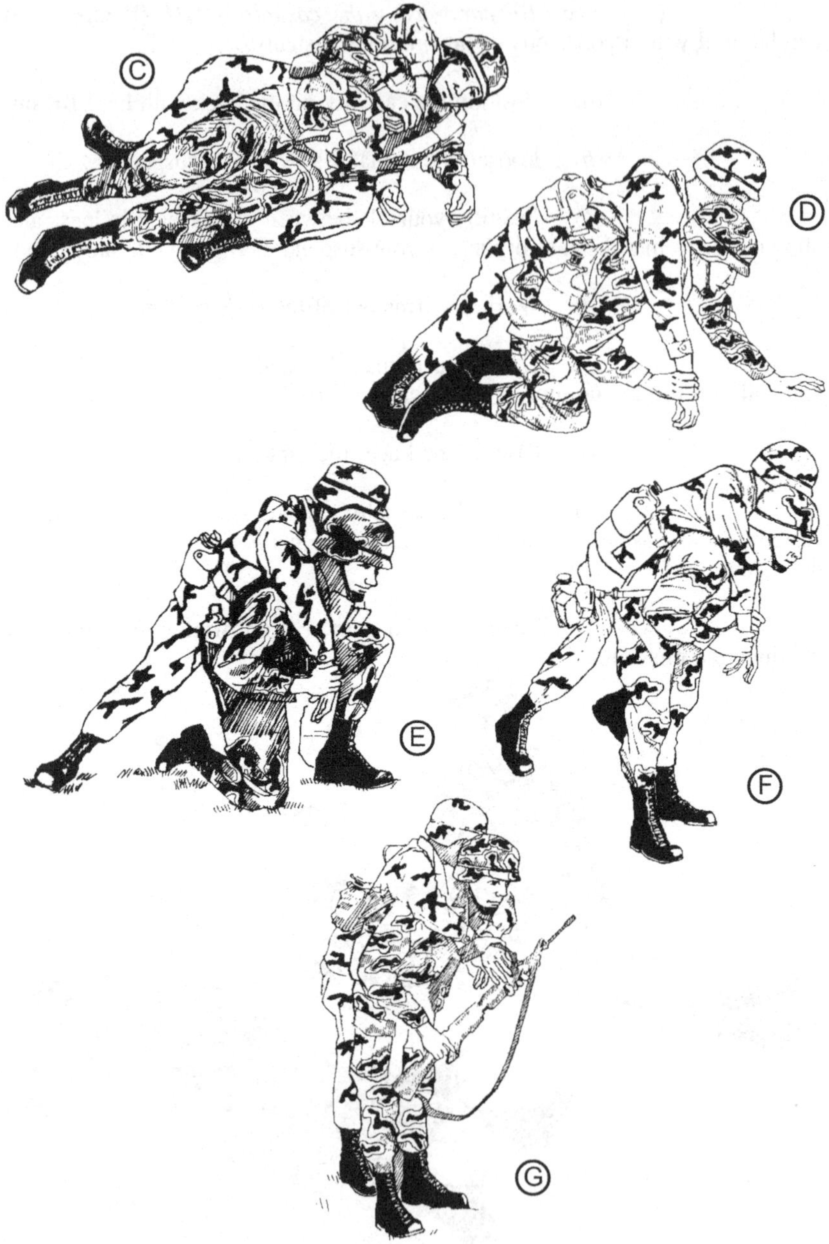

Figure B-15. Load bearing equipment carry using casualty's LBE (Illustrated A—G) (Continued).

b. *Two-man Carries*. These carries should be used whenever possible. They provide more casualty comfort, are less likely to aggravate injuries, and are less tiring for the bearers. Five different two-man carries can be used.

(1) The *two-man support carry* (Figure B-16) can be used in transporting either conscious or unconscious casualties. If the casualty is taller than the bearers, it may be necessary for the bearers to lift the casualty's legs and let them rest on their forearms. The bearers—

(*a*) Help the casualty to his feet and support him with their arms around his waist.

(*b*) Grasp the casualty's wrists and draw his arms around their necks.

Figure B-16. Two-man supporting carry.

(2) The *two-man arms carry* (Figure B-17) is useful in carrying a casualty for a moderate distance (50 to 300 meters) and placing him on a litter. To lessen fatigue, the bearers should carry the casualty high and as close to their chests as possible. In extreme emergencies when there is no time to obtain a spine board, this carry is the safest one for transporting a casualty with a back injury. If possible, two additional bearers should be used to keep the casualty's head and legs in alignment with his body. The bearers—

(*a*) Kneel at one side of the casualty; then they place their arms beneath the casualty's back, waist, hips, and knees.

(*b*) Lift the casualty while rising to their knees.

(*c*) Turn the casualty toward their chests, while rising to a standing position. Carry the casualty high to lessen fatigue.

Figure B-17. Two-man arms carry (Illustrated A—D).

(3) The *two-man fore-and aft-carry* (Figure B-18) is a useful *two-man carry* for transporting a casualty for a long distance (over 300 meters). The taller of the two bearers should position himself at the casualty's head. By altering this carry so that both bearers face the casualty, it is useful for placing a casualty on a litter.

(*a*) The shorter bearer spreads the casualty's legs and kneels between them with his back to the casualty. He positions his hands behind the casualty's knees. The other bearer kneels at the casualty's head, slides his hands under the arms, across the chest, and locks his hands together.

(*b*) The two bearers rise together, lifting the casualty.

Figure B-18. Two-man fore-and-aft carry (Illustrated A—B).

(4) Only a conscious casualty can be transported with the *four-hand seat carry* (Figure B-19) because he must help support himself by placing his arms around the bearers' shoulders. This carry is especially useful in transporting a casualty with a head or foot injury for a moderate distance (50 to 300 meters). It is also useful for placing a casualty on a litter.

(*a*) Each bearer grasps one of his wrists and one of the other bearer's wrists, thus forming a packsaddle.

(*b*) The two bearers lower themselves sufficiently for the casualty to sit on the packsaddle; then, they have the casualty place his arms around their shoulders for support. The bearers then rise to an upright position.

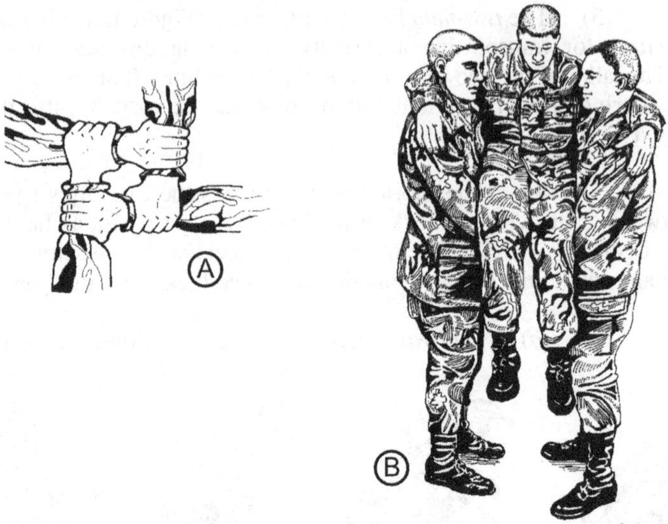

Figure B-19. Four-hand seat carry (Illustrated A—B).

(5) The *two-hand seat carry* (Figure B-20) is used when carrying a casualty for a short distance or for placing him on a litter. With the casualty lying on his back, a bearer kneels on each side of the casualty at his hips. Each bearer passes his arms under the casualty's thighs and back, and grasps the other bearer's wrists. The bearers rise lifting the casualty.

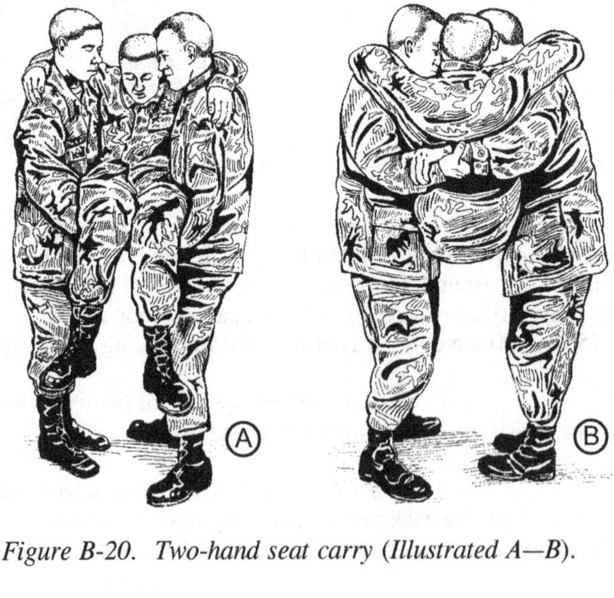

Figure B-20. Two-hand seat carry (Illustrated A—B).

B-9. Improvised Litters

Two men can support or carry a casualty without equipment for only short distances. By using available materials to improvise equipment, the casualty can be transported greater distances by two or more rescuers.

a. There are times when a casualty may have to be moved and a standard litter is not available. The distance may be too great for manual carries or the casualty may have an injury (such as a fractured neck, back, hip, or thigh) that would be aggravated by manual transportation. In these situations, litters can be improvised from materials at hand. Improvised litters must be as well constructed as possible to avoid risk of dropping or further injuring the casualty. Improvised litters are emergency measures and must be replaced by standard litters at the first opportunity.

b. Many different types of litters can be improvised, depending upon the materials available. A satisfactory litter can be made by securing poles inside such items as a blanket, poncho, shelter half, tarpaulin, mattress cover, jacket, shirt, or bed ticks, bags, and sacks (Figure B-18). Poles can be improvised from strong branches, tent supports, skis, lengths of pipe or other objects. If objects for improvising poles are not available, a blanket, poncho, or similar item can be rolled from both sides toward the center so the rolls can be gripped for carrying a patient. Most flat-surface objects of suitable size can be used as litters. Such objects include doors, boards, window shutters, benches, ladders, cots, and chairs. If possible, these objects should be padded for the casualty's comfort.

(1) To improvise a litter using a blanket and poles (Figure B-21), the following steps should be used.

Figure B-21. Litter made with blanket and poles.

(*a*) Open the blanket and lay one pole lengthwise across the center; then fold the blanket over the pole.

(*b*) Place the second pole across the center of the folded blanket.

(*c*) Fold the free edges of the blanket over the second pole and across the first pole.

(2) To improvise a litter using shirts or jackets (Figure B-22), button the shirt or jacket and turn it inside out, leaving the sleeves inside, (more than one shirt or jacket may be required), then pass the pole through the sleeves.

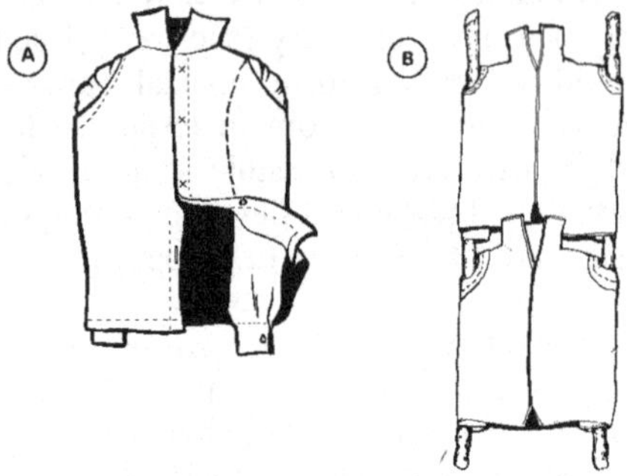

Figure B-22. Litter improvised from jackets and poles (Illustrated A—B).

(3) To improvise a litter from bed sacks and poles (Figure B-23), rip open the corners of bed ticks, bags, or sacks; then pass the poles through them.

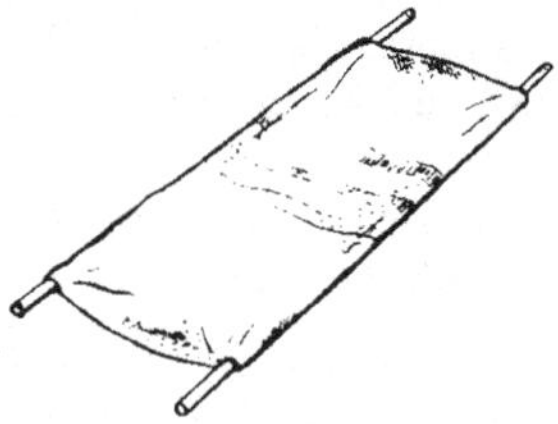

Figure B-23. Litter improvised from bed sacks and poles.

(4) If no poles are available, roll a blanket, shelter half, tarpaulin, or similar item from both sides toward the center (Figure B-24). Grip the rolls to carry the casualty.

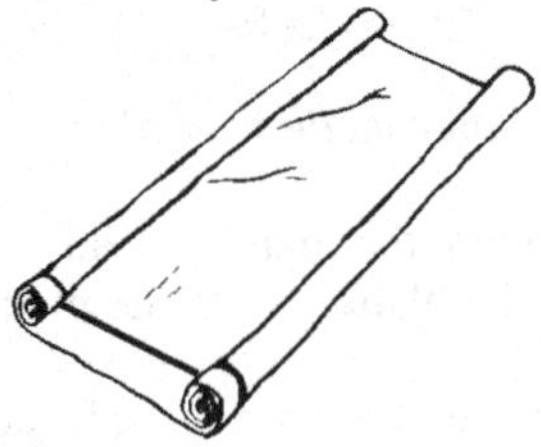

Figure B-24. Rolled blanket used as a litter.

c. Any of the appropriate carries may be used to place a casualty on a litter. These carries are:

- The one-man arms carry (Figure B-6).
- The two-man arms carry (Figure B-17).
- The two-man fore-and-aft carry (Figure B-18).
- The two-hand seat carry (Figure B-20).
- The four-hand seat carry (Figure B-19).

WARNING

Unless there is an immediate life-threatening situation (such as fire, explosion), DO NOT move a casualty with a suspected back or neck injury. Seek medical personnel for guidance on how to transport.

d. Either two or four service members (head/foot) may be used to lift a litter. To lift the litter, follow the procedure below.

(1) Raise the litter at the same time as the other carriers/ bearers.

(2) Keep the casualty as level as possible.

NOTE

Use caution when transporting on a sloping incline/hill.

GLOSSARY

ACRONYMS, ABBREVIATIONS, AND DEFINITIONS

AC hydrogen cyanide
AFMAN Air Force Manual
AOC area of concentration
AR Army regulation
ATM advanced trauma management
ATNAA Antidote Treatment, Nerve Agent, Autoinjector
attn attention

BDO battle dress overgarment
BDU battle dress uniform
BZ anticholinergic drugs

C Celsius
CANA Convulsant Antidote for Nerve Agent
CASEVAC casualty evacuation
cc cubic centimeter
CG phosgene
CHS combat health support
CK cyanogen chloride
Cl chlorine
CLS Combat Lifesaver
CNS central nervous system
CO_2 carbon dioxide
COSR combat and operational stress reactions
CSR combat stress reaction
CTA common table of allowance
CX phosgene oxime

DA Department of the Army
DD Department of Defense
DM diphenylaminochloroarsine (adamsite)
DNBI disease and nonbattle injury
DOD Department of Defense
DP diphosgene
DS direct support

EMT emergency medical treatment

F Fahrenheit
FM field manual

H mustard

HD mustard
HM Hospital Corpsman
HN nitrogen mustard
HSS health service support

IPE individual protective equipment
IV intravenous

JSLIST Joint Services Light Weight Integrated Suit Technology

L lewisite
lasers laser means Light Amplification by Stimulated Emission of Radiation and sources include range finders, weapons/guidance, communication systems, and weapons simulations such as MILES [Multiple Integrated Laser Engagement System].
LBE load bearing equipment
LX lewisite and mustard

MCRP Marine Corps Reference Publication
MILES Multiple Integrated Laser Engagement System
ml milliliter
MOPP mission-oriented protective posture
MOS military occupational specialty
MTF medical treatment facility

NAPP Nerve Agent Pyridostigmine Pretreatment
NATO North Atlantic Treaty Organization
NBC nuclear, biological, and chemical
NCO noncommissioned officer
NTRP Navy Tactical Reference Publication

occlusive dressing air tight transparent dressing used to seal and cover wounds
oz ounce

PAM pamphlet
PS chloropicrin
PTSD post-traumatic stress disorder

QSTAG Quadripartite Standardization Agreement

SOP standing operating procedure
STANAG standardization agreement
STP soldier training publication

2 PAM Cl pralidoxime chloride
TB MED technical bulletin medical
TM technical manual
TSOP tactical standing operating procedure

US United States

WP white phosphorus

REFERENCES

DOCUMENTS NEEDED

These documents must be available to the intended users of this publication.

NATO STANAGs

These agreements are available on request using DD Form 1425 from Standardization Document Order Desk, 700 Robin Avenue, Building 4, Section D, Philadelphia, Pennsylvania 19111-5094.

2122. *Medical Training in First Aid, Basic Hygiene and Emergency Care.* 10 December 1975.
2126. *First Aid Kits and Emergency Medical Care Kits.* 27 September 1983.
2358. *First Aid and Hygiene Training in NBC Operations.* 3 March 1989.
2871. *First Aid Material for Chemical Injuries.* 10 March 1986.

ABCA QSTAGs

These agreements are available on request using DD Form 1425 from Standardization Document Order Desk, 700 Robin Avenue, Building 4, Section D, Philadelphia, Pennsylvania 19111-5094.

535. *Medical Training in First Aid, Basic Hygiene and Emergency Care.* 12 November 1979.

Joint and Multiservice Publications

FM 21-10. *Field Hygiene and Sanitation.* MCRP 4-11.1D. 21 June 2000.

Army Publications

AR 350-41. *Training In Units.* 19 March 1993.
DA PAM 350-59. *Army Correspondence Course Program Catalog.* 26 October 2001.
FM 3-4. *NBC Protection.* FMFM 11-9. 29 May 1992 (Reprinted with basic including Change 1, 28 October 1992; Change 2, 26 February 1996.)
FM 3-5. *NBC Decontamination.* MCWP 3-37.3. 28 July 2000. (Change 1, 31 January 2002.)

FM 3-100. *Chemical Operations Principles and Fundamentals.* MCWP 3-3.7.1. 8 May 1996.

FM 4-02.33 (8-33). *Control of Communicable Diseases Manual* (17th Edition). 3 January 2000.

FM 8-10-6 (4-02.2). *Medical Evacuation in a Theater of Operations—Tactics, Techniques, and Procedures.* 14 April 2000.

FM 4-02.7 (8-10-7). *Health Service Support in a Nuclear, Biological, and Chemical Environment.* 1 October 2002.

FM 8-284 (4-02.284). *Treatment of Biological Warfare Agent Casualties.* NAVMED P-5042; AFMAN (I) 44-156; MCRP 4-11.1C. 17 July 2000. (Change 1, 8 July 2002.)

FM 8-285 (4-02.285). *Treatment of Chemical Agent Casualties and Conventional Military Chemical Injuries.* NAVMED P-5041; AFJMAN 44-149; FMFM 11-11. 22 December 1995.

FM 22-51 (4-02.22). *Leaders' Manual for Combat Stress Control.* 29 September 1994.

CTA 8-100. *Army Medical Department Expendable/Durable Items.* 31 August 1994.

CTA 50-900. *Clothing and Individual Equipment.* 1 September 1994.

STP 21-1-SMCT. *Soldier's Manual of Common Tasks Skill Level I.* 1 October 2001.

Department of Defense Forms

DD Form 1425. *Specifications and Standards Requisition.* March 1986.

READINGS RECOMMENDED

These readings contain relevant supplemental information.

Joint and Multiservice Publications

FM 8-9 (4-02.11). *NATO Handbook on the Medical Aspects of NBC Defensive Operations AMEDP-6 (B), Part I—Nuclear, Part II—Biological, Part III—Chemical.* NAVMED P-5059; AFJMAN 44-151V1V2V3. 1 February 1996.

TB MED 81. *Cold Injury.* NAVMED P-5052-29; AFP 161-11. 30 September 1976.

TB MED 507. *Occupational and Environmental Health Prevention, Treatment and Control of Heat Injury.* NAVMED P-5052-5; AFP 160-1. 25 July 1980.

Army Publications

AR 310-25. *Dictionary of United States Army Terms (Short Title: AD)*. 15 October 1983. (Reprinted with basic including Change 1, 21 May 1986.)

AR 310-50. *Authorized Abbreviations and Brevity Codes*. 15 November 1985.

TM 3-4230-216-10. *Operator's Manual for Decontaminating Kit, Skin: M258A1 (NSN 4230-01-101-3984) and Training Aid, Skin Decontaminating: M58A1 (6910-01-101-1768)*. 17 May 1985. (Change 1, 22 January 1997.)

INDEX

References are to paragraph numbers except where specified otherwise.

RANGER CREED

Recognizing that I volunteered as a Ranger, fully knowing the hazards of my chosen profession, I will always endeavor to uphold the prestige, honor, and high esprit de corps of the Rangers.

Acknowledging the fact that a Ranger is a more elite Soldier who arrives at the cutting edge of battle by land, sea, or air, I accept the fact that as a Ranger my country expects me to move further, faster, and fight harder than any other soldier.

Never shall I fail my comrades I will always keep myself mentally alert, physically strong, and morally straight and I will shoulder more than my share of the task whatever it may be, one hundred percent and then some.

Gallantly will I show the world that I am a specially selected and well trained Soldier. My courtesy to superior officers, neatness of dress, and care of equipment shall set the example for others to follow.

Energetically will I meet the enemies of my country. I shall defeat them on the field of battle for I am better trained and will fight with all my might. Surrender is not a Ranger word. I will never leave a fallen comrade to fall into the hands of the enemy and under no circumstances will I ever embarrass my country.

Readily will I display the intestinal fortitude required to fight on to the Ranger objective and complete the mission, though I be the lone survivor.

STANDING ORDERS, ROGERS' RANGERS

1. Don't forget nothing.
2. Have your musket clean as a whistle, hatchet scoured, sixty rounds powder and ball, and be ready to march at a minute's warning.
3. When you're on the march, act the way you would if you was sneaking up on a deer. See the enemy first.
4. Tell the truth about what you see and what you do. There is an army depending on us for correct information. You can lie all you please when you tell other folks about the Rangers, but don't never lie to a Ranger or officer.
5. Don't never take a chance you don't have to.
6. When we're on the march we march single file, far enough apart so one shot can't go through two men.
7. If we strike swamps, or soft ground, we spread out abreast, so it's hard to track us.
8. When we march, we keep moving till dark, so as to give the enemy the least possible chance at us.
9. When we camp, half the party stays awake while the other half sleeps.
10. If we take prisoners, we keep' em separate till we have had time to examine them, so they can't cook up a story between' em.
11. Don't ever march home the same way. Take a different route so you won't be ambushed.
12. No matter whether we travel in big parties or little ones, each party has to keep a scout 20 yards ahead, 20 yards on each flank, and 20 yards in the rear so the main body can't be surprised and wiped out.
13. Every night you'll be told where to meet if surrounded by a superior force.
14. Don't sit down to eat without posting sentries.
15. Don't sleep beyond dawn. Dawn's when the French and Indians attack.
16. Don't cross a river by a regular ford.
17. If somebody's trailing you, make a circle, come back onto your own tracks, and ambush the folks that aim to ambush you.
18. Don't stand up when the enemy's coming against you. Kneel down, lie down, hide behind a tree.
19. Let the enemy come till he's almost close enough to touch, then let him have it and jump out and finish him up with your hatchet.

--MAJOR ROBERT ROGERS, 1759

RANGER HISTORY

The history of the American Ranger is a long and colorful saga of courage, daring and outstanding leadership. It is a story of men whose skills in the art of fighting have seldom been surpassed. Only the highlights of their numerous exploits are told here.

Rangers primarily performed defensive missions until Benjamin Church's Company of Independent Rangers from Plymouth Colony proved successful in raiding hostile Indians during King Phillip's War in 1675. In 1756, Major Robert Rogers, a native of New Hampshire, recruited nine companies of American colonists to fight for the British during the French and Indian War. Ranger techniques and methods of operation were an inherent characteristic of the American frontiersmen; however, Major Rogers was the first to capitalize on them and incorporate them into the fighting doctrine of a permanently organized fighting force.

The method of fighting used by the first Rangers was further developed during the Revolutionary War by Colonel Daniel Morgan, who organized a unit known as "Morgan's Riflemen". According to General Burgoyne, Morgan's men were "....the most famous corps of the Continental Army, all of them crack shots."

Francis Marion, the "Swamp Fox," organized another famous Revolutionary War Ranger element known as "Marion's Partisans." Marion's Partisans, numbering anywhere from a handful to several hundred, operated both with and independent of other elements of General Washington's Army. Operating out of the Carolina swamps, they disrupted British communications and prevented the organization of loyalists to support the British cause, substantially contributing to the American victory.

The American Civil War was again the occasion for the creation of special units such as Rangers. John S. Mosby, a master of the prompt and skillful use of cavalry, was one of the most outstanding Confederate Rangers. He believed that by resorting to aggressive action he could compel his enemies to guard a hundred points. He would then attack one of the weakest points and be assured numerical superiority.

With America's entry into the Second World War, Rangers came forth to add to the pages of history. Major William O. Darby organized and activated the 1st Ranger Battalion on June19, 1942 at Carrickfergus, North Ireland. The members were all hand-picked volunteers; 50 participated in the gallant Dieppe Raid on the northern coast of France with British and Canadian commandos. The 1st, 3rd, and 4th Ranger Battalions participated with distinction in the North African, Sicilian and Italian campaigns. Darby's Ranger Battalions spearheaded the Seventh Army landing at Gela and Licata during the Sicilian invasion and played a key role in the subsequent campaign which culminated in the capture of Messina. They infiltrated German lines and mounted an attack against Cisterna, where they virtually annihilated an entire German parachute regiment during close in, night, bayonet, and hand-to-hand fighting.

The 2nd and 5th Ranger Battalions participated in the D-Day landings at Omaha Beach, Normandy. It was during the bitter fighting along the beach that the Rangers gained their official motto. As the situation became critical on Omaha Beach, the division commander of the 29th Infantry Division stated that the entire force must clear the beach and advance inland. He then turned to Lieutenant Colonel Max Schneider, Commander of the 5th Ranger Battalion, and said, "Rangers, lead the way." The 5th Ranger Battalion spearheaded the breakthrough and thus enabled the allies to drive inland away from the invasion beaches.

The 6th Ranger Battalion, operating in the Pacific, conducted Ranger type missions behind enemy lines which involved reconnaissance and hard-hitting, long-range raids. They were the first American contingent to return to the Philippines, destroying key coastal installations prior to the invasion. A reinforced company from the 6th Ranger Battalion formed the rescue force which liberated American and allied prisoners of war from the Japanese prison camp at Cabanatuan.

Another Ranger-type unit was the 5307th Composite Unit (Provisional), organized and trained as a long-range penetration unit for employment behind enemy lines in Japanese occupied Burma. The unit commander was Brigadier General (later Major General) Frank D. Merrill. Its 2,997 officers and men became popularly known as "Merrill's Marauders."

The men composing Merrill's Marauders were volunteers from the 5th, 154th, and 33rd Infantry Regiments and from other Infantry regiments engaged in combat in the southwest and South Pacific. These men responded to a call from then Chief of Staff, General George C. Marshall, for volunteers for a hazardous mission. These volunteers were to have a high state of physical ruggedness and stamina and were to come from jungle-trained and jungle-tested units.

Prior to their entry into the Northern Burma Campaign, Merrill's Marauders trained in India under the overall supervision of Major General Orde C. Wingate, British Army. There, they were trained from February to June 1943 in long-range penetration tactics and techniques of the type developed and first employed by General Wingate. The operations of the Marauders were closely coordinated with those of the Chinese 22nd and 38th Divisions in a drive to recover northern Burma and clear the way for the construction of Ledo Road, which was to link the Indian railhead at Ledo with the old Burma Road to China. The Marauders marched and fought through jungle and over mountains from Hukwang Valley in northwest Burma to Myitkyina and the Irrawaddy River. In 5 major and 30 minor engagements, they met and defeated the veteran soldiers of the Japanese 18th Division. Operating in the rear of the main force of the Japanese, they prepared the way for the southward advances of the Chinese by disorganizing supply lines and communications. The climax of the Marauder's operations was the capture of Myitkyina Airfield, the only all-weather strip in northern Burma. This was the final victory of "Merrill's Marauders," which disbanded in August 1944. Remaining personnel were consolidated into the 475th Infantry Regiment, which fought its last battle on February 3, and 4,1945, at Loi-Kang Ridge, China. This Infantry Regiment would serve as the forefather of today's 75th Ranger Regiment.

Shortly after the outbreak of the Korean War in June 1950, the 8th Army Ranger Company was formed of volunteers from American units in Japan. The Company was trained in Korea and distinguished itself in combat during the drive to the Yalu River, performing task force and spearhead operations. In November 1950 during the massive Chinese intervention, this small unit, though vastly outnumbered, withstood five enemy assaults on its position.

In September 1950, a Department of the Army message called for volunteers to be trained as Airborne Rangers. In the 82nd Airborne Division, five thousand regular Army paratroopers volunteered, and from that number nine hundred men were selected to form the initial eight Airborne Ranger Companies. An additional nine companies were formed from volunteers of regular Army and National Guard Infantry Divisions. These seventeen Airborne Ranger companies were activated and trained at Fort Benning, Georgia, with most receiving additional training in the mountains of Colorado.

In 1950 and 1951, some 700 men of the 1st, 2nd, 3rd, 4th, 5th, and 8th Airborne Ranger companies fought to the front of every American Infantry Division in Korea. Attacking by land, water, and air, these six Ranger companies conducted raids, deep penetrations, and ambush operations against North Korean and Chinese forces. They were the first Rangers in history to make a combat jump. After the Chinese intervention, these Rangers were the first Americans to re-cross the 38th parallel. The 2nd Airborne Ranger Company was the only African American Ranger unit in the history of the American Army. The men of the six Ranger companies who fought in Korea paid the bloody price of freedom. One in nine of this gallant brotherhood died on the battlefields of Korea.

Other Airborne Ranger companies led the way while serving with infantry divisions in the United States, Germany, and Japan. Men of these companies volunteered and fought as members of line infantry units in Korea. One Ranger, Donn Porter, would be posthumously awarded the Medal of Honor. Fourteen Korean War Rangers became general officers and dozens became colonels, senior noncommissioned officers, and leaders in civilian life. They volunteered for the Army, the Airborne, the Rangers, and for combat. The first men to earn and wear the coveted Ranger Tab, these men are the original Airborne Rangers.

In October 1951, the Army Chief of Staff, General J. Lawton Collins directed, "Ranger training be extended to all combat units in the Army." The Commandant of the Infantry School was directed to establish a Ranger Department. This new department would develop and conduct a Ranger course of instruction. The objective was to raise the standard of training in all combat units. This program was built upon what had been learned from the Ranger Battalions of World War II and the Airborne Ranger companies of the Korean conflict.

During the Vietnam Conflict, fourteen Ranger companies consisting of highly motivated volunteers served with distinction from the Mekong Delta to the DMZ. Assigned to separate brigade, division, and field force units, they conducted long-range reconnaissance and exploitation operations into enemy-held areas providing valuable combat intelligence. Initially designated at LRRP, then LRP companies, these units were later designated as C, D,E,F,G,H,I,K,L,M,N,O and P (Ranger) 75th Infantry.

Following Vietnam, recognizing the need for a highly trained and highly mobile reaction force, the Army Chief of Staff, General Abrams directed the activation of the first battalion-sized Ranger units since World War II, the 1st and 2nd Battalions (Ranger), 75th Infantry. The 1st Battalion was trained at Fort Benning, Georgia and was activated February 8, 1974 at Fort Stewart, Georgia with the 2nd Battalion being activated on October 3, 1974. The 1st Battalion is now located at Hunter Army Airfield, Georgia and the 2nd Battalion at Fort Lewis, Washington.

The farsightedness of General Abrams' decision, as well as the combat effectiveness of the Ranger battalions, was proven during the United States' invasion of the island of Grenada in October 1983 to protect American citizens there, and to restore democracy. As expected, Rangers led the way! During this operation, code named "Urgent Fury," the Ranger battalions conducted a daring, low level airborne assault (from 500 feet) to seize the airfield at Point Salines, and then continued operations for several days to eliminate pockets of resistance, and rescue American medical students.

As a result of the demonstrated effectiveness of the Ranger battalions, the Department of the Army announced in 1984 that it was increasing the strength of Ranger units to its highest level in 40 years. To do this, it activated another Ranger battalion as well as a Ranger Regimental Headquarters. These new units, the 3rd Battalion (Ranger), 75th Infantry, and Headquarters Company (Ranger), 75th Infantry, have increased the Ranger strength of the Army to over 2,000 soldiers actually assigned to Ranger units. On February 3, 1986, the 75th Infantry was renamed the 75th Ranger Regiment.

On December 20, 1989, the 75th Ranger Regiment was once again called upon to demonstrate its effectiveness in combat. For the first time since its reorganization in 1984, the Regimental Headquarters and all three Ranger battalions were deployed on Operation "Just Cause" in Panama. During this operation, the 75th Ranger Regiment spearheaded the assault into Panama by conducting airborne assaults onto Torrijos/Tocumen Airport and Rio Hato Airfield to facilitate the restoration of democracy in Panama, and protect the lives of American citizens. Between December 20, 1989 and January 7, 1990, numerous follow-on missions were performed in Panama by the Regiment.

Early in 1991, elements of the 75th Ranger Regiment deployed to Saudi Arabia in support of Operation Desert Storm.

In August 1993, elements of the 75th Ranger Regiment deployed to Somalia in support of Operation Restore Hope, and returned November 1993.

In 1994, elements of the 75th Ranger Regiment deployed to Haiti in support of Operation Uphold Democracy.

In 2000 – 2001, elements of the 75th Ranger Regiment deployed to Kosovo in support of Operation Joint Guardian.

Since September 11, 2001, the 75th Ranger Regiment has led the way in the Global War on Terrorism. In October 2001, elements of the 75th Ranger Regiment deployed to Afghanistan in support of Operation Enduring Freedom. In March 2003, elements of the Regiment deployed in support of Operation Iraqi Freedom.

The performance of these Rangers significantly contributed to the overall success of these operations and upheld the Ranger tradition of the past. As in the past, the Regiment stands ready to execute its mission to conduct special operations in support of the United States' policies and objectives.

RANGER MEDAL OF HONOR WINNERS

Millett, Lewis L. Sr	Captain	Feb 7 1951	Co. E 2/27th Infantry
Porter, Donn F.*	Sergeant	Sept 7 1952	Co. G 2/14th Infantry
Mize, Ola L.	Sergeant	June 10-11 1953	Co. K 3/15th Infantry
Dolby, David C.	Staff Sergeant	May 21 1966	Co. B 1/8th (ABN) Calvary
Foley, Robert F.	Captain	Nov 5 1966	Co. A 2/27th Infantry
Zabitosky, Fred M.	Staff Sergeant	Feb 19 1968	5th Special Forces
Bucha, Paul W.	Captain	May 16-19 1968	Co. D 3/187 Infantry
Rabel, Laszlo*	Staff Sergeant	Nov 13 1968	74th Infantry (LRRP)
Howard, Robert L.	Sergeant First Class	Dec 30 1968	5th Special Forces
Law, Robert D. *	Specialist 4	Feb 22 1969	Co. I 75th Infantry (Ranger)
Kerrey, J. Robert	Lieutenant	Mar 14 1969	Seal Team 1
Doane, Stephen H.*	1st Lieutenant	Mar 25 1969	Co. B 1/5th Infantry
Pruden, Robert J.*	Staff Sergeant	Nov 22 1969	Co. G 75th Infantry (Ranger)
Littrell, Gary L.	Sergeant First Class	April 4-8 1970	Advisory Team 21 (Ranger)
Lucas, Andre C.*	Lt Colonel	Jul 1-23 1970	HHC 2/506 Infantry
Gordon, Gary I. *	Master Sergeant	Oct 3 1993	Task Force Ranger
Shughart, Randall D. *	Sergeant First Class	Oct 3 1993	Task Force Ranger

*Awarded posthumously

SUMMARY OF CHANGES

THROUGHOUT	UPDATED doctrine, technology, terms; clarified verbiage
CHAPTER 1	UPDATED duties and responsibilities ADDED FO
CHAPTER 2	UPDATED WARNO ADDED sample WARNO UPDATED OPORD ADDED terrain model
CHAPTER 3	ADDED PLOT-CR ADDED risk estimate distances
CHAPTER 4	IDENTIFIED TL positions during movement
CHAPTER 5	REORGANIZED combat operations
CHAPTER 6	UPDATED battle drills IAW FM 3-21.8 CONSOLIDATED building clearing techniques under Battle Drill 5 (previously in Chapter 14)
CHAPTER 7	ADDED section on antennas
CHAPTER 8	UPDATED aircraft
CHAPTER 9	UPDATED rope bridge equipment UPDATED sequence and responsibilities
CHAPTER 10	UPDATED equipment and descriptions UPDATED climbing commands
CHAPTER 11	NA
CHAPTER 12	ADDED first aid techniques
CHAPTER 13	ADDED figure 'Lighting a time fuse with a match' ADDED expedient explosives ADDED table of US demolitions ADDED breaching and timber cutting charges
CHAPTER 14	MOVED building clearing techniques to Battle Drill 5 in Chapter 6 ADDED marking SOP
CHAPTER 15	NEW chapter covering vehicle convoy operations
APPENDIX A	NEW - Resources
APPENDIX B	NEW - Quick Reference Cards

TABLE OF CONTENTS

Chapter 1
LEADERSHIP

Leadership, the most essential element of combat power, gives purpose, direction, and motivation in combat. The leader balances and maximizes maneuver, firepower, and protection against the enemy. This chapter discusses how he does this by exploring the principles of leadership (BE, KNOW, DO); the duties, responsibilities, and actions of an effective leader; and the leader's assumption of command.

1-1. **PRINCIPLES**. The principles of leadership are BE, KNOW, and DO.

BE • Technically and tactically proficient • Able to accomplish to standard all tasks required for the wartime mission. • Courageous, committed, and candid. • A leader with integrity.
KNOW • The four major factors of leadership and how they affect each other are-- ---Led ---Leader ---Situation ---Communications • Yourself, and the strengths and weaknesses in your character, knowledge, and skills. Seek continual self-improvement, that is, develop your strengths and work to overcome your weaknesses. • Your Rangers, and look out for their well-being by training them for the rigors of combat, taking care of their physical and safety needs, and disciplining and rewarding them.
DO-- • Seek responsibility and take responsibility for your actions; exercise initiative; demonstrate resourcefulness; and take advantage of opportunities on the battlefield that will lead to you to victory; accept fair criticism, and take corrective actions for your mistakes. • Assess situations rapidly, make sound and timely decisions, gather essential information, announce decisions in time for Rangers to react, and consider the short- and long-term effects of your decision. • Set the example by serving as a role model for your Rangers. Set high but attainable standards; be willing do what you require of your Rangers; and share dangers and hardships with them. • Keep your subordinates informed to help them make decisions and execute plans within your intent, encourage initiative, improve teamwork, and enhance morale. • Develop a sense of responsibility in subordinates by teaching, challenging, and developing them. Delegate to show you trust them. This makes them want more responsibility. • Ensure the Rangers understand the task; supervise them, and ensure they accomplish it. Rangers need to know what you expect: when and what you want them to do, and to what standard. • Build the team by training and cross-training your Rangers until they are confident in their technical and tactical abilities. Develop a team spirit that motivates them to go willingly and confidently into combat. • Know your unit's capabilities and limitations, and employ them accordingly.

1-2. **DUTIES, RESPONSIBILITIES, AND ACTIONS**. To complete all assigned tasks, every Ranger in the patrol must do his job. Each must accomplish his specific duties and responsibilities and be a part of the team.

PLATOON LEADER (PL)

- Is responsible for what the patrol does or fails to do. This includes tactical employment, training, administration, personnel management, and logistics. He does this by planning, making timely decisions, issuing orders, assigning tasks, and supervising patrol activities. He must know his Rangers and how to employ the patrol's weapons. He is responsible for positioning and employing all assigned or attached crew-served weapons and employment of supporting weapons.
- Establishes time schedule using backwards planning. Considers time for execution, movement to the objective, and the planning and preparation phase of the operation.
- Takes the initiative to accomplish the mission in the absence of orders. Keeps higher informed by using periodic situation reports (SITREP).
- Plans with the help of the platoon sergeant (PSG), squad leaders, and other key personnel (team leaders, FO, attachment leaders).
- Stays abreast of the situation through coordination with adjacent patrols and higher HQ; supervises, issues FRAGOs, and accomplishes the mission.
- If needed to perform the mission, requests more support for his patrol from higher headquarters.
- Directs and assists the platoon sergeant in planning and coordinating the patrol's sustainment effort and casualty evacuation (CASEVAC) plan.
- During planning, receives on-hand status reports from the platoon sergeant and squad leaders.
- Reviews patrol requirements based on the tactical plan.
- Ensures that all-round security is maintained at all times.
- Supervises and spot-checks all assigned tasks, and corrects unsatisfactory actions.
- During execution, positions himself where he can influence the most critical task for mission accomplishment; usually with the main effort, to ensure that his platoon achieves its decisive point
- Is responsible for positioning and employing all assigned and attached crew-served weapons.
- Commands through his squad leaders IAW the intent of the two levels higher commanders.
- Conducts rehearsals.

PLATOON SERGEANT (PSG)

The PSG is the senior NCO in the patrol and second in succession of command. He helps and advises the patrol leader, and leads the patrol in the leader's absence. He supervises the patrol's administration, logistics, and maintenance, and he prepares and issues paragraph 4 of the patrol OPORD.

DUTIES

- Organizes and controls the patrol CP IAW the unit SOP, patrol leader's guidance, and METT-TC factors.
- Receives squad leader's requests for rations, water, and ammunition. Work with the company first sergeant or XO to request resupply. Directs the routing of supplies and mail.
- Supervises and directs the patrol medic and patrol aid-litter teams in moving casualties to the rear.
- Maintains patrol status of personnel, weapons, and equipment; consolidates and forwards the patrol's casualty reports (DA Forms 1155 and 1156); and receives and orients replacements.
- Monitors the morale, discipline, and health of patrol members.
- Supervises task-organized elements of patrol:
  - — Quartering parties.
  - — Security forces during withdrawals.
  - — Support elements during raids or attacks.
  - — Security patrols during night attacks.
- Coordinates and supervises patrol resupply operations.
- Ensures that supplies are distributed IAW the patrol leader's guidance and direction.
- Ensures that ammunition, supplies, and loads are properly and evenly distributed (a critical task during consolidation and reorganization).
- Ensures the casualty evacuation plan is complete and executed properly.

- Ensures that the patrol adheres to the platoon leader's time schedule.
- Assists the platoon leader in supervising and spot-checking all assigned tasks, and corrects unsatisfactory actions.

ACTIONS DURING MOVEMENT AND HALTS

- Takes actions necessary to facilitate movement.
- Supervises rear security during movement.
- Establishes, supervises, and maintains security during halts.
- Knows unit location.
- Performs additional tasks as required by the patrol leader and assists in every way possible. Focuses on security and control of patrol.

ACTIONS AT DANGER AREAS

- Directs positioning of near-side security (usually conducted by the trail squad or team).
- Maintains accountability of personnel.

ACTIONS ON THE OBJECTIVE AREA

- Assists with ORP occupation.
- Supervises, establishes, and maintains security at the ORP.
- Supervises the final preparation of men, weapons, and equipment in the ORP IAW the patrol leader's guidance.
- Assists the patrol leader in control and security.
- Supervises the consolidation and reorganization of ammunition and equipment.
- Establishes, marks, and supervises the planned CCP, and ensures that the personnel status (to include WIA/KIA) is accurately reported to higher.
- Performs additional tasks assigned by the patrol leader and reports status to platoon leader.

ACTIONS IN THE PATROL BASE

- Assists in patrol base occupation.
- Assist in establishing and adjusting perimeter.
- Enforces security in the patrol base.
- Keeps movement and noise to a minimum.
- Supervises and enforces camouflage.
- Assigns sectors of fire.
- Ensures designated personnel remain alert and equipment is maintained to a high state of readiness.
- Requisitions supplies, water, and ammunition, and supervises their distribution.
- Supervises the priority of work and ensures its accomplishment.
  - Security plan.
    - Ensures crew-served weapons have interlocking sectors of fire.
    - Ensures Claymores are emplaced to cover dead space.
    - Ensures range cards and sector sketch are complete.
  - Alert plan.
  - Evacuation plan.
  - Withdrawal plan.
  - Alternate patrol base.
  - Maintenance plan.
  - Hygiene plan.
  - Messing plan.
  - Water plan.
  - Rest plan.
- Performs additional tasks assigned by the patrol leader and assists him in every way possible.

SQUAD LEADER (SL)

Is responsible for what the squad does or fails to do. He is a tactical leader who leads by example.

DUTIES

- Completes casualty feeder reports and reviews the casualty reports completed by squad members.
- Directs the maintenance of the squad's weapons and equipment.

- Inspects the condition of Rangers' weapons, clothing, and equipment.
- Keeps the PL and PSG informed on status of squad.
- Submits ACE report to PSG.

ACTIONS THROUGHOUT THE MISSION

- Obtains status report from team leaders and submits reports to the PL and PSG.
- Makes a recommendation to the PL/PSG when problems are observed.
- Delegates priority task to team leaders, and supervises their accomplishment IAW squad leader's guidance.
- Uses initiative in the absence of orders.
- Follows the PL's plan and makes recommendations.

ACTIONS DURING MOVEMENT AND HALTS

- Ensures heavy equipment is rotated among members and difficult duties are shared.
- Notifies PL of the status of the squad.
- Maintains proper movement techniques while monitoring route, pace, and azimuth.
- Ensures the squad maintains security throughout the movement and at halts.
- Prevents breaks in contact.
- Ensures subordinate leaders are disseminating information, assigning sectors of fire, and checking personnel.

ACTION IN THE OBJECTIVE AREA

- Ensures special equipment has been prepared for actions at the objective.
- Maintains positive control of squad during the execution of the mission.
- Positions key weapons systems during and after assault on the objective.
- Obtains status reports from team leaders and ensures ammunition is redistributed and reports status to PL.

ACTIONS IN THE PATROL BASE

- Ensures patrol base is occupied according to the plan.
- Ensures that his sector of the patrol base is covered by interlocking fires; makes final adjustments, if necessary.
- Sends out LP or OPs in front of assigned sector. (METT-TC dependent).
- Ensures priorities of work are being accomplished, and reports accomplished priorities to the PL and PSG.
- Adheres to time schedule.
- Ensures personnel know the alert and evacuation plans, and the locations of key leaders, OPs, and the alternate patrol base.

WEAPONS SQUAD LEADER

Is responsible for all that the weapons squad does or fails to do. His duties are the same as those of the squad leader. Also, he controls the machine guns in support of the patrol's mission. He advises the PL on employment of his squad.

DUTIES

- Supervises machine gun teams to ensure they follow priorities of work.
- Inspects machine gun teams for correct range cards, fighting positions, and understanding of fire plan.
- Supervises maintenance of machine guns, that is, ensures that maintenance is performed correctly, that deficiencies are corrected and reported, and that the performance of maintenance does not violate the security plan.
- Assists PL in planning.
- Positions at halts and danger areas and according to the patrol SOP any machine guns not attached to squads.
- Rotates loads. Machine gunners normally get tired first.
- Submits ACE report to PSG.
- Designates sectors of fire, principal direction of fire (PDF), and secondary sectors of fire for all guns.
- Gives fire commands to achieve maximum effectiveness of firepower:
  - — Shifts fires.
  - — Corrects windage or elevation to increase accuracy.
  - — Alternates firing guns.
  - — Controls rates of fire and fire distribution.
- Knows locations of assault and security elements, and prevents fratricide.
- Reports to the PL.

TEAM LEADER (TL)

Controls the movement of his fire team and the rate and placement of fire. To do this, leads from the front and uses the proper commands and signals. Maintains accountability of his Rangers, weapons, and equipment. Ensures his Rangers maintain unit standards in all areas, and are knowledgeable of their tasks and the operation. The following checklist outlines specific duties and responsibilities of team leaders during mission planning and execution. The TL leads by example:

ACTIONS DURING PLANNING AND PREPARATION

- Warning Order.
  - — Assists in control of the squad.
  - — Monitors squad during issuance of the order.
- OPORD Preparation.
  - — Posts changes to schedule.
  - — Posts and updates team duties on warning order board.
  - — Submits ammunition and supply requests.
  - — Picks up ammunition and supplies.
  - — Distributes ammunition and special equipment.
  - — Performs all tasks given in the SL's special instructions paragraph.
- OPORD Issuance and Rehearsal.
  - — Monitors squad during issuance of the order.
  - — Assists SL during rehearsals.
- Takes actions necessary to facilitate movement.
- Enforces rear security.
- Establishes, supervises, and maintains security at all times.
- Performs additional tasks as required by the SL, and assists him in every way possible, particularly in control and security.

ACTION IN THE ORP

- Assists in the occupation of the ORP.
- Helps supervise, establish, and maintain security.
- Supervises the final preparation of Rangers, weapons, and equipment in the ORP IAW the SL's guidance.
- Assists in control of personnel departing and entering the ORP.
- Reorganizes perimeter after the reconnaissance party departs.
- Maintains communication with higher headquarters.
- Upon return of reconnaissance party, helps reorganize personnel and redistribute ammunition and equipment; ensures accountability of all personnel and equipment is maintained.
- Disseminates PIR to his team.
- Performs additional tasks assigned by the SL.

ACTIONS IN THE PATROL BASE

- Inspects the perimeter to ensure team has interlocking sectors of fire; prepares team sector sketch.
- Enforces the priority of work and ensures it is properly accomplished.
- Performs additional tasks assigned by the SL and assists him in every way possible.

MEDIC

Assists the PSG in directing aid and litter teams; monitors the health and hygiene of the platoon.

DUTIES

- Treats casualties, conducts triage, and assists in CASEVAC under the control of the PSG.
- Aids the PL or PSG in field hygiene matters. Personally checks the health and physical condition of platoon members.
- Requests Class VIII (medical) supplies through the PSG.
- Provides technical expertise to and supervision of combat lifesavers.
- Ensures casualty feeder reports are correct and attached to each evacuated casualty.
- Carries out other tasks assigned by the PL or PSG.

RADIO OPERATOR Is responsible for establishing and maintaining communications with higher headquarters and within the patrol. *DUTIES DURING PLANNING* • Enters the net at the specified time. • Ensures that all frequencies, COMSEC fills, and net IDs, are preset in squad/platoon radios. • Informs SL and PL of changes to call signs, frequencies, challenge and password, and number combination based on the appropriate time in the ANCD. • Ensures the proper function of all radios and troubleshoots and reports deficiencies to higher. • Weatherproofs all communications equipment. *DUTIES DURING EXECUTION* • Serves as en route recorder during all phases of the mission. • Records all enemy contact and reports it to higher in a SALUTE format. • Reports all OPSKEDs to higher. • Consolidates and records all PIR.
FORWARD OBSERVER (FO) Works for the PL. Serves as the eyes of the FA and mortars. Is mainly responsible for locating targets, and for calling for and adjusting indirect fire support. Knows the terrain where the platoon is operating; knows the tactical situation. Knows the mission, the concept, and the unit's scheme of maneuver and priority of fires. *DUTIES DURING PLANNING* • Selects targets to support the platoon's mission based on the company OPORD, platoon leader's guidance, and analysis of METT-TC factors. • Prepares and uses situation maps, overlays, and terrain sketches. *DUTIES DURING EXECUTION* • Informs the FIST headquarters of platoon activities and of the fire support situation • Selects new targets to support the platoon's mission based on the company OPORD, the platoon leader's guidance, and an analysis of METT-TC factors. • Calls for and adjusts fire support. • Operates as a team with the radio operator. • Selects OPs. • Maintains communications as prescribed by the FSO. • Maintains the current 8-digit coordinate of his location at all times.

1-3. **ASSUMPTION OF COMMAND**. Any platoon/squad member might have to take command of his element in an emergency, so every Ranger must be prepared to do so. During an assumption of command, situation permitting, the Ranger assuming command accomplishes the following tasks (not necessarily in this order) based on METT-TC:

- *INFORMS* the unit's subordinate leaders of the command and notifies higher.
- *CHECKS* security.
- *CHECKS* crew-served weapons.
- *PINPOINTS* location.
- *COORDINATES* and *CHECKS* equipment.
- *CHECKS* personnel status.
- *ISSUES* FRAGO (if required).
- *REORGANIZES* as needed, maintaining unit integrity when possible.
- *MAINTAINS* noise and light discipline.
- *CONTINUES* patrol base activities, especially security, if assuming command in a patrol base.
- *RECONNOITERS* or, at the least, conducts a map reconnaissance.
- *FINALIZES* plan.
- *EXECUTES* the mission.

NOTES

Chapter 2
OPERATIONS

This chapter provides techniques and procedures used by Infantry platoons and squads throughout the planning and execution phases of tactical operations. Specifically, it discusses the troop-leading procedures, combat intelligence, combat orders, and planning techniques and tools needed to prepare a platoon to fight. These topics are time sensitive and apply to all combat operations. When they have time, leaders can plan and prepare in depth. If they have less time, they must rely on previously rehearsed actions, battle drills, and standing operating procedures (SOPs).

2-1. TROOP-LEADING PROCEDURES

Troop leading procedures comprise the following steps. They are what a leader does to prepare his unit to accomplish a tactical mission. The TLP starts when the leader is alerted for a mission or receives a change or new mission. He can perform Steps 3 through 8 in any order, or at the same time. (He can also use the tools of the tactician shown in Figure 2-1):

1. Receive the mission.	5. Reconnoiter.
2. Issue a warning order.	6. Complete the plan.
3. Make a tentative plan.	7. Issue the complete order.
4. Initiate movement.	8. Supervise.

Figure 2-1. TOOLS OF THE TACTICIAN RELATIONSHIP

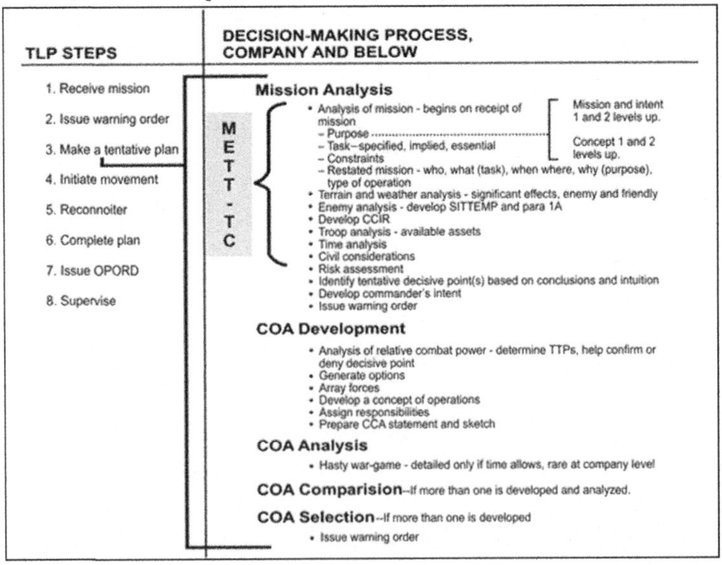

a. **STEP 1--RECEIVE THE MISSION.** The leader may receive the mission in a warning order, an operation order (OPORD), or a fragmentary order (FRAGO). He should use no more than one third of the available time for his own planning and for issuing his OPORD. The remaining two thirds is for subordinates to plan and prepare for the operation. Leaders should also consider other factors such as available daylight and travel time to and from orders and rehearsals.

b. **STEP 2--ISSUE A WARNING ORDER.** The leader provides initial instructions in a warning order. The warning order contains enough information to begin preparation as soon as possible. The warning order mirrors the five-paragraph OPORD format. A warning order may include--

- The mission or nature of the operation (mission statement).
- Time and place for issuance of the operation (coordinating instructions).
- Who is participating in the operation (coordinating instructions).
- Time of the operation (timeline).

c. **STEP 3--MAKE A TENTATIVE PLAN.** The leader uses METT-TC to develop an estimate of the situation, which he will use as the basis for his tentative plan. This set of actions is referred to as the leader's mission analysis:

(1) **Conduct a detailed mission analysis.**

(a) *Concept and Intent.* Higher commanders' concept and intent two levels up. This information is found in paragraph 1b for two levels up and in paragraphs 2 and 3 for one higher.

(b) *Unit Tasks.* Tasks that are clearly stated in the order (Specified Tasks) or tasks that become apparent as the OPORD is analyzed (Implied Tasks).

EXAMPLE SPECIFIED TASKS

- Retain hill 545 to prevent envelopment of B Co.
- Provide one squad to the 81-mm platoon to carry ammo.
- Establish an OP vic GA124325 NLT 301500 NOV 89.

EXAMPLE IMPLIED TASKS

- Provide security during movement.
- Conduct resupply operations.
- Coordinate with adjacent units.

(c) *Unit Limitations.* The leader next determines all control measures or instructions in the OPORD that restrict his freedom of action; these are called limitations. In every operation, there are some limitations on the unit.

COMMON LIMITATIONS

Graphic control measures.
Cross the LD at 100030 OCT 94 (time).
MOPP4 in effect (uniform and environmental considerations).
ADA weapons status, tight; warning status, yellow (rules of engagement).

(d) *Mission-Essential Task(s).* After reviewing all the above factors, the leader identifies his mission-essential task(s). Failure to accomplish a mission-essential task results in the unit's failure to accomplish its primary purpose for that operation. The mission essential task should be found in the maneuver paragraph.

(e) *Restated Mission.* The restated mission statement becomes the focus for the remainder of the estimate process. This is a clear, concise statement of the mission essential task(s) to be accomplished by the unit and the purpose to be achieved. The mission statement will state WHO, WHAT (the task), WHEN (the critical time), WHERE (usually a grid coordinate), and WHY (the purpose the unit must achieve). Some examples of restated missions follow:

EXAMPLE RESTATED MISSIONS
(WHO) 1st Platoon attacks (WHAT) to seize (WHERE) HILL 482 vic NB 457371 (OBJ BLUE) (WHEN) NLT 09050OZ Dec 92 L 482 (WHY) to enable the company's main effort to destroy enemy command bunker.
(WHO) 1st Platoon, C Company defends (WHAT) to destroy from (WHERE) AB163456 to AB163486 to AB123486 to AB123456 (WHEN) NLT 28153OZ Oct 97 (WHY) to prevent enemy forces from enveloping B Company, 1-66 Infantry (L) from the south.

(2) Analyze the situation and develop a course of action. Each COA must be--suitable, acceptable, feasible, distinguishable, and complete.

(3) Upon developing a COA, the unit leader will assign C2 headquarters, complete generic task organization assigning all organic and attached elements, and prepare COA statement and sketch.

(4) With the restated mission from Step 1 to provide focus, the leader continues the estimate process using the remaining factors of METT-TC.

(a) What is known about the **ENEMY**?

Composition	This is an analysis of the forces and weapons that the enemy can bring to bear. Determine what weapons systems they have available, and what additional weapons and units are supporting him.
Disposition	The enemy's disposition is how he is arrayed on the terrain, such as in defensive positions, in an assembly area, or moving in march formation.
Strength	Percentage strength.
Recent Activities	Identify recent and significant enemy activities that may indicate future intentions.
Reinforcement Capabilities	Determine positions for reserves and estimated time to counterattack or reinforce.
Possible Courses of Action	Determine the enemy's possible COAs. This will help the leader determine how to best array his own forces against the enemy and fight the battle.

(b) How will **TERRAIN** and **WEATHER** affect the operation? Analyze terrain using OAKOC.

- **Observation and Fields of Fire**. Determine locations that provide the best observation and fields of fire along friendly and enemy avenues of approach, near the objective, and on key terrain. The analysis of fields of fire is mainly concerned with the ability to cover the terrain with direct fire.
- **Avenues of Approach**. Analyze and identify friendly and enemy avenues of approach. Identify avenues of approach en route to the objective, on and around the objective, and for possible enemy counterattack. Also, consider aerial and subterranean avenues.

Offensive Considerations (Friendly)	• How can these avenues support my movement? • What are the advantages/disadvantages of each? (Consider enemy, speed, cover, and concealment.) • What are the likely enemy counterattack routes?

Offensive Considerations (Enemy)	• How can the enemy use these approaches? • Which avenue is most dangerous? Least? (Prioritize each approach.) • Which avenues would support a counterattack?

- **Cover and Concealment**. The analysis of cover and concealment is often inseparable from the fields of fires and observation. Weapon positions must have both to be effective and to be survivable. Infantry units are capable of improving poor cover and concealment by digging in and camouflaging their positions. When moving, the terrain is used to provide cover and concealment.
- **Obstacles**. Identify the existing and reinforcing obstacles and hindering terrain that will affect mobility.
- **Key Terrain**. Key terrain is any location or area that the seizure, retention, or control of affords a marked advantage to either combatant. Using the map, aerial photos, and information already gathered, look for key terrain that dominates avenues of approach or the objective area. Next, look for decisive terrain that if held or controlled will have an extraordinary impact on the mission.

(5) Analyze courses of action (war-game). This analysis is conducted by war-gaming friendly courses of action against the enemy's most probable courses of action. The leader can--and should--war-game with his subordinates.

(6) Compare courses of action. The leader compares the COAs and selects the one that is most likely to accomplish the assigned mission. He considers the advantages and disadvantages for each COA. He also considers how the critical events impact on COAs.

(7) Make a decision. The leader selects the COA that he believes has the best chance of accomplishing the mission.

d. **STEP 4--START NECESSARY MOVEMENT**. The unit may need to begin movement while the leader is still planning or forward reconnoitering. This step may occur anytime during the TLP.

e. **STEP 5--RECONNOITER**. If time allows, the leader makes a personal reconnaissance. When time does not allow, the leader must make a map or aerial photo reconnaissance. Sometimes the leader must rely on others, such as scouts, to conduct the reconnaissance.

f. ***STEP 6--COMPLETE THE PLAN***. The leader completes his plan based on the reconnaissance and any changes in the situation.

g. **STEP 7--ISSUE THE COMPLETE ORDER**. Platoon and squad leaders normally issue oral operations orders to aid subordinates in understanding the concept of the mission. Leaders may require subordinates to repeat part of the order, demonstrate it on a terrain model, or sketch their understanding of the operation. Leaders should also quiz their Rangers to ensure that all Rangers understand the mission.

h. **STEP 8--SUPERVISE AND REFINE**. The leader supervises the unit's preparation for combat by conducting rehearsals and inspections.

(1) ***Rehearsals***. Rehearsals include the practice of having squad leaders brief their planned actions in execution sequence to the platoon leader. The leader should conduct rehearsals on terrain that resembles the actual ground, and in similar light conditions. At a minimum, the leader should conduct a backbrief (least preferred), full force rehearsal (most preferred), or reduced force rehearsal (key leaders), and a rock drill.

(a) *Purpose*. The leader uses rehearsals to--

- Practice essential tasks (improve performance).
- Reveal weaknesses or problems in the plan.
- Coordinate the actions of subordinate elements.
- Improve Ranger understanding of the concept of the operation (foster confidence in Rangers).

(b) *Times and Tasks*. The platoon may begin rehearsals of battle drills and other SOP items before the receipt of the operation order. Once the order has been issued, it can rehearse mission specific tasks. Some important tasks to rehearse include--

- Actions on the objective.
- Assaulting a trench, bunker, or building.
- Actions at the assault position.
- Breaching obstacles (mine and wire).
- Using special weapons or demolitions.
- Actions on unexpected enemy contact.

(c) *Types*.

- Backbrief.
  - Key leaders sequentially brief the actions required during operation.
  - Patrol leader controls.
  - Conducted twice: right after FRAGO (confirmation brief), and again after subordinates develop own plan.
- Reduced force.
  - Conducted when time is key constraint.
  - Conducted when security must be maintained.
  - Key leaders normally attend.
  - Mock-ups, sand tables, and small scale replicas used.
- Full force.
  - Most effective type.
  - First executed in daylight and open terrain.
  - Secondly conduct in same conditions as operation.
  - All Rangers participate.
  - May use force on force.
- Techniques.
  - Force on force.
  - Map (limited value and limited number of attendees).
  - Radio (cannot mass leaders; confirms communications).
  - Sand table or terrain model (key leaders; includes all control measures).
  - Rock drill (similar to sand table/terrain model; subordinates actually move themselves). Very effective for coordinating timing and sequence of events--especially when air assets are used.

(d) *Inspections*. Squad leaders should conduct initial inspections shortly after receipt of the warning order. The platoon sergeant spot-checks throughout the unit's preparation for combat. The platoon leader and platoon sergeant make a final inspection of--

- Weapons and ammunition.
- Uniforms and equipment.
- Mission-essential equipment.
- Soldier's understanding of the mission and their specific responsibilities.
- Communications.
- Rations and water.
- Camouflage.
- Deficiencies noted during earlier inspections.

2-2. **COMBAT INTELLIGENCE**. Gathering information is one of the most important aspects of conducting patrolling operations. The following details what information to collect and how to report it:

a. **Reports**. All information must be quickly, completely, and accurately reported. Use the SALUTE report format for reporting and recording information.

SIZE	– Seven enemy personnel
ACTIVITY	– Traveling SW
LOCATION	– GA123456
UNIT/UNIFORM	– OD uniforms with red six-point star on left shoulder
TIME	– 210200JAN99
EQUIPMENT	– Carry one machine gun and one rocket launcher

b. **Field Sketches**. Try to include a sketch with each report. Include only any aspects of military importance such as targets, objectives, obstacles, sector limits, or troop dispositions and locations (use symbols from FM 1-02). Use notes to explain the drawing, but they should not clutter the sketch.

c. **Captured Documents**. The leader collects documents and turns them in with his reports. He marks each document with the time and place of capture.

d. **Prisoners**. If prisoners are captured during a patrolling operation, they should be treated IAW the Geneva Convention and handled by the 5-S rule:

(1) Search
(2) Silence
(3) Segregate
(4) Safeguard
(5) Speed to rear

e. **Debriefs**. Immediately upon return from a mission, the unit is debriefed. The intelligence officer will generally have a unit-specific format for debriefing a patrol.

2-3. **WARNING ORDER**. Warning orders give subordinates advance notice of upcoming operations. This gives them time to prepare. A warning order must be brief, but complete. *A warning order only authorizes execution when it clearly says so.* (Figure 2-2 shows an example format; Figure 2-3 shows an example warning order.)

Figure 2-2. WARNING ORDER FORMAT

WARNING ORDER ______

*Roll call, pencil/pen/paper, RHB, map, protractor, leader's monitor, hold all questions till the end
References: Refer to higher headquarters' OPORD, and identify map sheet for operation.
Time Zone Used throughout the Order: (Optional)
Task Organization: (Optional; see paragraph 1c.)

1. **SITUATION** (Higher's OPORD para 1a[1]--[3])

a. **Enemy Forces**. Include significant changes in enemy composition, dispositions, and courses of action. Information not available for inclusion in the initial WARNO can be included in subsequent warning orders. (who, what, where)

b. **Friendly Forces**. (Optional) only address if essential to the WARNO.

(1) Higher commander's mission (who, what, where, why).
(2) Higher commander's intent. (Higher's [go to mapboard] OPORD para 1b[2], give task and purpose)

c. **Attachments and Detachments**. Initial task organization, only address major unit changes.

(1) Orient relative to each point on the compass.
(2) Box in the entire AO with grid lines.
(3) Trace each Zone using boundaries.
(4) Familiarize by identifying 3 natural (terrain) and 3 man-made features in each zone.
(5) Point out the enemy and friendly locations on the map board.

2. **MISSION**. State mission twice (who, what, when, where, and why) and concisely state task and purpose.

3. **EXECUTION**.

a. **Concept of Operation**. Provide as much information as available. The concept should describe the form of maneuver or defensive technique, critical events, decisive point of the operation and why it is decisive, task and purpose for the main and supporting efforts, purposes of the warfighting systems, such as engineers, fire support, intelligence etc), acceptable risk, and the desired endstate. After the desired endstate is stated, leaders should talk through the concept of the operation like a story. Use a simple sketch to assist in relaying the information.

b. **Tasks to Maneuver Units**. Provide specified tasks to subordinate units. These are tactical instructions on how to execute the mission for each element in task organization. Planning guidance consists of tasks assigned to subordinate units, special teams, and key individuals.

c. **Tasks to Combat Support Units**. See paragraph 3b.

d. **Coordinating Instructions**. Include any information available at that time. If you know it, then at least cover--

- Uniform and equipment common to all. Consider the factors of METT-TC and tailor the load for each Ranger.
- Time line. (State 4 W's and specified and implied times. Reverse plan. Use 1/3-2/3 rule.)
- CCIR.
- Risk guidance.
- Deception guidance.
- Specific priorities, in order of completion.
- Guidance on orders and rehearsals.
- Orders group meeting (attendees, location, and time).
- Earliest movement time and degree of notice.

4. **SERVICE SUPPORT**. Include any known logistics preparation for the operation.

a. Special equipment. Identify requirements and any coordination measures the unit needs to take to transfer the special equipment. (State the equipment you will use, need, or want for the mission.)

b. Transportation. Identify requirements and any coordination needed to pre-position assets. (State unit's means of infil/exfil.)

5. **COMMAND AND SIGNAL**.

a. **Command**. State the chain of command if different from unit SOP.

b. **Signal**. Identify current SOI edition, and pre-position signal assets to support operation.

- Give subordinates guidance on tasks to complete for preparation of the OPORD and the mission.
- Give time, place, and uniform of the OPORD.
- Give a time hack and ask for questions.

NOTES

Figure 2-3. EXAMPLE WARNING ORDER

WARNING ORDER (SQUAD)

ROLL CALL, CAMP DARBY SPECIAL (CDS) 1:50,000 MAP, PEN, PAPER, PENCIL, PROTRACTOR, RANGER HANDBOOK (RHB), HOLD ALL QUESTIONS, AND TLs MONITOR AND TASK-ORGANIZE

1. **SITUATION**.

a. **Enemy**.

WHO: The Aragon Liberation Front (ALF).
WHAT: Has been conducting aggressive offensive operations.
WHERE: In Zones A, B, C.

b. **Friendly**:

(1) **Higher's Mission**.

WHO: 2nd PLT, C CO.
WHAT: Conducts area ambushes to destroy enemy personnel and equipment.
WHERE: Zone D.
WHY: To prevent enemy logistical resupply.

(2) **Higher Commander's Intent**.

- Rapid infiltration into Zone D.
- Find, fix, and finish enemy forces in Zone D.
- Enemy personnel and equipment located and targeted.
- Squads ready to conduct future combat operations.

c. **Attachments/Detachments**. MG Tm #1/2 PLT 160630-190630.

2. **MISSION**.

WHO: 1st SQD, 2nd PLT
WHAT: Conducts a point ambush to destroy enemy personnel and equipment.
WHERE: GA 1805 7878.
WHEN: NLT 25 1830 DEC 2007.
WHY: To prevent enemy logistical resupply and gather PIR.

3. **EXECUTION**.

a. **Concept of Operation**. (Orient Rangers to sketch or terrain model). We are currently located at Camp Darby, GA 1962 7902, which is on a ridge. Our ground tactical plan will begin when we will move generally southwest to our ORP at Grid GA 187778. Terrain feature there is a draw for 1,000 meters, by foot and it will take 1 hour. Here will we finalize the preparing of M, W, and E. We will then move generally southwest to our objective at Grid GA 184774. Terrain feature there is a ridge for 200 meters, by foot, and it will take us 30 minutes to an hour. The reason for this is stealth on the objective while we occupy our positions.

b. **Tasks to Maneuver Units**.

(1) HQs: HQ is responsible for command and control and communications. Commo/HQ is 2nd in the OOM. M240 will be placed in a position to provide supporting fires in the kill zone. RTO will be the recorder en route and during actions on the objective. SL will be located where he can best control his squad and observe the kill zone. SL is responsible for dissemination of information.

(2) ATM: ATM will be 1st OOM/responsible for navigation, and for clearing and securing the far side at all linear danger areas. ATM is flank security for AOO, will provide early warning and seal off objective/1-2 man EPW, and search TM/1-2 man aid and litter TM/1-2 man demo TM/1 man (P), 1 man (A) ORP clearing TM/ 2-2 man flank security. TM for AOO/ 1 compassman, 1 pace man, or ATL will be the sec. TM LDR along the most likely avenue of approach.

(3) BTM: BTM 3RD OOM/responsible for securing the near side of all linear danger areas/BTM. Is assault for AOO. Must provide suppressive fire and assault thru kill zone/1-2 man EPW and search team/1-2 man aid and litter team/1-2 man demolitions team; 1 man (P); 1 man (A) ORP clearing team; 1-2 man S/O team; 1-2 man security team for reentry; 1 compassman; 1 pace man; and BTL. You will be the assault element leader for AOO.

c. **Tasks to Combat Support Units**. None.

d. **Coordinating Instructions**.

(1) Uniform and equipment common to all. At a minimum, each Ranger wears or carries--

(a) A ballistic helmet w/camouflaged band and luminous tape; clean and serviceable ACUs, camelback, black belt w/buckle, blackened combat boots, serviceable OD or black socks, serviceable OD/brown T-shirt, ID tags w/breakaway lanyard around neck (silenced). Right cargo pocket; (RHB, CDS [map], protractor, pencil, pen, and paper) waterproofed, left cargo pocket; patrol cap w/ luminous tape, ID card in right breast pocket, SL will carry sterile fire support overlay in left breast pocket.

(b) LCE, consisting of pistol belt, suspenders, 2 ammo pouches, seven 30-round magazines, one first aid dressing with pouch, one compass and pouch, two 1-quart canteens w/covers and cups, one whistle, and four M67 fragmentation grenades. All are tied down IAW 4th Bn SOP, except that the pace cord is tied to suspenders.

(c) Rucksack, complete; one set serviceable ACUs, two 2-quart canteens w straps; E-tool with cover; two pair socks; two brown T-shirts; weapons cleaning kit; one waterproof bag; a sewing kit; a patrol cap; MREs; poncho and liner; wet weather top; personal hygiene kit; one brown towel; and camouflage stick. RTO also packs and ties off the ANCD.

(d) Optional bug juice, 550 cord, foot powder, black tape, knife, all other items will be approved by SL 15 minutes prior to completion time of OPORD.

(2) Time schedule.

WHEN	WHAT	WHERE	WHO	REF
*0630	WARNING ORDER	BAY AREA	All	3D1
0700	INITIAL INSPECTION	BAY AREA	All	
*0730	Required Ammo/Supplies	CO TOC	BTL	4B1
*0745	P/U-Ammo/Supplies	CO TOC	BTL/Detail	4B1
0800	Test Fire	T/F AREA	All	
*0830	S-2/S-3 Coord	CO BAY AO	SL/RTO	3D1
*0830	Fire Spt Coord	BN TOC	SL/ RTO	3A2
0850	Adj Unit Coord	BN TOC	ATL/compass man	3D1
*0900	Enter Net	BAY AREA	RTO	5B1
0930	Complete Plan	BAY AREA	All	
*1000	Issue OPORD	BAY AREA	All	3D1
1300	Rehearsal	BAY AREA	All	
1400	Final Inspection	BAY AREA	All	
1430	Conduct Move	BAY AREA	All	
1800	In ORP	GA	All	
2000	In Position	GA 184774	All	
*2300	Mission Complete	GA 186874	All	3D1
*0200	Reenter FFU	GA	All	
0500	S-2 Debrief	BN TOC	All	
* Specified time				

4. **SERVICE AND SUPPORT**.

a. **Special Equipment**.

(1) HQs: One hand-held radio and extra batteries; one AN/PRC-119 radio and extra batteries, one ANCD, one AN/PSN-11 and extra batteries, one M-22 binos, one AN/PVS-14 and extra batteries; one sheets overlay material, three sheets carbon paper; two HE smoke, two red smoke, one M240-B MG complete w/tripod, flex mount, T&E, and Pintle, one AN/PVS-4 and extra batteries, one AN/PVS-7D, and extra batteries.

(2) ATM: One hand-held radio and extra batteries, one M-24 binos, two AN/PVS-14 and extra batteries, two HE smoke, two yellow smoke, one thermite grenade, two gags, two flex cuffs, two sandbags, one CLS bag, one pole-less litter, two M18A1 claymore mines, one AT-4, one LAW, one M-203 vest, two M-249 SAW pouches, one small metal flashlight, and extra batteries secured to rifle.

(3) BTM: One hand-held radio and extra batteries, one M-24 binos, two AN/PVS 14 and extra batteries, two HE smoke, two yellow smoke, one thermite grenade; two each gags, flex cuffs, and sandbags; one CLS bag, one pole-less Litter, two M18A1 claymore mines, one AT-4, one LAW, one M-203 vest, two M249 SAW pouches, one small metal flashlight, and extra batteries secured to rifle.

b. **Ammo**. (per weapon system).

M16/M4	**M249**	**M203**	**M240**	**SL AND TLs**
210 Rounds 5.56-MM Ball	625 Rounds 5.56-MM (4:1 Mix)	16 Rounds 40-MM HEDP 5 Rounds 40-MM Shot 4 Rounds 40-MM Illumination 2 Rounds 40-MM TP-T	825 Rounds 7.62-MM (4:1 Mix)	42 Rounds 5.56-MM Tracer

c. **Transportation**. NA

5. **COMMAND AND SIGNAL**.

a. **Command**.

PERSONNEL	**POS**	**SOC**
HQ/SUPPORT		
BARNETT	SL	1
MARENGO	RTO	4
CASELY	MG	11
WADE	AG	12
ATM/SECURITY		
TRUITT	ATL	3
SILCOX	GRN	5
WISE	R	6
EDMUNDS	R	7
BTM/ASSAULT		
SPENCER	BTL	2
CHRISTENSEN	GRN	8
RUFUS	R	9
ROBINSON	R	10

SUBORDINATE GUIDANCE:	
ATL:	Write paragraph 1 of squad OPORD, and for drawing all sketches, terrain models, routes, and nonsterile fire support overlays.
BTL:	You are second in the chain of command and in charge at all times during my absence. Write paragraph 4 of squad OPORD Write the supply, DX, and ammo lists, draw and issue all items, ensure that everyone conducts a test fire, and all equipment is tied down IAW 4th RTBn SOP, and update the squad status card and hand receipt.
RTO:	Write paragraph 5 of squad OPORD, ensure all radios are operational with proper freqs loaded, and ensure we enter the net on time.
TL:	TLs: You will update the Warning Order board with all the correct information. As a task is accomplished, you will line it out. Also, post your succession of command (SOC), duty positions, and job descriptions (specialty teams and key individuals). Come and see me for further guidance at the conclusion of this WARNO.

b. **Signal**.

TACTICAL FREQUENCY	62,000
MEDEVAC FREQUENCY	35,000
CHALLENGE	BAYONET
PASSWORD	MILLETT
RUNNING PASSWORD	RANGER
NUMBER COMBO	7
TIME, PLACE, AND UNIFORM FOR OPORD. TIME HACK. ANY QUESTIONS?	

NOTES

2-4. **OPERATIONS ORDER**. An operations order (OPORD) is a directive issued by a leader to his subordinates in order to effect the coordinated execution of a specific operation. A five-paragraph format (shown below) is used to organize the briefing, to ensure completeness, and to help subordinate leaders understand and follow the order. Use a terrain model or sketch along with a map to explain the order. When possible, such as in the defense, give the order while observing the objective. The platoon/squad leader briefs his OPORD orally off notes that follow the five-paragraph format. Before the issuance of the OPORD, the leader ensures that the following resources are in place: pencil, pen, paper; RHB; map; protractor; leader's monitor. Then he calls roll and says "*Please hold all questions till the end.*" Figure 2-4 shows an example OPORD format. Figure 2-5 shows an OPORD shell.

Figure 2-4. EXAMPLE OPORD FORMAT

OPERATIONS ORDER

[Plans and orders normally contain a code name and are numbered consecutively within a calendar year.]

References: [The heading of the plan or order includes a list of maps, charts, datum, or other related documents the unit will need to understand the plan or order. The user need not reference the SOP, but may refer to it in the body of the plan or order. He refers to a map by map series number (and country or geographic area, if required), sheet number and name, edition, and scale, if required. "*Datum*" refers to the mathematical model of the earth that applies to the coordinates on a particular map. It is used to determine coordinates. Different nations use different datum for printing coordinates on their maps. The datum is usually referenced in the marginal information of each map.]

Time zone used throughout the order: The time zone used throughout the order (including annexes and appendixes) is the time zone applicable to the operation. Operations across several time zones use Zulu time.

Task organization: Describe the allocation of forces to support the commander's concept. Task organization may be shown in one of two places: preceding paragraph one, or in an annex, if the task organization is long and complicated.

**Weather and light data, and general forecast (Only discuss what will affect your patrol. Note effects on friendly and enemy personnel and equipment.):*

High
Moonrise
Sunrise
Low
Moonset
Sunset
Wind Speed
Moon Phase
BMNT
Wind Direction
Percent Illumination
EENT

**This is the information the leader received from higher's OPORD.*

Terrain: Make a statement concerning the effects terrain will have on both friendly and enemy forces in the area of operation using the OAKOC format.

1. **SITUATION**.

a. **Enemy Forces**. The enemy situation in higher headquarters' OPORD (paragraph 1a) is the basis for this, but the leader refines this to provide the detail required by his subordinates.

(1) Enemy's composition, disposition, strength.

(2) Recent activities.

(3) Known/suspected locations and capabilities.

(4) Describe the enemy's most likely and most dangerous course of action on a map.

b. **Friendly Forces**. This information is in paragraphs 1b, 2, and 3 of higher headquarters' OPORD.

(1) State the mission of the Higher Unit (2 levels up).

(2) State the mission of the Higher Unit (1 level up).

(3) State intent 2 levels up.

(4) State locations of units to the left, right, front, and rear. State those units' tasks and purposes; and say how those units will influence your unit's patrol.

(a) Show other units locations on map board.

(b) Include statements about the influence each of the above patrols will have on your mission, if any.

- Obtain this information from higher's OPORD. It gives each leader an idea of what other units are doing and where they are going. This information is in paragraph 3a (1) (execution, concept of the operation, maneuver).
- Also include any information obtained when the leader conducts adjacent unit coordination.

c. **Attachments and Detachments**. Do not repeat information already listed under Task Organization. Try to put all information in the Task Organization. However, when not in the Task Organization, list units that are attached or detached to the headquarters that issues the order. State when attachment or detachment is to be effective if different from when the OPORD is effective, such as on order, on commitment of the reserve). Use the term "*remains attached*" when units will be or have been attached for some time.

2. **MISSION**. State the mission derived during the planning process. There are no subparagraphs in a mission statement. Include the 5 W's: Who, What (task), Where, When, and Why (purpose).

- Make it a clear and concise statement and read it twice.
- Go to map and point out the exact location of the OBJ and point out the unit's present location.

3. **EXECUTION**. State the commander's intent.

a. **Concept of Operation**. The concept statement should be concise and clear. It describes how the unit will accomplish its mission from start to finish. The number of subparagraphs, if any, is based on what the leader considers appropriate for the level of leadership and complexity of the operation. Generally, the concept should describe the form of maneuver or defensive technique and critical events; identify what the decisive point of the operation is and why; describe the task and purpose for the main effort and supporting efforts; state the purpose of the warfighting functions, such as engineers, fire support, or intelligence; define acceptable risk; and identify the desired endstate. After stating the desired endstate, leaders should talk through the concept of the operation like a story, using a simple sketch to help them convey the information. Figure 1 shows the six warfighting functions (WFF):

Figure 1. Warfighting functions.

- Fire support
- Movement and Maneuver
- Protection
- Command and Control
- Intelligence
- Sustainment (formerly called "CSS")

(1) ***Maneuver***. The maneuver paragraph details the mechanics of the operations. This subparagraph specifically addresses all subordinate units and attachments by name, giving each its mission in the form of a task and purpose. The main effort's mission must be supported by the subordinate unit's missions. Actions on the objective will comprise the majority of this paragraph and should include a detailed plan for engagement/disengagement criteria, an alternate plan in the event of compromise or unplanned movement of enemy forces, and a withdrawal plan.

(2) ***Fires***. State scheme of fires to support the overall concept. This paragraph should state which maneuver unit has priority of fires. Fires should be planned using the PLOT-CR format (purpose, location, observer, trigger, communication method, resources). A target list worksheet and overlay are referenced here, if applicable. Specific targets are discussed and pointed out on the terrain model (see Chapter 3, Fire Support).

(3) ***Casualty Evacuation***.

(a) *Plan*. State a detailed CASEVAC plan during each phase of the operation. Include CCP locations, tentative extraction points, and methods of extraction.

b. **Tasks to Maneuver Units**. Clearly state the missions or tasks for each maneuver unit that reports directly to the headquarters issuing the order. List the units in the task organization, including reserves. Use a separate subparagraph for each maneuver unit. Only state the tasks that are necessary for comprehension, clarity, and emphasis. Place tactical tasks that affect two or more units in Coordinating Instructions (subparagraph 3d). Platoon leaders task their subordinate squads. Those squads may be tasked to provide any of the following special teams: reconnaissance and security, assault, support, aid and litter, EPW and search, clearing, and demolitions. Detailed instructions may also be given to the platoon sergeant, RTO, compassman, and paceman.

(1) ***Tasks to Combat Support Units***. Add subparagraphs here only as necessary. List CS units in subparagraphs from the task organization. Use CS subparagraphs to list only those specific tasks that CS units must accomplish and that are not specified or implied elsewhere. Include organization for combat if not clear from task organization.

c. **Coordinating Instructions**. This is always the last subparagraph under paragraph 3. List only the instructions that apply to two or more units, and which are seldom covered in unit SOPs. Refer the user to an annex for more complex instructions. The following information is required:

1. Time schedule (state time, place, uniform, and priority of rehearsals, backbriefs, inspections, and movement).

2. Give the commander's critical information requirements (CCIR), which includes priority intelligence requirements (PIR), essential elements of friendly information (EEFI), and friendly force information requirements (FFIR):

a. PIR is intelligence that the commander must have for planning and decision making.

b. EEFI are critical aspects of friendly operations that, if known by the enemy, would compromise, lead to failure, or limit success of the operation.

c. FFIR include what the commander needs to know about friendly forces available for the operation. It can include personnel status, ammunition status, and leadership capabilities.

3. Risk reduction control measures. These are measures unique to the operation. They supplement the unit SOP and can include mission-oriented protective posture, operational exposure guidance, vehicle recognition signals, and fratricide prevention measures.

4. Rules of engagement (ROE).

5. Environmental considerations.

6. Force protection

7. Movement plan. Use terrain model, sketch, or both. State azimuths, directions, and grid coordinates. Use names of real Rangers on the sketches.

a. Sketch out the order of movement, the formation, and the movement technique.

b. Sketch actions at halts (long and short).

c. Routes (use terrain model):

1) Orient the terrain model (North, South, East, and West).

2) Identify the grid lines, both North-South and East-West.

3) Brief the legend.

4) Box in the AO with distinguishable natural or man-made terrain features such as roads, ridgelines, rivers, and streams.

5) *Brief all the routes depicted on the terrain model from start to finish using specific azimuths and distance.*

a) *The primary route from the start point to the ORP, from the ORP to the objective, and from the objective to the patrol base while describing the terrain over which the squad/platoon will travel. (Include near/far side rally points for danger areas.)*

b) *The alternate route from the start point to the ORP, from the ORP to the objective, and from the objective to the patrol base while describing the terrain over which the squad/platoon will travel. (Include near/far side rally points for danger areas.)*

6) *Brief the fire support plan (if not given in fires paragraph). When covering fires the leader should cover the following (PLOT-CR):*

- *Purpose.*
- *Location and terrain feature to the targets.*
- *Observer.*
- *Trigger.*
- *Communication method.*
- *Resources.*

7) *State where the sterile fire support overlay is being carried.*

a) *Rally points and actions at all rally points (include grid location and terrain reference, and use sketch).*

b) *Actions at danger areas. This is a general plan for unknown linear, small open areas and large open areas; specific plan for all known danger areas that unit will encounter along the route (use sketch).*

c) *Actions on enemy contact; battle drills (use sketch).*

4. **SERVICE SUPPORT**. Address service support in the areas shown below as needed to clarify the service support concept. Subparagraphs can include:

a. **General**. Refer to the SOPs that govern the sustainment operations of the unit. Provide current and proposed company trains locations, casualty, and damaged equipment collection points and routes.

b. **Materiel and Services**.

(1) ***Supply***.

Class I--Rations plan.
Class III--Petroleum.
Class V--Ammunition.
Class VII--Major end items.
Class VIII--Medical.
Class IX--Repair parts.
Distribution Methods.

(2) ***Transportation***.

(3) ***Services***. (Laundry and showers.)

(4) ***Maintenance***. (Weapons and equipment.)

(a) *Medical Evacuation and Hospitalization*. Method of evacuating dead and wounded, friendly and enemy personnel. Include priorities and location of CCP.

(b) *Personnel Support*. Method of handling EPWs and designation of the EPW collection point.

5. **COMMAND AND SIGNAL**.

This paragraph states where command and control facilities and key leaders are located during the operation. Use the OPORD shell to save time and organize your thoughts while preparing an OPORD.

a. **Command**.

- Location of the higher unit commander and CP.
- Location of key personnel and CP during each phase of the operation.

- Succession of command.
- Adjustments to the patrol SOP.

b. **Signal**.
- SOI index in effect.
- Methods of communication in priority.
- Pyrotechnics and signals, to include arm and hand signals (demonstrate).
- Code words such as OPSKEDs.
- Challenge and password (use behind friendly lines).
- Number combination (use forward of friendly lines).
- Running password.
- Recognition signals (near/far and day/night).

c. **Special Instructions to RTOs**.

d. **Actions after Issuance of OPORD--**
(1) Issue annexes.
(2) Give time hack.
(3) ASK for questions.

Figure 2-5. OPORD SHELL

OPORD

DTG:

REFERENCE:

TASK ORGANIZATION:

1ST PLT   2ND PLT   TM CONTROL:

1. **SITUATION**.

BMNT: SR: SS: EENT: MR: MS: HI: LO: ILLUM:

a. **Enemy**.

Terrain | End state

(1) Higher mission:

Cdr's intent:

(2) Units to N, S, E, W.

Terrain | End state

2. **MISSION**:

3. **EXECUTION**:

a. **Concept of Operations**.

Cdr's intent:

(ME): T: P:	(SE1): T: P:	(SE2): T: P:	(SE3): T: P:	(SE4): T: P:	(SE5): T: P:

b. **Maneuver**

PHASE 1:
ME:
SE1:
SE2:
SE3:
SE4:
SE5:
FIRES:
MTRS:
C2:

PHASE 2:
ME:
SE1:
SE2:
SE3:
SE4:
SE5:
FIRES:
MTRS:
C2:

PHASE 3:
ME:
SE1:
SE2:
SE3:
SE4:
SE5:
FIRES:
MTRS:
C2:

PHASE 4:
ME:
SE1:
SE2:
SE3:
SE4:
SE5:
FIRES:
MTRS:
C2:

c. **Tasks to Maneuver Units**.

(1) Risk-reduction control measures.
Rules of engagement.
(2) Environmental considerations.
(3) Non-Airborne personnel.

(ME): (SE1): (SE2):

(SE3): (SE4): (SE5):

d. **Tasks to Subordinate Units**

Mortars

Coordinating instructions:

PIR:

EEFI:

FFIR:

Timeline.

Coordinating Instructions.

WHEN	WHO	WHAT	WHERE

4. **SERVICE SUPPORT**.

a. **General**.

b. **Material and Services**.

(1) Supply:

Class I.

Class III:

Class V:

Class VIII:

Class IX:

(2) LOGSTAT.

(3) The chain of command is responsible to ensure that ammo lasts between resupplies.

c. **Medical Evacuation**.

(1) Air MEDEVAC is on call through [this] frequency. JP capable; use standard 9-line, 24-hr coverage.

(2) Establish friendly CCPs in AAs. Adjust CCPs as mission progresses.

(3) Troop medical clinic located vic [this location].

(4) Precedence for evacuation of WIA is US military, host nation military, host nation civilian, and enemy military.

(5) Precedence for evacuation of KIA is same. Evacuation of KIA will be as per mission constraints.

d. **Personnel**.

(1) Method of handling EPWs:

ANNEXES:

5. **COMMAND AND SIGNAL**.

a. **Command**.

(1) TF CP:

(2) Succession of Command:

b. **Signal**.

(1) Challenge and Password:

(2) Number Combination:

(3) Running Password:

(4) Recognition:

(a) Far:

(b) Near:

(7) Code word:

(8) OPSKED:

(9) Reports. SALUTE to higher on enemy contact.

2-5. **FRAGMENTARY ORDER**. A FRAGO is an abbreviated form of an operations order, usually issued on a day-to-day basis that eliminates the need for restating information contained in a basic operations order. It is issued after an OPORD to change or modify that order or to execute a branch or sequel to that order. Figure 2-6 shows an annotated FRAGO format:

Figure 2-6. ANNOTATED FRAGO FORMAT

FRAGMENTARY ORDER____________

Time Zone referenced throughout order:

Task Organization:

Weather and Light Data:

High: BMNT: Moonrise:

Low: Sunrise: Moonset:

Wind Speed: Sunset: % Illum:

Wind Direction: EENT:

Forecast:

Terrain (changes that will effect operation in new area of operations):

OAKOC: Observation and fields of fire, avenues of approach, key terrain, observation, and cover and concealment.

1. **SITUATION**. (Brief changes from base OPORD specific to this day's operation.)

a. **Enemy Situation**.

(1) Composition, disposition and strength.

(2) Capabilities.

(3) Recent activities.
(4) Most likely COA.

b. **Friendly Situation.**
(1) Higher mission.
(2) Adjacent patrols task/purpose.
(3) Adjacent patrol objective/route (if known).

2. **MISSION** (Who, what [task], when, where, why [purpose]—from higher HQ's maneuver paragraph).

3. **EXECUTION**

a. **Concept of Operation.** (Explain how unit will accomplish mission in general terms. Identify mission essential task, designate the main effort and how the supporting efforts support the main effort.)

(1) ***Maneuver***. (Assign task/purpose and discuss actions on the objective in detail. Use a sketch or terrain model to brief.)

(2) ***Fires***. (In support of today's mission, portions that do not change are not briefed.) PLOT-CR:
(a) Purpose.
(b) Location.
(c) Observer.
(d) Trigger.
(e) Communication method.
(f) Resource.

(3) ***CASEVAC***.

b. **Tasks to Maneuver Units.** (List unique tasks that apply to each squad.)

c. **Coordinating Instructions.**

(1) ***Timeline***.
(a) Hit time:
(b) ORP time:
(c) Movement time from PB:
(d) Final inspection:
(e) Rehearsals:
(f) FRAGO complete:

(2) ***Movement Plan***. (Use a sketch to brief.)
(a) Routes (primary and alternate):
(b) OOM:
(c) Formations:
(d) Movement technique:

(3) ***PIR***. (Specific to this mission):

(4) ***Rehearsal Plan***.

(5) ***Patrol Base Plan***. (If not IAW SOP):
(a) Teams:
(b) Occupation plan:
(c) Operations plan (security plan, alert plan, Black and Gold):
(d) Priorities of work:

(6) ***Air Assault Plan***. (Provided with higher's FRAGO, if applicable):
(a) Number, type, and ACL aircraft:
(b) PZ grid, DOL, and PZ posture time:
(c) Load time/lift time/flight time:
(d) Number of lifts and composition:
(e) Air checkpoints en route:
(f) LZ grid/DOL/LZ time:
(g) Actions after getting off aircraft:
(h) Actions on contact on LZ:

(7) ***Linkup Plan***. (If applicable):

(a) Time of linkup:
(b) Location of linkup site:
(c) Stationary element:
(d) Moving element:
(e) Rally points:
(f) Actions at linkup point:
(g) Near/far recognition signals (day and night):

4. **SUPPORT**. (Only cover changes from base order that apply for today's mission).

a. **Material and Services**.

(1) Any changes in classes of supply:
(2) Resupply plan:
(3) Water resupply plan:
(4) Aerial resupply plan (if applicable) :
(5) Truck plan:
(6) Maintenance issues specific to plan:

b. **Medical Evacuation Plan**. (Specific to mission):

(1) CCP point and markings:
(2) Aid and litter duties (if not SOP):

5. **COMMAND AND SIGNAL**.

a. **Command**.

(1) Location of Company CP:
(2) Location of PL:
(3) Location of key leaders:
(4) Succession of command:

b. **Signal**.

(1) Location of radios:

(a) During movement:
(b) During actions on the objective:

(2) SOI in effect:

(a) Bn command freq:	Co CP call sign:
(b) Bn MEDEVAC freq:	PL call sign:
(c) Retrans freq:	PSG call sign:
(d) Company freq:	1 SL call sign:
(e) Platoon freq:	2 SL call sign:
	3 SL call sign:
	WSL call sign:

(3) Pyro signals in use:
(4) Running password/number combination:
(5) OPSKEDs in effect:

(a) The field FRAGO should take no more than 40 minutes to issue, with 30 minutes for the target. The proposed planning guide is as follows:

Paragraphs 1 and 2:	5 minutes
Paragraph 3:	20 to 30 minutes
Paragraphs 4 and 5:	5 minutes

(b) The FRAGO should focus on actions on the objective. The PL may use subordinates to prepare para 1, 4, 5 and routes and fires for the FRAGO. It is acceptable for subordinates to brief the portions of the FRAGO they prepare.

(c) Use of sketches and a terrain model are critical to allow rapid understanding of the operation/FRAGO.

(d) Rehearsals are critical as elements of the constrained planning model. The FRAGO used in conjunction with effective rehearsals reduces preparation time and allows the PL more time for movement and recon.

(e) Planning in a field environment will necessarily reduce the amount of time leaders have for in-depth mission planning. The TLP gives leaders a framework to plan missions and produce orders when time is short.

2-6. **ANNEXES**. Operation order annexes are necessary to complete the plan and to provide greater clarity and understanding during complex or critical aspects of the operation. Information issued in annex form include the: aerial re-supply, truck movement, air assault, patrol base, small boat, linkup, and stream crossing annexes. Annexes are prepared only if the subject is not addressed thoroughly enough in the OPORD; brevity remains the standard. Annexes are always issued after the operation order.

AIR MOVEMENT ANNEX

1. **SITUATION**.
   a. **Enemy Situation**.
      (1) Enemy air capability.
      (2) Enemy ADA capability.
      (3) Include in weather: % Illum, illum angle, NVG window, ceiling, and visibility.
   b. **Friendly Situation**.
      (1) Unit(s) supporting operation.
      (2) Friendly ADA status.
2. **MISSION**.
3. **EXECUTION**.
   a. **Concept of Operation**.
   b. **Subunit Missions**.
   c. **Coordinating instructions**.
      (1) ***Pickup Zone***.
         (a) Name/Number/
         (b) Coordinates/
         (c) Load Time/
         (d) Takeoff Time/
         (e) Markings/
         (f) Control/
         (g) Landing Formation/
         (h) Approach/Departure Direction.
         (i) Alternate PZ Name/Number.
         (j) Penetration Points.
         (k) Extraction Points.
      (2) ***Landing Zone***.
         (a) Name/Number.
         (b) Coordinates.
         (c) H-Hour.
         (d) Markings.
         (e) Control.
         (f) Landing Formation/Direction.
         (g) Alt LZ Name/Number.
         (h) Deception Plan.
         (i) Extraction LZ.
      (3) ***Laager Site***.
         (a) Communications.
         (b) Security Force.
      (4) ***Flight Routes and Alternates***.
      (5) ***Abort Criteria***.
      (6) ***Down Aircraft/Crew*** (designated area of recovery [DAR]).
      (7) ***Special Instructions***.
      (8) ***Cross-FLOT Considerations***.
      (9) ***Aircraft Speed***.

(10) ***Aircraft Altitude***.
(11) ***Aircraft Crank Time***.
(12) ***Rehearsal Schedule/Plan***.
(13) ***Actions on Enemy Contact*** (en route and on the ground).

4. **SERVICE SUPPORT**.
   a. **Forward Area Refuel/Rearm Points**.
   b. **Classes I, III, and V** (specific).

5. **COMMAND AND SIGNAL**.
   a. **Command**.
   b. **Signal**.
      (1) Air/ground call signs and frequencies.
      (2) Air/ground emergency code.
      (3) Passwords/number combinations.
      (4) Fire net/Quickfire net.
      (5) Time zone.
      (6) Time hack.

AERIAL RESUPPLY ANNEX

1. **SITUATION**.
   a. **Enemy Forces** (include weather).
   b. **Friendly Forces**.
   c. **Attachments and Detachments**.

2. **MISSION**.

3. **EXECUTION**.
   a. **Concept of Operation**.
      (1) ***Maneuver***.
      (2) ***Fires***.
   b. **Tasks to Combat Units**.
      (1) ***Command and Control***.
      (2) ***Security***.
      (3) ***Marking***.
      (4) ***Recovery/Transport***.
   c. **Tasks to Combat Support Units**.
   d. **Coordinating Instructions**.
      (1) Flight Route.
         (a) General.
         (b) Checkpoints.
         (c) Communication checkpoint (CCP).
            - Marking of CCP.
            - Report time.
         (d) Heading from CCP.
      (2) Landing/Drop Zone.
         (a) Location.
            - Primary.
            - Alternate.
         (b) Marking.
            - Near.
            - Far.

(3) Drop Information.
   (a) Date/time of resupply (and alternates).
   (b) Code letter at DZ/LZ.
   (c) Length of DZ in seconds or dimensions of LZ.
   (d) Procedures for turning off DZ/LZ.
   (e) Formation, altitude, and air speed.
   - En route.
   - At DZ/LZ.

(4) Actions on enemy contact during resupply.
(5) Abort Criteria: En route and at DZ/LZ.
(6) Actions at DZ/LZ.
   - Rehearsals.

4. **SERVICE SUPPORT.**

5. **COMMAND AND SIGNAL.**
   a. **Command.**
   (1) Location of patrol leader.
   (2) Location of assistant patrol leader.
   (3) Location of members not involved in resupply.
   b. **Signal.**
   (1) Air to ground call-signs and frequencies (primary and alternate).
   (2) Long range visual signals.
   (3) Short range visual signals.
   (4) Emergency procedures and signals.
   (5) Air drop communication procedures.
   (6) Code words.

PATROL BASE ANNEX

1. **SITUATION**
   a. Enemy Forces
   b. Friendly Forces
   c. Attachments and Detachments

2. **MISSION.**

3. **EXECUTION.**
   a. **Concept of Operation.**
   (1) ***Maneuver.***
   (2) ***Fires.***
   b. **Tasks to Combat Units.**
   (1) ***Teams.***
   - Security
   - Recon
   - Surveillance
   - LP/OPs

   (2) ***Individuals.***
   c. **Tasks to Combat Support Units.**
   d. **Coordinating instructions.**
   (1) Occupation plan.
   (2) Operations plan.
   - Security Plan.
   - Alert Plan.

- Priority of work.
- Evacuation plan.
- Alternate patrol base (used when primary is unsuitable or compromised).

4. **SERVICE SUPPORT**.
   a. Water plan.
   b. Maintenance plan.
   c. Hygiene plan.
   d. Messing plan.
   e. Rest plan.

5. **Command and Signal**.
   a. **Command**.
      (1) Location of patrol leader.
      (2) Location of assistant patrol leader.
      (3) Location of patrol CP.
   b. **Signal**.
      (1) Call signs and frequencies.
      (2) Code words.
      (3) Emergency signals.

SMALL BOAT ANNEX

1. **SITUATION**.
   a. **Enemy Forces**.
      (1) **Weather**.
         (a) Tide.
         (b) Surf.
         (c) Wind.
      (2) **Terrain**.
         (a) River width.
         (b) River depth and water temperature.
         (c) Current.
         (d) Vegetation.
      (3) **Identification, Location, Activity, and Strength**.
   b. **Friendly Forces** (unit furnishing support).
   c. **Attachments and Detachments**.
   d. **Organization for Movement**.

2. **MISSION**.

3. **EXECUTION**.
   a. **Concept of Operation**.
      (1) ***Maneuver***.
      (2) ***Fires***.
   b. **Tasks to Combat Units**.
      (1) Security.
      (2) Tie-down teams.
         (a) Load equipment.
         (b) Secure equipment.
      (3) Designation of coxswains and boat commanders.
      (4) Selection of navigator(s) and observer(s).

c. **Coordinating Instructions**.
(1) Formations and order of movement.
(2) Route and alternate route.
(3) Method of navigation.
(4) Actions on enemy contact.
(5) Rally points.
(6) Embarkation plan.
(7) Debarkation plan.
(8) Rehearsals.
(9) Time schedule.

4. **SERVICE SUPPORT**.
a. **Ration Plan**.
b. **Weapons and Ammunition**.
c. **Uniform and Equipment**.
(1) Method of distribution of paddles and life jackets.
(2) Disposition of boat, paddles and life jackets upon debarkation.

5. **COMMAND AND SIGNAL**.
a. **Command**.
(1) Location of patrol leader.
(2) Location of assistant patrol leader.
b. **Signal**.
(1) Signals used between boats and in boats.
(2) Code words.

STREAM CROSSING ANNEX

1. **SITUATION**.
a. **Enemy Forces**.
(1) ***Weather***.
(2) ***Terrain***.
(a) River width.
(b) River depth and water temperature.
(c) Current.
(d) Vegetation.
(e) Obstacles.
(3) ***Enemy*** (location, identification, activity).
b. **Friendly Forces**.
c. **Attachments and Detachments**.

2. **MISSION**.

3. **EXECUTION**.
a. **Concept of Operation**.
(1) ***Maneuver***.
(2) ***Fires***.
b. **Tasks to Combat Units**.
(1) ***Elements***.
(2) ***Teams***.
(3) ***Individuals***.

c. **Tasks to Combat Support Units.**
d. **Coordinating Instructions.**
(1) Crossing procedure/techniques.
(2) Security.
(3) Order of crossing.
(4) Actions on enemy contact.
(5) Alternate plan.
(6) Rallying points.
(7) Rehearsal plan.
(8) Time schedule.

4. **SERVICE SUPPORT.**

5. **COMMAND AND SIGNAL.**
a. **Command.**
(1) Location of patrol leader.
(2) Location of assistant patrol leader.
(3) Location of CP.
b. **Signal.**
(1) Emergency signals.
(2) Signals.

TRUCK ANNEX

1. **SITUATION.**
a. **Enemy Forces.**
b. **Friendly Forces.**
c. **Attachments and Detachments.**

2. **MISSION.**

3. **EXECUTION.**
a. **Concept of Operation.**
(1) ***Maneuver.***
(2) ***Fires.***
b. **Tasks to Combat Units.**
c. **Tasks to Combat Support Units.**
d. **Coordinating Instructions.**
(1) Time of departure and return.
(2) Loading plan and order of movement.
(3) Route (primary and alternate).
(4) Air Guards.
(5) Actions on enemy contact (vehicle ambush) during movement, loading, and downloading.
(6) Actions at the detrucking point.
(7) Rehearsals.
(8) Vehicle speed, separation, and recovery plan.
(9) Broken vehicle instructions.

4. **SERVICE SUPPORT**.

5. **COMMAND AND SIGNAL**.
   a. **Command**. Location of PL and PSG
   b. **Signal**.
      (1) Radio call signs and frequencies
      (2) Code words

2-7. **COORDINATION CHECKLISTS**. The following checklists are items that a platoon/squad leader must check when planning for a combat operation. In some cases, he will coordinate directly with the appropriate staff section, in most cases this information will be provided by the company commander or platoon leader. The platoon/squad leader, to keep him from overlooking anything that may be vital to his mission, may carry copies of these checklists:

Figure 2-7. INTELLIGENCE COORDINATION CHECKLIST

INTELLIGENCE COORDINATION CHECKLIST

In this coordination, the leader is informed of any changes in the situation as given in the operation order of mission briefing. He must keep himself constantly updated to ensure the plan is sound:

1. Identification of enemy unit.
2. Weather and light data.
3. Terrain update.
   a. Aerial photos.
   b. Trails and obstacles not on map.
4. Known or suspected enemy locations.
5. Weapons.
6. Probable course of action.
7. Recent enemy activities.
8. Reaction time of reaction forces.
9. Civilians on the battlefield.
10. Update to CCIR.

Figure 2-8. OPERATIONS COORDINATION CHECKLIST

OPERATIONS COORDINATION CHECKLIST

This coordination occurs with the platoon leader/company commander so that the platoon/squad leader can confirm his mission and operational plan, receive any last-minute changes to his mission or plan, and to update his subordinates or issue a FRAGO, if required:

1. Mission backbrief.
2. Identification of friendly units.
3. Changes in the friendly situation.
4. Route selection, LZ/PZ/DZ selection.
5. Linkup procedures.
6. Transportation/movement plan.

7. Resupply (in conjunction with S4).
8. Signal plan.
9. Departure and re-entry of forward units.
10. Special equipment requirements.
11. Adjacent units operating in the area of operations.
12. Rehearsal areas.
13. Method of insertion/extraction.

Figure 2-9. FIRE SUPPORT COORDINATION CHECKLIST

FIRE SUPPORT COORDINATION CHECKLIST

The platoon/squad leader will normally coordinate the following with the platoon forward observer (FO):

1. Mission backbrief.
2. Identification of supporting unit.
3. Mission and objective.
4. Route to and from the objective (include alternate routes).
5. Time of departure and expected time of return.
6. Unit target list (from fire plan).
7. Type of available support (artillery, mortar, naval gunfire, and aerial support to include Army, Navy, and Air Force) and their locations.
8. Ammunition available (to include different fuses).
9. Priority of fires.
10. Control measures.
    a. Checkpoints.
    b. Boundaries.
    c. Phase lines.
    d. Fire support coordination measures.
    e. Priority targets (target list).
    f. RFA (restrictive fire area).
    g. RFL (restrictive fire line).
    h. NFA (no-fire areas).
    i. Precoordinated authentication.
11. Communication (include primary and alternate means, emergency signals and code words).

Figure 2-10. COORDINATION WITH FORWARD UNIT CHECKLIST

COORDINATION WITH FORWARD UNIT CHECKLIST

A platoon/squad that requires foot movement through a friendly forward unit must coordinate with that unit's commander for a safe and orderly passage. If no time and place has been designated for coordination with the forward unit, the platoon/squad leader should set a time and place when he coordinates with the S-3. He must talk with someone at the forward unit who has the authority to commit that unit in assisting the platoon/squad during departure. Coordination entails a two-way exchange of information:

1. Identification (yourself and your unit).
2. Size of platoon/squad.
3. Time(s) and place(s) of departure and return, location(s) of departure point(s), ERRP, and detrucking points.
4. General area of operations.
5. Information on terrain and vegetation.
6. Know or suspected enemy positions or obstacles.
7. Possible enemy ambush sites.
8. Latest enemy activity.
9. Detailed information on friendly positions such as crew-served weapons, FPF.

10. Fire and barrier plan.
    a. Support the unit can furnish. How long and what can they do?
        (1) Fire support.
        (2) Litter teams.
        (3) Navigational signals and aids.
        (4) Guides.
        (5) Communications.
        (6) Reaction units.
        (7) Other.
    b. Call signs and frequencies.
    c. Pyrotechnic plan.
    d. Challenge and password, running password, number combination.
    e. Emergency signals and code words.
    f. If the unit is relieved, pass the information to the relieving unit.
    g. Recognition signals.

Figure 2-11. ADJACENT UNIT COORDINATION CHECKLIST

ADJACENT UNIT COORDINATION CHECKLIST

Immediately after the operation order of mission briefing, the platoon/squad leader should check with other platoon/squad leaders who will be operating in the same areas. If the leader is not aware of any other units operating is his area, he should check with the S3 during the operations coordination. The S3 can help arrange this coordination if necessary. The platoon/squad leaders should exchange the following information with other units operating in the same area:

1. Identification of the unit.
2. Mission and size of unit.
3. Planned times and points of departure and reentry.
4. Route(s).
5. Fire support and control measures.
6. Frequencies and call signs.
7. Challenge and password, running password, number combination.
8. Pyrotechnic plan.
9. Any information that the unit may have about the enemy.
10. Recognition signals.

Figure 2-12. REHEARSAL AREA COORDINATION CHECKLIST

REHEARSAL AREA COORDINATION CHECKLIST

This coordination is conducted with the platoon leader/company commander to facilitate the unit's safe, efficient and effective use of rehearsal area prior to its mission:

1. Identification of your unit.
2. Mission.
3. Terrain similar to objective site.
4. Security of the area.
5. Availability of aggressors.
6. Use of blanks, pyrotechnics, and ammunition.
7. Mock-ups available.
8. Time the area is available (preferably when light conditions approximate light conditions of patrol).
9. Transportation.
10. Coordination with other units using area.

Figure 2-13. ARMY AVIATION COORDINATION CHECKLIST

ARMY AVIATION COORDINATION CHECKLIST

This coordination is conducted with the platoon leader/company commander and/or S3 Air to facilitate the time, detailed and effective use of aviation assets as they apply to your tactical mission:

1. **SITUATION.**
   a. **Enemy Situation.**
      (1) Enemy air capability.
      (2) Enemy ADA capability.
      (3) Include in Weather: Percent illum, illum angle, NVG window, ceiling, and visibility.
   b. **Friendly Situation.**
      (1) Unit(s) supporting operation, Axis of movement/corridor/routes.
      (2) Friendly ADA status.

2. **MISSION.**

3. **EXECUTION.**
   a. **Concept of the Operation.** Overview of what requesting unit wants to accomplish with the air assault/air movement.
   b. **Tasks to Combat Units.**
      (1) Infantry.
      (2) Attack aviation.
   c. **Tasks to Combat Support Units.**
      (1) Artillery.
      (2) Aviation (lift).
   d. **Coordinating Instructions.**
      (1) ***Pickup Zone.***
      - Direction of landing.
      - Time of landing/flight direction.
      - Location of PZ/alternate PZ.
      - Loading procedures.
      - Marking of PZ (panel, smoke, SM, lights).
      - Flight route planned (SP, ACP, RP).
      - Formations: PZ, en route, LZ.
      - Code words:
        -- PZ secure (prior to landing), PZ clear (lead bird and last bird).
        -- Alternate PZ (at PZ, en route, LZ), names of PZ/alt PZ.
      - TAC air/artillery.
      - Number of pax per bird and for entire lift.
      - Equipment carried by individuals.
      - Marking of key leaders.
      - Abort criteria (PZ, en route, LZ).

      (2) ***Landing Zone.***
      - Direction of landing.
      - False insertion plans.
      - Time of landing (LZ time).
      - Location of LZ and Alternate LZ.
      - Marking of LZ (panel, smoke, SM, lights).
      - Formation of landing.
      - Code words, LZ name, alternate LZ name.
      - TAC air/artillery preparation, fire support coordination.
      - Secure LZ or not?

4. **SERVICE AND SUPPORT**.
   a. Number of aircraft per lift and number of lifts.
   b. Refuel/rearm during mission or not?
   c. Special equipment/aircraft configuration for weapons carried by unit personnel.
   d. Bump plan.
5. **COMMAND AND SIGNAL**.
   a. Frequencies, call signs and code words.
   b. Locations of air missions commander ground tactical commander, and air assault task force commander.

Figure 2-14. VEHICULAR MOVEMENT COORDINATION CHECKLIST

VEHICULAR MOVEMENT COORDINATION CHECKLIST

This is coordinated with the supporting unit through the platoon sergeant/first sergeant to facilitate the effective, detailed, and efficient use of vehicular support and/or assets:

1. Identification of the unit.
2. Supporting unit identification.
3. Number and type of vehicles and tactical preparation.
4. Entrucking point.
5. Departure/loading time.
6. Preparation of vehicles for movement.
   a. Driver responsibilities.
   b. Platoon/squad responsibilities.
   c. Special supplies/equipment required.
7. Availability of vehicles for preparation/rehearsals/inspection (time and location).
8. Routes.
   a. Primary.
   b. Alternate.
   c. Checkpoints.
9. Detrucking points.
   a. Primary.
   b. Alternate.
10. March internal/speed.
11. Communications (frequencies, call signs, codes).
12. Emergency procedures and signals.

2-8. DOCTRINAL TERMS

TACTICAL TASK--A clearly defined, measurable activity accomplished by individuals and organizations. Tasks are specific activities that contribute to the accomplishment of encompassing missions or other requirements. A task should be definable, measurable, and decisive (achieve the purpose).

Actions by Friendly Force		**Effects on Enemy Force**
Assault	Follow and Assume	Block
Attack-by-Fire	Follow and Support	Canalize
Breach	Linkup	Contain
Bypass	Occupy	Defeat
Clear	Reconstitution	Destroy
Combat Search and Rescue	Reduce	Disrupt
Consolidation and Reorganization	Retain	Fix
Control	Secure	Interdict
Counterreconnaissance	Seize	Isolate
Disengagement	Support-by-Fire	Neutralize
Exfiltrate	Suppress	Penetrate
		Turn

PURPOSE (in order to)--The desired or intended result of the tactical operation stated in terms related to the enemy or the desired situation. The why of the mission statement. The most important component of the mission statement.

Allow	Divert	Prevent
Cause	Enable	Protect
Create	Envelop	Support
Deceive	Influence	Surprise
Deny	Open	

OPERATION--A military action or the carrying out of a military action to gain the objectives of any battle or campaign.

Types of operations include:	Forms of Offensive Maneuver:	Types of Defensive Techniques:	Other Operations:
Movement to Contact Search and Attack Attack Ambush Demonstration Feint Raid Spoiling Attack Exploitation Pursuit	Envelopment Frontal Attack Infiltration Penetration Turning Movement	Area Defense Mobile Defense Retrograde Ops Delay Withdrawal Retirement	Reconnaissance Ops Security Ops Information Ops Combined Arms Breach Passage of Lines Relief in Place River Crossing Ops Troop Movement Admin Approach March Road March

2-9. **TERRAIN MODEL**. The terrain model (Figure 2-15) is an outstanding visual means used during the planning process to communicate the patrol routes and also detailed actions on the objective. At a minimum, the model is required to during planning to display routes to the objective and to highlight prominent terrain features the patrol will encounter during movement. A second terrain model is usually constructed of the objective area, enlarged to a sufficient size and detail to brief the patrol's actions on the objective.

a. **Checklist**.

(1) North-seeking arrow.

(2) Scale.

(3) Grid lines.

(4) Objective location.

(5) Exaggerated terrain relief, water obstacles.

(6) Friendly patrol locations.

(7) Targets (indirect fires, including grid and type of round).

(8) Routes, primary and alternate.

(9) Planned RPs (ORP, L/URP, RP).

(10) Danger areas (roads, trails, open areas).

(11) Legend.

(12) Blowup of objective area.

b. **Construction**. The following are some field-expedient techniques to aid in terrain model construction:

(1) Use a 3 x 5 card, MRE box, or piece of paper to label the objective or key sites.

(2) Grid lines can be made using string from the guts of 550 cord or colored tape. (Grid lines are identified by writing numbers on small pieces of paper.)

(3) Trees and vegetation are replicated by using moss, green or brown spray paint, pine needles, crushed leaves, or cut up grass.

(4) Water is designated by blue chalk, blue spray paint, blue yarn, tin foil, or MRE creamer.

(5) North seeking arrows are made from sharpened twigs, pencils, or colored yarn.

(6) Enemy positions are designated using red yarn, M-16 rounds, toy Rangers, or poker chips.

(7) Friendly positions such as security elements, support by fire, and assault elements are made using M-16 rounds, toy Rangers, poker chips, small MRE packets of sugar and coffee, or pre-printed acetate cards.

(8) Small pieces of cardboard or paper can identify target reference points and indirect fire targets. Ensure grids are shown for each point.

(9) Breach, support by fire, and assault positions are made using colored yarn or string so that these positions can be easily identified.

(10) Bunkers and buildings are constructed using MRE boxes or tongue depressors/sticks.

(11) Perimeter wire is constructed from a spiral notebook.

(12) Key phase lines are constructed with colored string or yarn.

(13) Trench lines are replicated by colored tape or yarn, by digging a furrow and coloring it with colored chalk or spray paint.

Note: All symbols used on the terrain model must be clearly identified in the legend.

Figure 2-15. TERRAIN MODEL

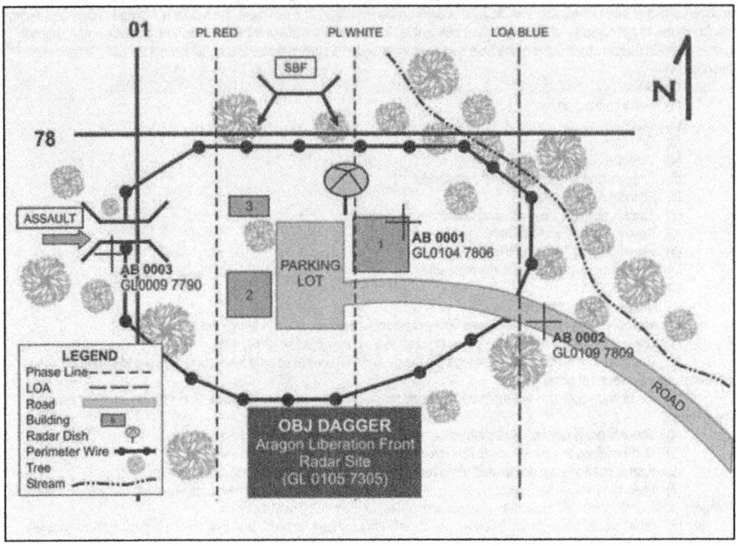

NOTES

Chapter 3
FIRE SUPPORT

Indirect fire support can greatly increase the combat effectiveness and survivability of any Infantry unit. The ability to plan for and effectively use this asset is a task that every Ranger and small unit leader should master. Fire support assets can assist a unit suppress, fix, destroy, or neutralize the enemy. Employment of indirect fire support should be considered throughout every offensive and defensive operation. This chapter discusses planning, tasks, capabilities, risk estimate distances, target overlays, close air support, elements and sequence of call for fire, and example call-for-fire transmissions.

3-1. **PLANNING**. Planning is the continual process of selecting targets on which fires are prearranged to support a phase of the commander's plan. The principles of planning follow:

- Consider what the commander wants to do.
- Plan early and continuously.
- Exploit all available targeting assets.
- Use all available lethal and nonlethal fire support means.
- Use the lowest echelon able to furnish effective support.
- Observe all fires.
- Use the most effective fire support asset available.
- Provide adequate fire support.
- Avoid unnecessary duplication.
- Provide for safety of friendly forces and installations.
- Provide for flexibility.
- Furnish the type of fire support requested.
- Consider the airspace.
- Provide rapid and effective coordination.
- Keep all fire support informed.

3-2. **TASKS**.

a. **All Operations**.
   (1) Locate targets.
   (2) Integrate all available assets.
   (3) Destroy, neutralize, or suppress all enemy direct and indirect fire systems.
   (4) Provide illumination and smoke.
   (5) Provide fires in support of JAAT and SEAD missions.
   (6) Deliver scatterable mines.
   (7) Prepare for future operations.
   (8) Provide positive clearance of fires.

b. **Offensive Operations**.
   (1) Support the movement to contact, chance contact.
   (2) Soften enemy defenses before the attack by arranging short, violent preparations, where required.
   (3) Provide support during the attack by attacking high payoff targets.
   (4) Plan for deep and flanking fires.
   (5) Plan fires during consolidation.
   (6) Provide counterfires.

c. **Defensive Operations**.
   (1) Execute all plans as the commander intends.
   (2) Disorganize, delay, and weaken the enemy before the attack.
   (3) Provide counterfire.
   (4) Provide fires in support of planned engagement areas (EAs).
   (5) Attack high payoff targets.
   (6) Plan fires in support of barriers and obstacle plans.
   (7) Plan for deep, flanking, and rear fires.

(8) Provide fires in support of counterattacks.
(9) Plan final protective fire (FPFs).

3-3. **CAPABILITIES**. Tables 3-1 and 3-2 show capabilities of field artillery and mortars.

Table 3-1. CAPABILITIES OF FIELD ARTILLERY

WEAPON	MAX RANGE (meters)	MIN RANGE (meters)	MAX RATE Rds per Min	Burst Radius	SUSTAINED RATE Rds per Min
105-mm Howitzer M119, Towed	14,000 m	0	6 for 2 min	35 m	3 Rds for 30 min, then 1 rd per min
155-mm Howitzer M198, Towed	18 m,100 m 30,000 m (RAP)	0	4 for 3 min 2 for 30 min	50 m	1 rd per min Temp Dependent
155-mm Howitzer M109A6 SP	18 m, 100 m 30,000 m (RAP)	0	4 for 3 min	50 m	1 for 60 min 0.5

Table 3-2. CAPABILITIES OF MORTARS

WPN	MUNITIONS AVAILABLE	MAX HE RANGE	MIN RANGE (meters)	MAX RATE (Rds per Min)	BURST RADIUS	SUSTAINED RATE (Rds per Min)
60mm	HE, WP, Illum	3,500 m	70 m (HE)	30 for 4 min	30 m	20
81mm	HE, WP, Illum	5,600 m	70 m (HE)	25 for 2 min	38 m	8
120mm	HE, Smoke, Illum	7,200 m	180 m (HE)	15 for 1 min	60 m	5

DANGER

DANGER CLOSE

1. **When the target is within 600 meters of any friendly troops (for mortars and field artillery), announce DANGER CLOSE in the Method of Engagement portion of the call for fire.**
2. **When adjusting 5-inch or smaller naval guns on targets within 750 meters, announce DANGER CLOSE. For larger naval guns, announce DANGER CLOSE for targets within 1,000 meters. Failure to adhere to this guidance can result in fratricide.**
3. **Avoid making corrections using the bracketing method of adjustment, because doing so can cause serious injury or death. Use only the creeping method of adjustment during DANGER CLOSE missions. Make corrections of no more than 100 meters by creeping the rounds to the target.**

3-4. **RISK ESTIMATE DISTANCES**. RED applies to combat only. Minimum safe distances (Table 3-3) apply to training IAW AR 350-1. RED takes into account the bursting radius of particular munitions and the characteristics of the delivery system and associates this combination with a percentage representing the likelihood of becoming a casualty, that is, the percentage of risk. RED is defined as the minimum distance friendly troops can approach the effects of friendly fires without suffering appreciable casualties of 0.1% PI or higher.

> **WARNING**
> **[Commanders] use risk-estimate distance formulas to determine acceptable risk levels in combat only. Specifically, use them to identify the risk to your Rangers at various distances from their targets. Risk estimate distances apply only in combat. In training, use minimum safe distances (MSD).**

a. **Casualty Criterion**. The casualty criterion is the 5-minute assault criterion for a prone soldier in winter clothing and helmet. Physical incapacitation means that a soldier is physically unable to function in an assault within a 5-minute period after an attack. A PI value of less than 0.1 percent can be interpreted as being less than or equal to one chance in one thousand.

Table 3-3. RISK ESTIMATE DISTANCES FOR MORTARS AND CANNON ARTILLERY

	Risk Estimate Distances (Meters)					
	10 % PI			0.1 % PI		
System	1/3 range	2/3 range	Max range	1/3 range	2/3 range	Max range
60-mm mortar	60	65	65	100	150	175
81-mm mortar	75	80	80	165	185	230
120-mm mortar	100	100	100	150	300	400
105-mm howitzer	85	85	90	175	200	275
155-mm howitzer	100	100	125	200	280	450
155-mm DPICM	150	180	200	280	300	475

b. **Risk**. Using echelonment of fires within the specified RED for a delivery system requires the unit to assume some risks. The maneuver commander determines by delivery system how close he will allow fires to fall in proximity to his forces. The maneuver commander makes the decision for this risk level, but he relies heavily on the FSO's expertise.

3-5. **TARGET OVERLAYS**.

a. Figure 3-1 shows contents of fire support overlay.

Figure 3-1. CONTENTS OF FIRE SUPPORT OVERLAY

- Unit and official capacity of person making overlay.
- Date the overlay was prepared.
- Map Sheet Number.
- Effective Period of Overlay (DTG).
- Priority target.
- ORP Location.
- Call signs and frequencies. (PRI/ALT)
- Routes--Primary/Alternate.
- Phase Lines/Checkpoints used by the patrol.
- Spares.
- Index marks to position overlay on map.
- Objective.
- Target Symbols.
- Description, location and remarks column, complete.

b. Using a PLOT-CR checklist helps ensure the leader's fire support plan is complete:
   (1) Purpose--Purpose of the planned indirect fires.
   (2) Location--Planned targets with an 8-digit grid (minimum).
   (3) Observer--Planned observer to see the impact of the rounds and adjust.
   (4) Trigger--Method to initiate fires.
   (5) Communication--Method to communicate between the observer and the supporting unit.
   (6) Resources--Planned allocated resource for each target.
c. Sterile overlay includes---
   (1) Index marks to position overlay on map.
   (2) Target symbols.
d. Target overlay symbols (Figure 3-2) include--
   (1) Point target:
   (2) Linear target:
   (3) Circular target:

Figure 3-2. TARGET OVERLAY SYMBOLS--POINT (TOP), LINEAR (CENTER), AND CIRCULAR (BOTTOM)

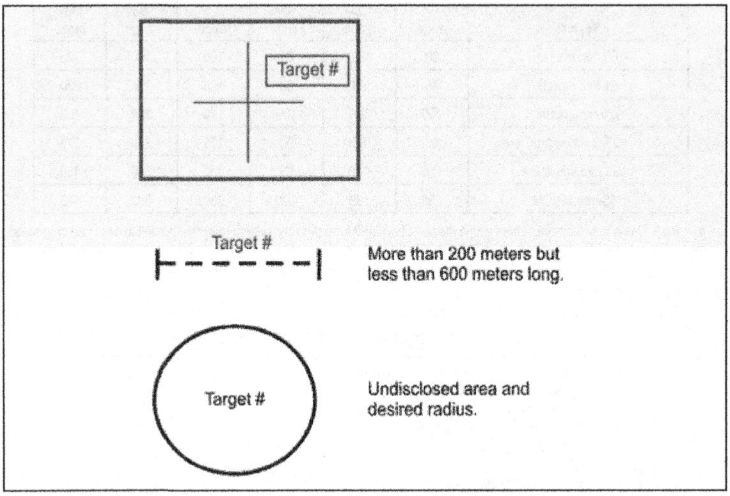

3-6. **CLOSE AIR SUPPORT**. The two types of close air support requests are planned and immediate. Planned requests are processed by the Army chain to corps for approval. Immediate requests are initiated at any level and processed by the battalion S-3 FSO, and air liaison officer.

a. Format for requesting immediate CAS:
   (1) Observer identification.
   (2) Warning order (request close air).
   (3) Target location (grid).

(4) Target description. (Description must include, as a minimum: type and number of targets, activity or movement; point or area targets, include desired results on target and time on target.

(5) Location of friendly forces.

(6) NAV details (elevation).

(7) Threats--ADA, small arms, and so on.

(8) Hazards--friendly aircraft in area.

(9) Wind direction.

b. Close air support capabilities are shown in Table 3-4.

Table 3-4. CLOSE AIR SUPPORT

AIRCRAFT	SERVICE	CHARACTERISTICS
A-10 *	AF	Specialized CAS aircraft; subsonic; typical load 6,000 lbs, max load 16,000 lbs; 30-mm gun.
F-16 *	AF	A multirole aircraft; complements the F-4 and F-15 in an air-to-air role. Most accurate ground delivery system in the inventory; supersonic; typical load 6,000 lbs, max load 10,600 lbs.
F-18 *	N/MC	A multirole fighter scheduled to replace the F-4. A wide variety of air-to-surface weapons. Typical load 7,000 lbs, max load 17,000 lbs, 20-mm gun mounted in the nose; air-to-air missiles.
AC-130	AF/R	A specialized CAS/RACO aircraft, propeller driven. Two models: The "A" model has two 40-mm guns, two 20-mm guns and two 7.62-mm miniguns. The similar "H" model has no 7.62-mm miniguns, and has a 105-mm howitzer in place of one of the 40-mm guns. Both models have advanced sensors and a target-acquisition system, including FLIR and low light TV. Very accurate. Vulnerable to enemy air defense systems, so must operate in a low-threat environment.
*FM capability.		

3-7. **CALL FOR FIRE.**

a. **Observer's Identification - Call Signs.**

b. **Warning Order.**

(1) ***Type of mission.***

(a) Adjust fire.

(b) Fire for effect.

(c) Suppress.

(d) Immediate suppression/immediate smoke.

(2) ***Size of element to Fire for Effect.*** When the observer does not specify what size element to fire, the battalion FDC will decide.

(3) ***Method of Target Location.***

(a) Polar plot.

(b) Shift from a known point.

(c) Grid.

(4) **Location of Target.**

(a) *Grid Coordinate - 6 Digit.* 8 digit if greater accuracy is required.

(b) *Shift from a Known Point.* Send observer-target (OT) direction:

- Mils (nearest 10).
- Degrees.
- Cardinal direction.
- Send lateral shift (right/left, nearest 10 mils).
- Send range shift (add/drop, nearest 100 mils).
- Send vertical shift (up/down). Use only if it exceeds 35 mils (nearest 5 mils).

(c) *Polar Plot.* Send direction. (Nearest 10 mils)

- Send distance (nearest 100 mils).
- Send vertical shift (nearest 5 mils).

(5) ***Description of Target.***

(a) Type.

(b) Activity.

(c) Number.

(d) Degree of protection.

(e) Size and shape (length/width or radius).

(6) ***Method of Engagement, Type of Adjustment***. When the observer does not request a specific type of fire control adjustment, area fire is issued.

(a) Area fire--moving target.

(b) Precision fire--point target.

(7) ***Danger Close***. Use when friendly troops are within--

(a) 600 mils for mortars.

(b) 600 mils for artillery.

(c) 750 mils for naval guns 5 inches or smaller.

(8) ***Mark***. Use to orient observer or to indicate targets.

(9) ***Trajectory***.

(a) Low angle (standard).

(b) High angle (mortars or if requested).

(10) ***Ammunition***. Use HE quick used unless observer specifies otherwise.

(a) Projectile (HE, illum, ICM, smoke, and so on).

(b) Fuse (quick, time, and so on).

(c) Volume of fire (observer may request the number of rounds to be fired).

(11) ***Distribution***.

(a) 100 mil sheaf (standard).

(b) Converged sheaf (used for small hard targets).

(c) Special sheaf (any length, width, and attitude).

(d) Open sheaf (separate bursts).

(e) Parallel sheaf (linear target).

(12) ***Method of Fire and Control***.

(a) *Method of Fire*. Specific guns and a specific interval between rounds. Normally adjust fire, one gun is used with a 5-second interval between rounds.

(b) *Method of Control.*

- AT MY COMMAND - FIRE--Remains in effect until observer announces, "*Cancel at my command*."
- CANNOT OBSERVE--Observer can't see the target.
- TIME ON TARGET--Observer tells FDC when he wants the rounds to impact.
- CONTINUOUS ILLUMINATION. Calculated by the FDC, otherwise observer indicates interval between rounds in seconds.
- COORDINATED ILLUMINATION. Observer may order the interval between illum and HE shells.
- CEASE LOADING--Stop loading rounds.
- CHECK FIRING--Stop firing at once.
- CONTINUOUS FIRE--Load and fire as fast as possible.
- REPEAT--Fire another round(s) with or without adjustments.

(13) ***Authentication***. Challenge and reply.

(14) ***Message to Observer***.

(a) Battery(ies) to fire for effect.

(b) Adjusting battery.

(c) Changes to the initial call for fire.

(d) Number of rounds (per tube) to be fired for effect.

(e) Target numbers.

(f) Additional information:

- Time of flight. Moving target mission.
- Probable error in range. 38 meters or greater (normal mission).
- Angle T. 500 mils or greater.

c. **Correction of Errors**. When FDC has made an error when reading back he fire support data, the observer announces "*CORRECTION*" and transmits the correct data in its entirety.

d. **Call-for-Fire Transmissions**. Example call-for-fire transmissions are shown in Table 3-5.

Table 3-5. EXAMPLE CALL-FOR-FIRE TRANSMISSIONS.

GRID MISSION	
OBSERVER	**FIRING UNIT**
F24, this is J42, ADJUST FIRE, OVER.	J42, this is F24, AJUST FIRE, OUT.
GRID WM180513, DIRECTION 0530, OVER.	GRID WM180513, DIRECTION 0530, OUT.
Infantry platoon dug in, OVER.	Infantry platoon dug in, OUT.
	SHOT OVER.
SHOT OUT.	
	SPLASH, OVER.
SPLASH OUT.	
End of mission, 15 casualties, platoon dispersed, OVER.	End of mission, 15 casualties, platoon dispersed, OUT.
SHIFT FROM KNOWN POINT	
OBSERVER	**FIRING UNIT**
J42, THIS IS F24, ADJUST FIRE, SHIFT AB1001, OVER.	F24, THIS IS J42, ADJUST FIRE, SHIFT AB1001, OUT.
DIRECTION 2420, RIGHT 400, ADD 400, OVER.	DIRECTION 2420, RIGHT 400, ADD 400, OUT.
FIVE T-72 TANKS AT POL SITE, OVER.	FIVE T-72 TANKS AT POL SITE, AUTHENTICATE JULIET NOVEMBER, OVER.
I AUTHENTICATE TANGO, OVER.	
	SHOT, OVER.
SHOT OUT.	SPLASH, OVER.
SPLASH OUT.	
END OF MISSION, 2 TANKS DESTROYED, 3 IN WOODLINE, OVER	END OF MISSION, 2 TANKS DESTROYED, 3 IN WOODLINE, OUT

POLAR	
OBSERVER	**FIRING UNIT**
J42, this is F24, adjust fire, polar OVER.	F24, this is J42, adjust fire, polar, OUT
Direction 2300, distance 4,000, OVER.	Direction 2300, distance 4,000, OUT.
Infantry platoon dug in, OVER	Infantry platoon dug in, OUT
	Shot OVER.
Shot OUT.	Splash, OVER.
Splash OUT.	
End of mission, 15 casualties, platoon dispersed, OVER.	End of mission, 15 casualties, platoon dispersed, OUT.

NOTES

Chapter 4
MOVEMENT

To survive on the battlefield, stealth, dispersion, and security must be enforced in all tactical movements. The leader must be skilled in all movement techniques.

4-1. **FORMATIONS**. Movement formations are comprised of elements and Soldiers in relation to each other. Fire teams, squads and platoons use several movement formations. Formations provide the leader control based on a METT-TC analysis. Leaders position themselves where they can best command and control formations. The formations below allow the fire team leader to lead by example, "*Follow me, and do as I do.*" All Soldiers in the team must be able to see their leader. The formations in Figure 4-1 reflect fire team formations. Squad formations are very similar with more Soldiers. Squads can operate in lines and files similar to fire teams. When squads operate in wedges or in echelon, the fire teams use those formations and simply arrange themselves in column or one team behind the other. Squads may also use the vee, with one team forming the lines of the vee, and the SL at the apex for command and control. Platoons work on the same basic formations as the Squads. When operating as a platoon, the platoon leader must carefully select the location for his machine guns in the movement formation.

Figure 4-1. FORMATIONS

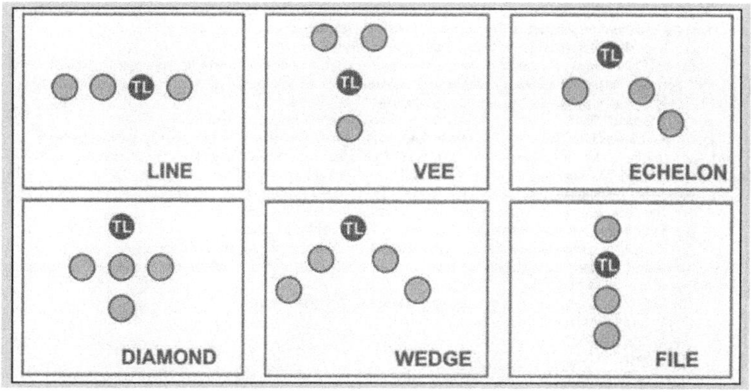

a. **Techniques**. A movement technique is the method a unit uses to traverse terrain. There are three movement techniques: traveling, traveling overwatch, and bounding overwatch. The selection of a movement technique is based on the likelihood of enemy contact and the need for speed. Factors to consider for each technique are control, dispersion, speed, and security. Movement techniques are not fixed formations. They refer to the distances between Soldiers, teams, and squads that vary based on METT-TC. Soldiers must be able to see their fire team leaders. The platoon leader should be able to see his lead squad leader. Leaders control movement with hand-and-arm signals and use radios only when needed.

b. **Standards**.

(1) Unit moves on designated route or arrives at specified location IAW OPORD maintaining accountability of all assigned/attached personnel.

(2) Unit uses movement formation and technique ordered by the leader based on METT-TC.

(3) Leaders remain oriented (within 200 meters) and follow planned route unless METT-TC dictates otherwise.

(4) Unit maintains 360 degree security and remains 100% alert during movement.

(5) Unit maintains 360 degree security and a minimum of 75% security during halts.

(6) If contact with the enemy is made, it is made with the smallest element possible.

(7) Control measures are used during movement such as head counts, rally points, or phase lines.

c. **Fundamentals.**

(1) ***Land Navigation.*** Mission accomplishment depends on successful land navigation. The patrol should use stealth and vigilance to avoid chance contact. Designate a primary and alternate compass and pace man per patrol.

Note: The point man will not be tasked to perform compass or pace duties. The point man's sole responsibility is forward security for the element.

(2) ***Avoidance of Detection.*** Patrols must use stealth, and use the cover and concealment of the terrain to its maximum advantage. Whenever possible, move during limited visibility in order to maximize technological advantages gained by night vision devices and to hinder the enemy's ability to detect the patrol. Exploit the enemy's weaknesses, and attempt to time movements to coincide with other operations that distract the enemy. The enemy threat and terrain determines which of the three movement techniques will be used:

(a) Fire teams maintain visual contact, but the distance between them is such that the entire patrol does not become engaged if contact is made. Fire teams can spread their formations as necessary to gain better observation to the flanks. Although widely spaced, men retain their relative position in their wedge and follow their team leader. Only in extreme situations should the file be used.

(b) The lead squad must secure the front along with assuming responsibility for navigation. For a long movement, the PL may rotate the lead squad responsibilities. The fire team/squad in the rear is charged with rear security.

(c) Vary movement techniques to meet the changing situation.

(d) With the exception of fire team leaders, leaders move inside their formations where they can maintain the best control.

(3) ***Security.*** The patrol must use both active and passive security measures constantly. Assign subunits responsibility for security at danger areas, patrol bases, and most importantly in the objective area.

(4) ***Fire Support.*** Plan fire support (mortars, artillery, tactical air, attack helicopter, naval gunfire).

(5) ***Three-Dimensional Battlefield.*** 360 Degree Security is achieved through high and low security. Within a fire team, squad, and so on, the leader must assign appropriate sectors of fire to their subordinate in order to ensure all aspects of the battlefield are covered. This includes trees, multiple storied structures, tunnels, sewers, ditches.

d. **Movement Techniques.**

(1) The traveling technique is used when enemy contact is not likely but speed is necessary.

(2) The traveling overwatch technique is used when enemy contact is possible.

(3) The bounding overwatch technique is used when enemy contact is likely, or when crossing a danger area.

e. **Traveling.** In the traveling technique, the distance between individuals is about 10 meters with 20 meters between squads. It has the following characteristics:

(1) More control than traveling overwatch but less than bounding overwatch.

(2) Minimum dispersion.

(3) Maximum speed.

(4) Minimum security.

f. **Traveling Overwatch.** The traveling overwatch technique is the basic movement technique. The distance between individuals is about 20 meters, between teams about 50 meters.

(1) In platoon traveling overwatch, the lead squad must be far enough ahead of the rest of the platoon to detect or engage any enemy before the enemy observes or fires on the main body. However, it must be close enough to be supported by the platoon's small arm's fires. This is normally between 50 to 100 meters, depending on terrain, vegetation, and light and weather conditions.

(2) In a column formation, only the lead squad should use the traveling overwatch; however, if greater dispersion is desired, all squads may use it.

(3) In other formations, all squads use traveling overwatch unless the platoon leader specifies otherwise.

(4) Traveling overwatch has the following characteristics:

- Good control.
- Good dispersion.
- Good speed.
- Good security forward.

g. **Bounding Overwatch**. In the bounding overwatch technique (Figure 4-2), the distance between men remains about 20 meters. The distance between teams and squads varies.

(1) The squad or platoon has a bounding element and an overwatch element. The bounding element moves while the overwatch element occupies an overwatch position that can cover the route of the bounding element by fire. Each bound is within supporting range of the overwatch element.

(2) There are two types of bounding, successive and alternating (Figure 4-2). Successive is nothing more than one squad moving to a position, then the overwatching squad moving to a position generally online with the first squad. Alternating bounding is when one squad moves into position, then the overwatching squad moves to a position in front of the first squad.

Figure 4-2. SQUAD BOUNDING OVERWATCH

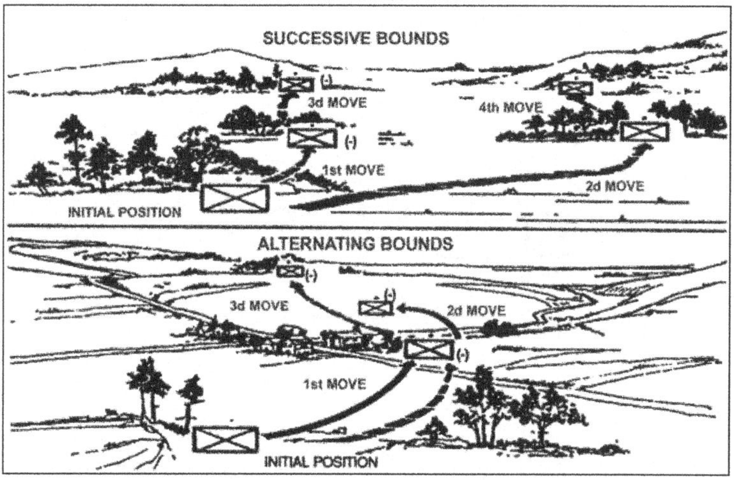

(3) The length of a bound depends on the terrain, visibility, and control.

(4) Before a bound, the leader gives the following instructions to his subordinates:

- Direction of the enemy if known
- Position of overwatch elements
- Next overwatch position
- Route of the bounding element
- What to do after the bounding element reaches the next position
- How the elements receive follow-on orders

(5) The characteristics of bounding overwatch are:

- Maximum control
- Maximum dispersion
- Minimum speed
- Maximum security

h. **Platoon Bounding Overwatch** (Figure 4-3).

(1) ***Method One***. When platoons use bounding overwatch, one squad bounds and one squad overwatches; the third squad awaits orders. Forward observers stay with the overwatching squad to call for fire. Platoon leaders normally stay with the overwatching squad who use machine guns and attached weapons to support the bounding squad.

(2) ***Method Two***. Another way is to have one squad use bounding overwatch and have the other two squads use traveling or traveling overwatch technique.

Figure 4-3. PLATOON BOUNDING OVERWATCH

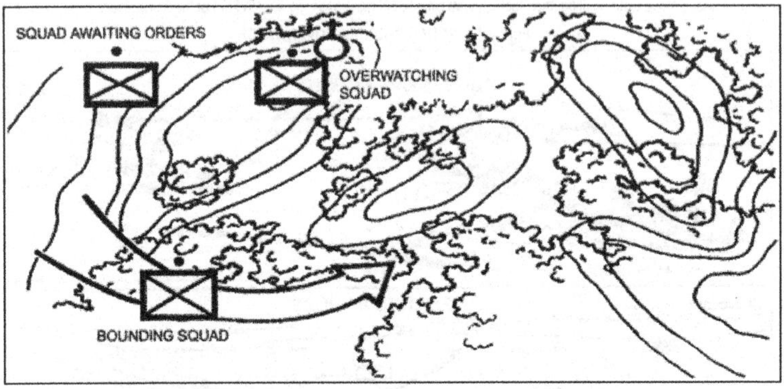

(3) ***Movement Considerations***. When deciding where to move the bounding element, consider--

- Where the enemy is likely to be.
- The mission.
- The routes to the next overwatch position.
- The weapons ranges of the overwatching unit.
- The responsiveness of the rest of the unit.
- The fields of fire at the next overwatch position.

4-2. **TACTICAL MARCHES**. Platoons conduct two types of marches with the company: foot marches and motor (road) marches.

a. **Purpose/General**. A successful foot march is when troops arrive at their destination at the prescribed time, physically able to execute their tactical mission.

b. **Standard**.

(1) The unit crosses the start point and release point at the time specified in the order.

(2) The unit follows the prescribed route, rate of march, and interval without deviation unless required otherwise by enemy action or higher headquarters action.

c. **Fundamentals**.

(1) Effective control.

(2) Detailed planning.

(3) Rehearsals.

d. **Considerations**.

(1) ***METT-TC***.

- Mission ....Task and purpose.
- Enemy ....Intentions, capabilities, and course of action.
- Terrain and weather.................Road condition/trafficability, and visibility.

- Troops/equipment....................Condition of Soldiers and their loads, numbers and types of weapons and radios.
- Time...Start time, release time, rate of march, time available.
- Civilians ....Movement through populated areas, refugees, OPSEC.

(2) ***Task Organization***:

- Security....................................Advance and trail teams.
- Main body ....Two remaining line squads and weapons squad.
- Headquarters............................Command and control.
- Control measures.

(3) ***Start Point and Release Point*** (given by higher):

(a) Checkpoints.............................At checkpoints report to higher and use checkpoint to remain oriented.
(b) Rally or rendezvous points ......Used when elements become separated.
(c) Location of leaders ....Where they can best control their elements.
(d) Commo plan ....Locations of radios, frequencies, call signs, and OPSKEDs.
(e) Dispersion between Soldiers--
- 3 to 5 meters / day.
- 1 to 3 meters / night.

(4) ***March Order***. May be issued as an OPORD, FRAGO, or annex to either (must use operational overlay or strip map)"

(a) Formations and order of movement
(b) Route of march........................Assembly area, start point, release point, rally points, checkpoints, break/halt points.
(c) Start point time, release point time, and rate of march.
(d) March interval for squads, teams, and individuals.
(e) Actions on enemy contact........Air and ground.
(f) Actions at halts.
(g) Fires ....Detailed plan of fire support for the march.
(h) Water supply plan.
(i) MEDEVAC plan.

e. **Duties and Responsibilities**.

(1) ***Platoon Leader***:

- Before ....Issues warning order and FRAGO; inspects and supervises.
- During ....Ensures unit makes movement time; ensures interval is maintained and that unit remains oriented; maintains security; checks condition of men; and enforces water discipline and field sanitation.
- After..Ensures men are prepared to accomplish their mission, supervises SLs, and ensures medical coverage is provided to men as needed.

(2) ***Platoon Sergeant***:

- Before......................................Assists PL, makes recommendations, and enforces uniform and packing lists.
- During......................................Controls stragglers, helps platoon leader maintain proper interval and security.
- At Halts ....Enforces security, ensures welfare of men, enforces field sanitation and litter discipline, and enforces use of preventive medicine.
- After...Coordinates for water, rations, and medical supplies. Recovers casualties.

(3) ***Squad Leaders***:

- Before......................................Provides detailed instruction to TLs, inspects boots and socks for serviceability and proper fit, inspects adjustment of equipment, inspects to ensure canteens are full, and ensures equal distribution of loads.
- During......................................Controls squad, maintains proper interval between men and equipment, enforces security, remains oriented.

- At Halts....................................Ensures security is maintained, provides men for water resupply as detailed. Physically checks the men in his squad, ensures they drink water, and that they change socks as necessary. Rotates heavy equipment. (Units should plan this in detail to avoid confusion before, during, and after halts.)
- After..Occupies squad sector of assembly area, conducts foot inspection, reports condition of men to PL, and prepares men to accomplish the mission.

(4) ***Security Squad***:
- Lead Team.
  - -- Serves as point element for platoon.
  - -- Recons route to SP.
  - -- Calls in checkpoints.
  - -- Provides early warning.
  - -- Maintains rate of march.
  - -- Moves 10 to 20 meters in front of main body.
- Trail Team.
  - -- Provides rear security.
  - -- Moves 10 to 20 meters behind main body.

(5) ***Medic***:
- Assesses and treats march casualties.
- Advise chain of command on evacuation and transportation requirements of casualties.

(6) ***Individual***:
- Maintains interval.
- Follows TL's examples.
- Relays hand and arm signals; remains alert during movement and at halts.
- Remains alert during movement and at halts.

4-3. **MOVEMENT IN LIMITED VISIBILITY CONDITIONS**. During hours of limited visibility, a platoon will use surveillance, target acquisition, and night observation (STANO) devices to enhance their effectiveness. Leaders must be able to control, navigate, maintain security, and move during limited visibility.

a. **Control**. When visibility is poor, the following methods aid in control.

(1) Use of night vision devices.
(2) Leaders move closer to the front.
(3) Platoon reduces speed.
(4) Use of luminescent tape on equipment.
(5) Reduce intervals between men and elements.
(6) Conducts headcount regularly.

b. **Navigation**. While navigating during limited visibility the same techniques are used as during the day; however, leaders exercise more care to keep the patrol oriented.

c. **Security**.

(1) Enforce strict noise and light discipline.
(2) Use radio-listening silence.
(3) Use camouflage.
(4) Use terrain to avoid detection by enemy surveillance or night vision devices.
(5) Make frequent listening halts; Conduct SLLS (Stop, Look, Listen, Smell).
(6) Mask the sounds of movement when possible. (Rain, wind, and flowing water will mask the sounds of movement).

d. **Rally Points**. Plan actions to be taken at rally points in detail. All elements must maintain communications at all time. There are two techniques for actions at rally points:

(1) ***Minimum Force***: Patrol members assemble at the rally point, and the senior leader assumes command. When the minimum force (designated in the OPORD) is assembled and organized the patrol will continue the mission.

(2) ***Time Available***: The senior leader determines if the patrol has enough time remaining to accomplish the mission.

e. **Actions at Halts**. During halts, post security and cover all approaches into the sector with key weapons.

(1) ***Short Halt***. Typically 1-2 minutes long. Soldiers seek immediate cover and concealment and take a knee. Leaders assign sectors of fire.

(2) ***Long Halt***. More than 2 minutes. Soldiers assume the prone position behind cover and concealment. (Ensure soldiers have clear fields of fire.) Leaders assign sectors of fire.

4-4. **DANGER AREAS**. A danger area is any place on a unit's route where the leader determines his unit may be exposed to enemy observation or fire. Some examples of danger areas are open areas, roads and trails, built-up areas, enemy positions, and natural and man-made obstacles. Bypass danger areas whenever possible.

a. **Standards**.

(1) The unit prevents the enemy from surprising the main body.
(2) The unit moves all personnel and equipment across the danger area.
(3) The unit prevents decisive engagement by the enemy.

b. **Fundamentals**.

(1) Designate near and far side rally points.
(2) Secure near side, left and right flank, and rear security.
(3) Recon and secure the far side.
(4) Cross the danger area.
(5) Plan for fires on all known danger areas.

c. **Technique for Crossing Danger Areas**.

(1) ***Linear Danger Area*** (LDA, Figure 4-4) ***Actions for a Squad***:

STEP 1: The alpha team leader (ATL) observes the linear danger area and sends the hand and arm signal to the SL who determines to bound across.

STEP 2: SL directs the ATL to move his team across the LDA far enough to fit the remainder of the squad on the far side of the LDA. Bravo team moves to the LDA to the right or left to provide an overwatch position prior to A team crossing.

STEP 3: SL receives the hand and arm signal that it is safe to move the rest of the squad across (B team is still providing overwatch).

STEP 4: SL moves himself, radio operator, and B team across the LDA. (A team provides overwatch for squad missions.)

STEP 5: A team assumes original azimuth at SL's command or hand and arm signal.

Figure 4-4. LINEAR DANGER AREA

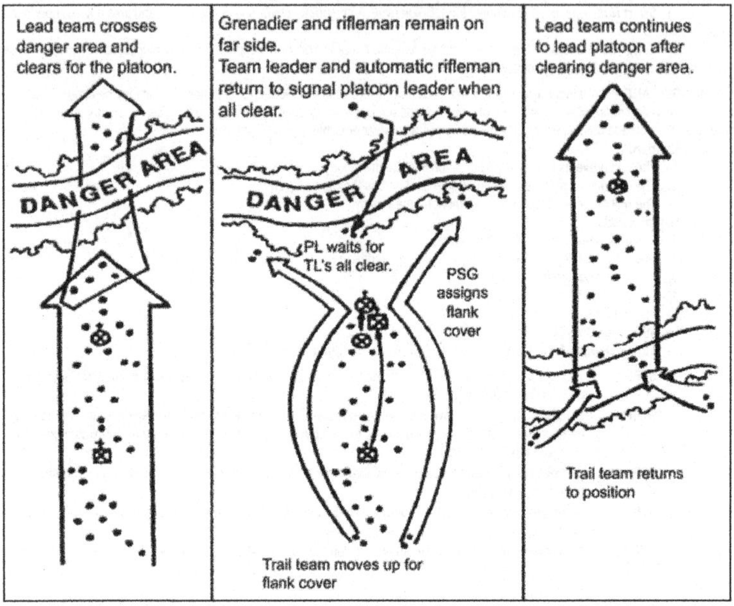

(2) ***LDA Crossing for a Platoon***:

(a) The lead squad halts the platoon, and signals danger area.

(b) The platoon leader moves forward to the lead squad to confirm the danger area and decides if current location is a suitable crossing site.

(c) The platoon leader confirms danger area/crossing site and establishes near and far side rally points.

(d) On the platoon leader's signal, the A team of the lead squad establishes an overwatch position to the left of the crossing site. Prior to crossing, the compassman with the lead two squads confirm azimuth and pace data.

(e) B team of the lead squad establishes an overwatch position to the right of the crossing site.

(f) Once overwatch positions are established, the platoon leader gives the second squad in movement the signal to bound across by fire team.

(g) Once across, the squad is now lead in movement and continues on azimuth

(h) Once Stop, Look, Listen, and Smell (SLLS) is conducted, squad leader signals platoon leader all clear.

- Day time—hand and arm signal such as thumbs up.
- Night time—Clandestine signal such as IR, red lens.

(i) Platoon leader receives all clear and crosses with radio operator, FO, WSL, and 2 gun teams.

(j) Once across, PL signals the 3rd squad in movement to cross.

(k) PSG with medic and one gun team crosses after 2nd squad is across (sterilizing central crossing site).

(l) PSG signals security squad to cross at their location.

Notes: Platoon leader will plan for fires at all known LDA crossing sites.
Nearside security in overwatch will sterilize signs of the patrol .

(3) ***Danger Area (Small/Open)***

(a) The lead squad halts the platoon and signals danger area.

(b) The PL moves forward to the lead squad to confirm the danger area.

(c) The platoon leader confirms danger area and establishes near and far side rally points.

(d) The PL designates lead squad to bypass danger area using the detour-bypass method.

(e) Paceman suspends current pace count and initiates an interim pace count. Alternate pace/compass man moves forward and offsets compass 90 degrees left or right as designated and moves in that direction until clear of danger area.

(f) After moving set distance (x-meters as instructed by PL). Lead squad assumes original azimuth, and primary pace man resumes original pace.

(g) After the open area, the alternate pace/compass man offsets his compass 90 degrees left or right and leads the platoon/squad the same distance (x-meters) back to the original azimuth.

(4) ***Danger Areas (Series)***: A series of danger areas is two or more danger areas within an area that can be either observed or covered by fire.

- Double linear danger area (use linear danger area technique and cross as one LDA).
- Linear/small open danger area (use by-pass/contour technique. Figure 4-5).
- Linear/large open danger area (use platoon wedge when crossing).

Note: A series of danger areas is crossed using the technique which provides the most security.

(5) ***Danger Area (Large)***:

(a) Lead squad halts the platoon, and signals danger area.

(b) The platoon leader moves forward with RTO and FO and confirms danger area.

(c) The platoon leader confirms danger area and establishes near and far side rally points.

(d) PL designates direction of movement.

(e) PL designates change of formation as necessary to ensure security.

Notes: Before point man steps into danger area, PL and FO adjust targets to cover movement.
If far side of danger area is within 250 meters, PL establishes overwatch, and designates lead squad to clear wood line on far side.

Figure 4-5. SMALL OPEN AREA

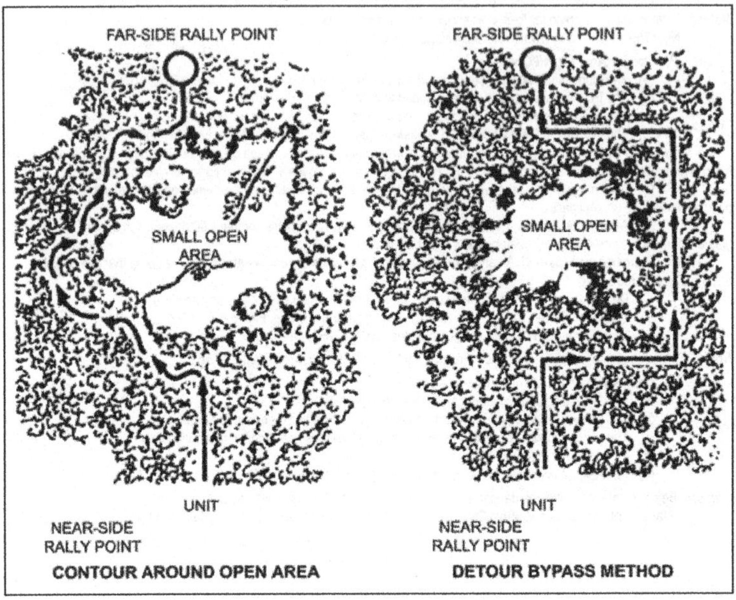

NOTES

Chapter 5

PATROLS

Infantry platoons and squads primarily conduct two types of patrols: reconnaissance and combat. This chapter describes the principles of patrolling, planning considerations, types of patrols, supporting tasks, patrol base, and movement to contact (FM 7-8, FM 3-0, and FM 101-5-1). In this chapter, the terms "*element*" and "*team*" refer to the squads, fire teams, or buddy teams that perform the tasks as described.

5-1. **PRINCIPLES**. All patrols are governed by five principles:

a. **Planning**. Quickly make a simple plan and effectively communicate it to the lowest level. A great plan that takes forever to complete and is poorly disseminated isn't a great plan. Plan and prepare to a realistic standard and rehearse everything.

b. **Reconnaissance**. Your responsibility as a Ranger leader is to confirm what you think you know, and to learn that which you do not already know.

c. **Security**. Preserve your force as a whole. Every Ranger and every rifle counts; anyone could be the difference between victory and defeat.

d. **Control**. Clear understanding of the concept of the operation and commander's intent, coupled with disciplined communications, to bring every man and weapon available to overwhelm the enemy at the decisive point.

e. **Common Sense**. Use all available information and good judgment to make sound, timely decisions.

5-2. **PLANNING**. This paragraph provides the planning considerations common to most patrols. It discusses task organization, initial planning and coordination, completion of the plan, and contingency planning.

a. **Task Organization**. A patrol is a mission, not an organization. To accomplish the patrolling mission, a platoon or squad must perform specific tasks, for example, secure itself, cross danger areas, recon the patrol objective, breach, support, or assault. As with other missions, the leader tasks elements of his unit in accordance with his estimate of the situation, identifying those tasks his unit must perform and designating which elements of his unit will perform which tasks. Where possible, in assigning tasks, the leader should maintain squad and fire team integrity. The chain of command continues to lead its elements during a patrol. Squads and fire teams may perform more than one task in an assigned sequence; others may perform only one task. The leader must plan carefully to ensure that he has identified and assigned all required tasks in the most efficient way. Elements and teams for platoons conducting patrols include--

(1) Elements common to all patrols:

(a) *Headquarters Element*. The headquarters consists of the platoon leader (PL), RTO, platoon sergeant (PSG), FO, RTO, and medic. It may include any attachments that the PL decides that he or the PSG must control directly.

(b) *Aid and Litter Team*. Aid and litter teams are responsible for buddy aid and evacuation of casualties.

(c) *Enemy Prisoner of War Team*. EPW teams control enemy prisoners using the five S's and the leader's guidance.

(d) *Surveillance Team*. The surveillance team keeps watch on the objective from the time that the leader's reconnaissance ends until the unit deploys for actions on the objective. They then rejoin their parent element.

(e) *En Route Recorder*. Part of the HQ element, maintains communications with higher and acts as the recorder for all CCIR collected during the mission.

(f) *Compass Man*. The compass man assists in navigation by ensuring the patrol remains on course at all times. Instructions to the compass man must include initial and subsequent azimuths. As a technique, the compass man should preset his compass on the initial azimuth before the unit moves out, especially if the move will be during limited visibility conditions. The platoon or squad leader should also designate an alternate compass man.

(g) *Point/Pace Man*. As required, the PL designates a primary and alternate point man and a pace man for the patrol. The pace man aids in navigation by keeping an accurate count of distance traveled. The point man selects the actual route through the terrain, guided by the compass man or team leader. In addition, the point man also provides frontal security.

(2) Elements common to all combat patrols:

(a) *Assault Element*. The assault element seizes and secures the objective and protects special teams as they complete their assigned actions on the objective.

(b) *Security Element*. The security element provides security at danger areas, secures the ORP, isolates the objective, and supports the withdrawal of the rest of the patrol once actions on the objective are complete. The security element may have separate security teams, each with an assigned task or sequence of tasks.

(c) *Support Element.* The support element provides direct and indirect fire support for the unit. Direct fires include machine guns, medium and light antiarmor weapons, small recoilless rifles. Indirect fires available may include mortars, artillery, CAS, and organic M203 weapon systems.

(d) *Demolition Team.* Demolition teams are responsible for preparing and detonating the charges to destroy designated equipment, vehicles, or facilities on the objective.

(e) *EPW and Search Teams.* The assault element may provide two-man (buddy teams) or four-man (fire team) search teams to search bunkers, buildings, or tunnels on the objective. These teams will search the objective or kill zone for any PIR that may give the PL an idea of the enemy concept for future operations. Primary and alternate teams may be assigned to ensure enough prepared personnel are available on the objective.

(f) *Breach Element.* The breach team conducts initial penetration of enemy obstacles to seize a foothold and allow the patrol to enter an objective. This is typically done IAW METT-TC and the steps outlined in the "*Conduct an Initial Breach of a Mined Wire Obstacle*" battle drill in Chapter 6 of this Handbook.

(3) Elements common to all reconnaissance patrols:

(a) *Reconnaissance Team.* Reconnaissance teams reconnoiter the objective area once the security teams are in position. Normally these are two-man teams (buddy teams) to reduce the possibility of detection.

(b) *Reconnaissance and Security Teams.* R&S teams are normally used in a zone reconnaissance, but may be useful in any situation when it is impractical to separate the responsibilities for reconnaissance and security.

(c) *Security Element.* When the responsibilities of reconnaissance and security are separate, the security element provides security at danger areas, secures the ORP, isolates the objective, and supports the withdrawal of the rest of the platoon once the recon is complete. The security element may have separate security teams, each with an assigned task or sequence of tasks.

b. **Initial Planning and Coordination**. Leaders plan and prepare for patrols using the troop-leading procedures and the estimate of the situation, as described in Chapter 2. Through an estimate of the situation, leaders identify required actions on the objective (mission analysis) and plan backward to departure from friendly lines and forward to reentry of friendly lines. Because patrolling units act independently, move beyond the direct-fire support of the parent unit, and operate forward of friendly units, coordination must be thorough and detailed. Coordination is continuous throughout planning and preparation. PLs use checklists to preclude omitting any items vital to the accomplishment of the mission.

(1) ***Coordination with Higher Headquarters***. This coordination includes Intelligence, Operations, and Fire Support IAW Chapter 2, *Coordination Checklists* (**page 2-34**). This initial coordination is an integral part of Step 3 of Troop-Leading Procedures, *Make a Tentative Plan.*

(2) ***Coordination with Adjacent Units***. The leader also coordinates his unit's patrol activities with the leaders of other units that will be patrolling in adjacent areas at the same time, IAW Chapter 2-7, *Coordination Checklists* (**page 2-34**).

c. **Completion of Plan**. As the PL completes his plan, he considers--

(1) ***Specified and Implied Tasks***. The PL ensures that he has assigned all specified tasks to be performed on the objective, at rally points, at danger areas, at security or surveillance locations, along the route(s), and at passage lanes. These make up the maneuver and tasks to maneuver units subparagraphs of the Execution paragraph.

(2) ***Key travel and Execution Times***. The leader estimates time requirements for movement to the objective, leader's reconnaissance of the objective, establishment of security and surveillance, completion of all assigned tasks on the objective, and passage through friendly lines. Some planning factors are--

- Movement: Average of 1 Kmph during daylight hours in Woodland Terrain; Average limited visibility ½ Kmph. Add additional time for restrictive, or severely restrictive terrain such as mountains, swamps, or thick vegetation.
- Leader's recon: NLT 1.5 hr.
- Establishment of security and surveillance: 0.5 hr.

(3) ***Primary and Alternate Routes***. The leader selects primary and alternate routes to and from the objective. The return routes should differ from the routes to the objective. The PL may delegate route selection to a subordinate, but is ultimately responsible for the routes selected.

(4) ***Signals***. The leader should consider the use of special signals. These include hand-and-arm signals, flares, voice, whistles, radios, and infrared equipment. Primary and alternate signals must be identified and rehearsed so that all Soldiers know their meaning.

(5) ***Challenge and Password Forward of Friendly Lines***. The challenge and password from the unit's ANCD must not be used beyond the FLOT.

(a) *Odd-Number System.* The leader specifies an odd number. The challenge can be any number less than the specified number. The password will be the number that must be added to it to equal the specified number, for example, the number is 7, the challenge is 3, and the password is 4.

(b) *Running Password.* ANCDs may also designate a running password. This code word alerts a unit that friendly Soldiers are approaching in a less than organized manner and possibly under pressure. The number of Soldiers approaching follows the running password. For example, if the running password is "*Ranger,*" and five friendly Soldiers are approaching, they would say "*Ranger five.*"

(6) ***Location of Leaders***. The PL considers where he and the PSG and other key leaders are located during each phase of the mission. The PL positions himself where he can best control the actions of the patrol. The PSG is normally located with the assault element during a raid or attack to help the PL control the use of additional assaulting squads, and will assist with securing the OBJ. The PSG will locate himself at the CCP to facilitate casualty treatment and evacuation. During a reconnaissance mission, the PSG will stay behind in the ORP to facilitate the transfer of Intel to the higher headquarters, and control the recon elements movement into and out of the ORP.

(7) ***Actions on Enemy Contact.*** Unless required by the mission, the unit avoids enemy contact. The leader's plan must address actions on chance contact at each phase of the patrol mission. The unit's ability to continue will depend on how early contact is made, whether the platoon is able to break contact successfully (so that its subsequent direction of movement is undetected), and whether the unit receives any casualties because of the contact. The plan must address the handling of seriously wounded Soldiers and KIAs. The plan must also address the handling of prisoners who are captured because of chance contact and are not part of the planned mission.

(8) ***Contingency Plans***. The leader leaves his unit for many reasons throughout the planning, coordination, preparation, and execution of his patrol mission. Each time the leader departs the patrol main body, he must issue a five-point contingency plan to the leader left in charge of the unit. The contingency plan is described by the acronym GOTWA, as follows. The patrol leader will additionally issue specific guidance stating what tasks are to be accomplished in the ORP in his absence:

- G: Where the leader is GOING.
- O: OTHERS he is taking with him.
- T: TIME he plans to be gone.
- W: WHAT to do if the leader does not return in time.
- A: The unit's and the leader's ACTIONS on chance contact while the leader is gone.

(9) ***Rally Points***. The leader considers the use and location of rally points. A rally point is a place designated by the leader where the unit moves to reassemble and reorganize if it becomes dispersed. Soldiers must know which rally point to move to at each phase of the patrol mission should they become separated from the unit. They must also know what actions are required there and how long they are to wait at each rally point before moving to another.

(a) *Criteria*. Rally points must be--

- Easily identifiable in daylight and limited visibility.
- Show no signs of recent enemy activity.
- Covered and concealed.
- Away from natural lines of drift and high-speed avenues of approach.
- Defendable for short periods of time.

(b) *Types*. The most common types of rally points are initial, en route, objective, and near-and-far-side rally points.

(10) ***Objective Rally Point***. The ORP is typically 200 to 400m from the objective, or at a minimum, one major terrain feature away. Actions at the ORP include--

- Conduct SLLS and pinpoint location.
- Leaders Recon of the Objective.
- Issuing a FRAGO, if needed.
- Making final preparations before continuing operations, for example, recamouflaging, preparing demolitions, lining up rucksacks for quick recovery. Preparing EPW bindings, first aid kits, litters, and inspecting weapons.
- Accounting for Soldiers and equipment after actions at the objective are complete.
- Reestablishing the chain of command after actions at the objective are complete.
- Disseminating information from reconnaissance, if contact was not made.

(11) ***Leader's reconnaissance of the objective***. The plan must include a leader's reconnaissance of the objective once the platoon or squad establishes the ORP. Before departing, the leader must issue a 5-point contingency plan. During his

reconnaissance, the leader pinpoints the objective, selects reconnaissance, security, support, and assault positions for his elements, and adjusts his plan based on his observation of the objective. Each type of patrol requires different tasks during the leader's reconnaissance. The platoon leader will bring different elements with him. (These are discussed separately under each type of patrol). The leader must plan time to return to the ORP, complete his plan, disseminate information, issue orders and instructions, and allow his squads to make any additional preparations. During the Leader's Reconnaissance for a Raid or Ambush, the PL will leave surveillance on the OBJ.

(12) ***Actions on the objective***. Each type of patrol requires different actions on the objective. Actions on the objective are discussed under each type of patrol.

5-3. **RECONNAISSANCE PATROLS**. Recon patrols are one of the two types of patrols. provide timely and accurate information on the enemy and terrain. They confirm the leader's plan before it is executed. Units on reconnaissance operations collect specific information (*priority intelligence requirements* [PIR]) or general information (*information requirements* [IR]) based on the instructions from their higher commander. The two types of recon patrols discussed here are area and zone. This section discusses the fundamentals of reconnaissance, task standards for the two most common types of recon, and actions on the objective for those types of recon.

a. **Fundamentals of Reconnaissance**. In order to have a successful area reconnaissance, the platoon leader must apply the fundamentals of the reconnaissance to his plan during the conduct of the operation.

(1) ***Gain all required information***. The parent unit tells the patrol leader (PL) what information is required. This is in the form of the IR and PIR. The platoon's mission is then tailored to what information is required. During the entire patrol, members must continuously gain and exchange all information gathered, but cannot consider the mission accomplished unless all PIR has been gathered.

(2) ***Avoid detection by the enemy***. A patrol must not let the enemy know that it is in the objective area. If the enemy knows he is being observed, he may move, change his plans, or increase his security measures. Methods of avoiding detection are--

(a) Minimize movement in the objective area (area reconnaissance).

(b) Move no closer to the enemy than necessary.

(c) If possible, use long-range surveillance or night vision devices.

(d) Camouflage, stealth, noise, and light discipline.

(e) Minimize radio traffic.

(3) ***Employ security measures***. A patrol must be able to break contact and return to the friendly unit with what information is gathered. If necessary, break contact and continue the mission. Security elements are emplaced so that they can overwatch the reconnaissance elements and suppress the enemy so the reconnaissance element can break contact.

(4) ***Task organization***. When the platoon leader receives the order, he analyzes his mission to ensure he understands what must be done. Then he task organizes his platoon to best accomplish the mission IAW METT-TC. Recons are typically squad-sized missions.

b. **Task Standards**.

(1) ***Area Reconnaissance***. The area recon patrol collects all available information on PIR and other intelligence not specified in the order for the area. The patrol completes the recon and reports all information by the time specified in the order. The patrol is not compromised.

(2) ***Zone Reconnaissance***. The zone recon patrol determines all PIR and other intelligence not specified in the order for its assigned zone. The patrol reconnoiters without detection by the enemy. The patrol completes the recon and reports all information by the time specified in the order.

c. **Actions on the Objective, Area Reconnaissance** (Figure 5-1).

(1) The element occupies the ORP as discussed in the section on occupation of the ORP. The RTO calls in spare for occupation of ORP. The leader confirms his location on map while subordinate leaders make necessary perimeter adjustments.

(2) The PL organizes the platoon in one of two ways: separate recon and security elements, or combined recon and security elements.

(3) The PL takes subordinates leaders and key personnel on a leader's recon to confirm the objective and plan.

(a) Issues a 5-point contingency plan before departure.

(b) Establishes a suitable release point that is beyond sight and sound of the objective if possible, but that is definitely out of sight. The RP should also have good rally point characteristics.

(c) Allow all personnel to become familiar with the release point and surrounding area.

(d) Identifies the objective and emplaces surveillance. Designates a surveillance team to keep the objective under surveillance. Issues a contingency plan to the senior man remaining with the surveillance team. The surveillance team is positioned with one man facing the objective, and one facing back in the direction of the release point.

(e) Takes subordinate leaders forward to pinpoint the objective, emplace surveillance, establish a limit of advance, and choose vantage points.

(f) Maintains commo with the platoon throughout the leader's recon.

(4) The PSG maintains security and supervises priorities of work in the ORP.

(a) Reestablishes security at the ORP.

(b) Disseminates the PLs contingency plan.

(c) Oversees preparation of recon personnel (personnel recamouflaged, NVDs and binos prepared, weapons on safe with a round in the chamber).

(5) The PL and his recon party return to the ORP.

(a) Confirms the plan or issues a FRAGO.

(b) Allows subordinate leaders time to disseminate the plan.

(6) The patrol conducts the recon by long-range observation and surveillance if possible.

(a) R&S elements move to observation points that offer cover and concealment and that are outside of small-arms range.

(b) Establishes a series of observation posts (OP) if information cannot be gathered from one location.

(c) Gathers all PIR using the SALUTE format.

(7) If necessary, the patrol conducts its recon by short-range observation and surveillance.

(a) Moves to an OP near the objective.

(b) Passes close enough to the objective to gain information.

(c) Gathers all PIR using the SALUTE format.

(8) R&S teams move using a technique such as the cloverleaf method to move to successive OP's. In this method, R&S teams avoid paralleling the objective site, maintain extreme stealth, do not cross the limit of advance, and maximize the use of available cover and concealment.

(9) During the conduct of the recon, each R&S team will return to the release point when any of the following occurs:

- They have gathered all their PIR.
- They have reached the limit of advance.
- The allocated time to conduct the recon has elapsed.
- Contact has been made.

(10) At the release point, the leader will analyze what information has been gathered and determine if he has met the PIR requirements.

(11) If the leader determines that he has not gathered sufficient information to meet the PIR requirements, or if the information he and the subordinate leader gathered differs drastically, he may have to send R&S teams back to the objective site. In this case, R&S teams will alternate areas of responsibilities. For example, if one team reconnoitered from the 6 – 3 – 12, then that team will now recon from the 6 – 9 – 12.

(12) The R&S element returns undetected to the ORP by the specified time.

(a) Disseminates information to all patrol members through key leaders at the ORP, or moves to a position at least one terrain feature or one kilometer away to disseminate. To disseminate, the leader has the RTO prepare three sketches of the objective site based on the leader's sketch and provides the copies to the subordinate leaders to assist in dissemination.

(b) Reports any information requirements and/or any information requiring immediate attention to higher headquarters, and departs for the designated area.

(13) If contact is made, move to the release point. The recon element tries to break contact and return to the ORP, secure rucksacks, and quickly move out of the area. Once they have moved a safe distance away, the leader will inform higher HQ of the situation and take further instructions from them.

(a) While emplacing surveillance, the recon element withdraws through the release point to the ORP, and follows the same procedures as above.

(b) While conducting the reconnaissance, the compromised element returns a sufficient volume of fire to allow them to break contact. Surveillance can fire an AT-4 at the largest weapon on the objective. All elements will pull off the objective and move to the release point. The senior man will quickly account for all personnel and return to the ORP. Once in the ORP, follow the procedures previously described.

Figure 5-1. ACTIONS ON THE OBJECTIVE, AREA RECONNAISSANCE

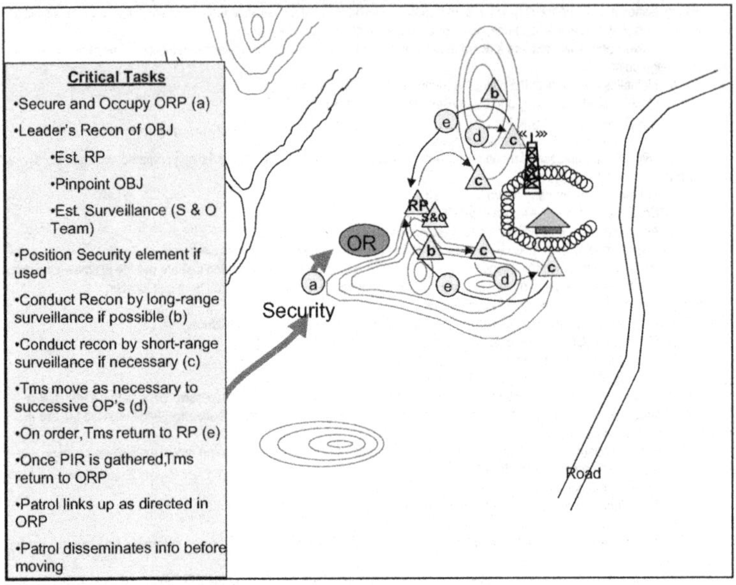

d. **Actions on the Objective, Zone Reconnaissance**.

(1) The element occupies the initial ORP as discussed in the section occupation of the ORP. The radio operator calls in spare for occupation of ORP. The leader confirms his location on map while subordinate leaders make necessary perimeter adjustments.

(2) The recon team leaders organize their recon elements.

(a) Designate security and recon elements.

(b) Assign responsibilities (point man, pace man, en route recorder, and rear security), if not already assigned.

(c) Designates easily recognizable rally points.

(d) Ensure local security at all halts.

(3) The patrol recons the zone.

(a) Moves tactically to the ORPs.

(b) Occupies designated ORPs.

(c) Follows the method designated by the PL (fan, converging routes, or box method, Figure 5-2).

Figure 5-2. COMPARISON OF ZONE RECONNAISSANCE METHODS

FAN METHOD	*CONVERGING ROUTES METHOD*	*BOX METHOD*
• Uses a series (fan) of ORPs. • Patrol establishes security at first ORP. • Each recon element moves from ORP along a different fan-shaped route. Route overlaps with that of other recon elements. This ensures recon of entire area. • Leader maintains reserve at ORP. • When all recon elements return to ORP, PL collects and disseminates all info before moving to next ORP.	• PL selects routes from ORP thru zone to a linkup point at the far side of the zone from the ORP. Each recon element moves and recons along a specified route. They converge (link up) at one time and place.	• PL sends recon elements from the first ORP along routes that form a box. He sends other elements along routes throughout the box. All teams link up at the far side of the box from the ORP.

(d) The recon teams reconnoiter.

- During movement, the squad will gather all PIR specified by the order.
- Recon team leaders will ensure sketches are drawn or digital photos are taken of all enemy hard sites, roads, and trails.
- Return to the ORP, or link up at the rendezvous point on time.
- When the squad arrives at new rendezvous point or ORP, the recon team leaders report to the PL with all information gathered.

(e) The PL continues to control the recon elements.

- PL moves with the recon element that establishes the linkup point.
- PL changes recon methods as required.
- PL designates times for the elements to return to the ORP or to linkup.
- PL collects all information and disseminates it to the entire patrol. PL will brief all key subordinate leaders on information gathered by other squads, establishing one consolidated sketch if possible, and allow team leaders time to brief their teams.
- PL and PSG account for all personnel.

(f) The patrol continues the recon until all designated areas have been reconned, and returns undetected to friendly lines.

5-4. **COMBAT PATROLS**. Combat patrols are the second type of patrol. Combat patrols are further divided into raids and ambushes Units conduct combat patrols to destroy or capture enemy Soldiers or equipment; destroy installations, facilities, or key points; or harass enemy forces. Combat patrols also provide security for larger units. This section describes overall combat patrol planning considerations, task considerations for each type of combat patrol, and finally actions on the objective for each type.

a. **Planning Considerations (General)**. In planning a combat patrol, the PL considers the following:

(1) ***Tasks to Maneuver Units***. Normally the platoon headquarters element controls the patrol on a combat patrol mission. The PL makes every try to maintain squad and fire team integrity as he assigns tasks to subordinates units.

(a) The PL must consider the requirements for assaulting the objective, supporting the assault by fire, and security of the entire unit throughout the mission.

- For the assault on the objective, the PL considers the required actions on the objective, the size of the objective, and the known or presumed strength and disposition of the enemy on and near the objective.
- The PL considers the weapons available, and the type and volume of fires required to provide fire support for the assault on the objective.
- The PL considers the requirement to secure the platoon at points along the route, at danger areas, at the ORP, along enemy avenues of approach into the objective, and elsewhere during the mission.
- The PL will also designate engagement/disengagement criteria.

(b) The PL assigns additional tasks to his squads for demolition, search of EPWs, guarding of EPWs, treatment and evacuation (litter teams) of friendly casualties, and other tasks required for successful completion of patrol mission (if not already in the SOP).

(c) The PL determines who will control any attachments of skilled personnel or special equipment.

(2) ***Leader's Reconnaissance of the Objective***. In a combat patrol, the PL has additional considerations for the conduct of his reconnaissance of the objective from the ORP.

(a) *Composition of the leader's reconnaissance party*. The platoon leader will normally bring the following personnel.

- Squad leaders to include the weapons squad leader.
- Surveillance team.
- Forward observer.
- Security element (dependent on time available).

(b) *Conduct of the leader's reconnaissance*. In a combat patrol, the PL considers the following additional actions in the conduct of the leader's reconnaissance of the objective.

- The PL designates a release point approximately half way between the ORP and this objective. The PL posts the surveillance team. Squads and fire teams separate at the release point, and then they move to their assigned positions.
- The PL confirms the location of the objective or kill zone. He notes the terrain and identifies where he can emplace claymores to cover dead space. Any change to his plan is issued to the squad leaders (while overlooking the objective if possible).
- If the objective is the kill zone for an ambush, the leader's reconnaissance party should not cross the objective; to do so will leave tracks that may compromise the mission.
- The PL confirms the suitability of the assault and support positions and routes from them back to the ORP.
- The PL issues a five-point contingency plan before returning to the ORP.

b. **Ambush**.

(1) ***Planning Considerations***. An ambush is a surprise attack from a concealed position on a moving or temporarily halted target. Ambushes are classified by category--hasty or deliberate; type--point or area; and formation--linear or L-shaped. The leader uses a combination of category, type, and formation in developing his ambush plan. The key planning considerations include--

(a) Coverage of entire kill zone by fire.

(b) METT-TC.

(c) Use of existing or reinforcing obstacles, including Claymores, to keep the enemy in the kill zone.

(d) Security teams (typically equipped with hand-held antitank weapons such as AT-4 or LAW; Claymores; and various means of communication.

(e) Protect the assault and support elements with claymores or explosives.

(f) Use security elements or teams to isolate the kill zone.

(g) Assault through the kill zone to the limit of advance (LOA). (The assault element must be able to move quickly through its own protective obstacles.)

(h) Time the actions of all elements of the platoon to preclude loss of surprise. In the event any member of the ambush is compromised, he may immediately initiate the ambush.

(i) When the ambush must be manned for a long time, use only one squad to conduct the entire ambush and determining movement time of rotating squads from the ORP to the ambush site.

(2) ***Categories***.

(a) *Hasty Ambush*. A unit conducts a hasty ambush when it makes visual contact with an enemy force and has time to establish an ambush without being detected. The actions for a hasty ambush must be well rehearsed so that Soldiers know what to do on the leader's signal. They must also know what action to take if the unit is detected before it is ready to initiate the ambush.

(b) *Deliberate Ambush*. A deliberate ambush is conducted at a predetermined location against any enemy element that meets the commander's engagement criteria. The leader requires the following detailed information in planning a deliberate ambush: size and composition of the targeted enemy, and weapons and equipment available to the enemy.

(3) ***Types***.

(a) Point ambush. In a point ambush, Soldiers deploy to attack an enemy in a single kill zone.

(b) Area ambush. In an area, Soldiers deploy in two or more related point ambushes.

(4) ***Formations***. (Figure 5-3).

(a) *Linear Ambush*. In an ambush using a linear formation, the assault and support elements deploy parallel to the enemy's route. This positions both elements on the long axis of the kill zone and subjects the enemy to flanking fire. This formation can be used in close terrain that restricts the enemy's ability to maneuver against the platoon, or in open terrain provided a means of keeping the enemy in the kill zone can be effected.

(b) *L-Shaped Ambush*. In an L-shaped ambush, the assault element forms the long leg parallel to the enemy's direction of movement along the kill zone. The support element forms the short leg at one end of and at right angles to the assault element. This provides both flanking (long leg) and enfilading fires (short leg) against the enemy. The L-shaped ambush can be used at a sharp bend in a trail, road, or stream. It should not be used where the short leg would have to cross a straight road or trail.

Figure 5-3. AMBUSH FORMATIONS

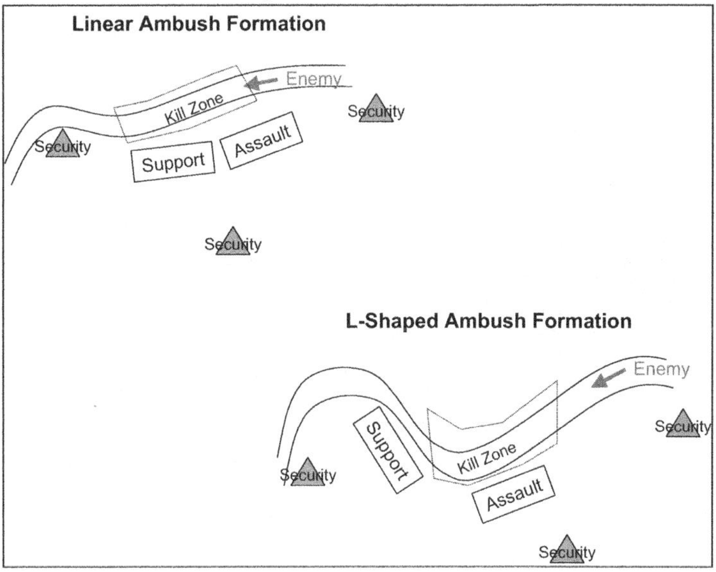

c. **Hasty Ambush**.

(1) ***Task Standards***. The platoon moves quickly to concealed positions. The ambush is not initiated until the majority of the enemy is in the kill zone. The unit does not become decisively engaged. The platoon surprises the enemy. The patrol captures, kills, or forces the withdrawal of all of the enemy within the kill zone. On order, the patrol withdraws all personnel and equipment in the kill zone from observation and direct fire. The unit does not become decisively engaged by follow-on elements. The platoon continues follow-on operations.

(2) ***Actions on the Objective (Hasty Ambush)***. See Figure 5-4.

(a) Using visual signals, any Soldier alerts the unit that an enemy force is in sight. The Soldier continues to monitor the location and activities of the enemy force until his team or squad leader relieves him, and gives the enemy location and direction of movement.

(b) The platoon or squad halts and remains motionless.

- The PL gives the signal to conduct a hasty ambush, taking care not to alert the enemy of the patrol's presence.
- The leader determines the best nearby location for a hasty ambush. He uses arm-and-hand signals to direct the unit members to covered and concealed positions.

(c) The leader designates the location and extent of the kill zone.

d) Teams and squads move silently to covered and concealed positions, ensuring positions are undetected and have good observation and fields of fire into the kill zone.

(e) Security elements move out to cover each flank and the rear of the unit. The leader directs the security elements to move a given distance, set up, and then rejoin the unit on order or, after the ambush (the sound of firing ceases). At squad level, the two outside buddy teams normally provide flank security as well as fires into the kill zone. At platoon level, fire teams make up the security elements.

(f) The PL assigns sectors of fire and issues any other commands necessary such as control measures.

(g) The PL initiates the ambush, using the greatest casualty-producing weapon available, when the largest percentage of enemy is in the kill zone. The PL--

- Controls the rate and distribution of fire.
- Employs indirect fire to support the ambush.
- Orders cease fire.
- (If the situation dictates) Orders the patrol to assault through the kill zone.

(h) The PL designates personnel to conduct a hasty search of enemy personnel and process enemy prisoners and equipment.

(i) The PL orders the platoon to withdraw from the ambush site along a covered and concealed route.

(j) The PL gains accountability, reorganizes as necessary, disseminates information, reports the situation, and continues the mission as directed.

d. **Deliberate (Point/Area) Ambush**.

(1) ***Task Standards***. The ambush is emplaced NLT the time specified in the order. The patrol surprises the enemy and engages the enemy main body. The patrol kills or captures all enemy in the kill zone and destroys equipment based on the commander's intent. The patrol withdraws all personnel and equipment from the objective, on order, within the time specified in the order. The patrol obtains all available PIR from the ambush and continues follow-on operations.

(2) ***Actions on the Objective (Deliberate Ambush).*** See Figure 5-5.

(a) The PL prepares the patrol for the ambush in the ORP.

(b) The PL prepares to conduct a leader's reconnaissance. He--

- Designates the members of the leader's recon party (typically includes squad leaders, surveillance team, FO, and possibly the security element.
- Issues a contingency plan to the PSG.

Figure 5-4. ACTIONS ON THE OBJECTIVE -- HASTY AMBUSH

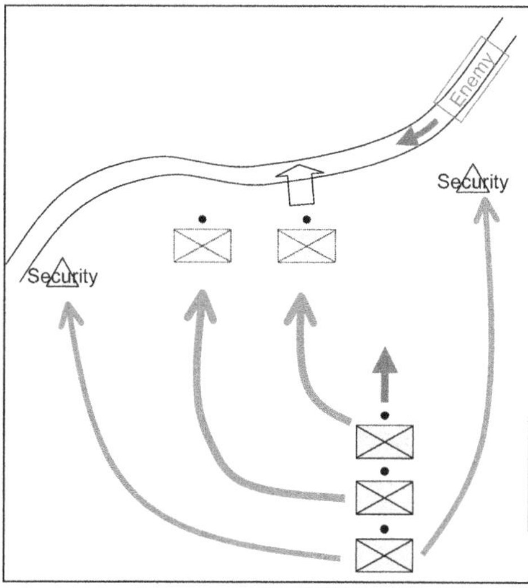

Critical Tasks

- Patrol detects an enemy unit; PL is notified
- Patrol halts and remains motionless
- PL gives signal for Hasty Ambush
- PL directs elements to covered and concealed positions
- Security Elements move to flanks of patrol
- PL establishes control measures
- PL initiates and controls ambush
- PL directs a hasty search
- Patrol consolidates, reorganizes, withdraws, reports, and continues mission

(c) The PL conducts his leader's reconnaissance. He--

- Ensures the leader's recon party moves undetected.
- Confirms the objective location and suitability for the ambush.
- Selects a kill zone.
- Posts the surveillance team at the site and issues a contingency plan.
- Confirms suitability of assault and support positions, and routes from them to the ORP.
- Selects the position of each weapon system in the support-by-fire position, and then designates sectors of fire.
- Identifies all offensive control measures to be used. Identifies the PLD, the assault position, LOA, any boundaries or other control measures. If available, the PL can use infrared aiming devices to identify these positions on the ground.

(d) The PL adjusts his plan based on info from the reconnaissance. He--

- Assigns positions.
- Designates withdrawal routes.

(e) The PL confirms the ambush formation.

(f) The security team(s) occupy first, securing the flanks of the ambush site, and providing early warning. The security element must be in position before the support and assault elements move forward of the release point. A security team remains in the ORP if the patrol plans to return to the ORP after actions on the objective. If the ORP is abandoned, a rear security team should be emplaced.

(g) Support element leader assigns sectors of fire. He--

- Emplaces claymores and obstacles as designated.
- Identifies sectors of fire and emplaces limiting stakes to prevent friendly fires from hitting other elements.
- Overwatches the movement of the assault element into position.

(h) Once the support element is in position, or on the PLs order, the assault element--

- Departs the ORP and moves into position.
- Upon reaching the PLD, the assault element transitions from the movement formation to the battle formation.
- Identifies individual sectors of fire as assigned by the PL. Emplaces aiming stakes or uses metal-to-metal contact with the machine gun tripods to prevent fratricide on the objective.
- Emplaces claymores to help destroy the enemy in the kill zone.
- Camouflages positions.

(i) The security element spots the enemy and notifies the PL, and reports the direction of movement, size of the target, and any special weapons or equipment carried. The security element also keeps the platoon leader informed if any enemy forces are following the lead force.

(j) The PL alerts other elements, and determines if the enemy force is too large, or if the ambush can engage the enemy successfully.

(k) The PL initiates the ambush using the highest casualty-producing device. He may use a command-detonated claymore. He must also plan a backup method for initiating the ambush, in case his primary means fails. This should also be a casualty-producing device such as his individual weapon. He passes this information to all Rangers, and practices it during rehearsals.

(l) The PL ensures that the assault and support elements deliver fire with the heaviest, most accurate volume possible on the enemy in the kill zone. In limited visibility, the PL may use infrared lasers to further define specific targets in the kill zone.

(m) Before assaulting the target, the PL gives the signal to lift or shift fires.

(n) The assault element--

- Assaults before the remaining enemy can react.
- Kills or captures enemy in the kill zone.
- Uses individual movement techniques or bounds by fire teams to move.
- Upon reaching the limit of advance, halts and establishes security. If needed, it reestablishes the chain of command and remains key weapon systems. All Soldiers will load a fresh magazine or drum of ammunition using the buddy system. ACE reports will be submitted through the chain of command. The PL will submit an initial contact report to higher.

(o) The PL directs special teams (EPW search, aid and litter, demo) to accomplish their assigned task once the assault element has established its LOA.

- Once the kill zone had been cleared, collect and secure all EPWs and move them out of the kill zone before searching bodies. Coordinate for an EPW exchange point to link up with higher to extract all EPWs and treat them IAW the five S's.
- Search from one side to the other and mark bodies that have been searched to ensure the area is thoroughly covered. Units should use the "*clear out, search in*" technique, clear from the center of the objective out ensuring the area is clear of all enemy combatants; then search all enemy personnel towards the center of the objective. Search all dead enemy personnel using two-man search techniques.
  - -- As the search team approaches a dead enemy Soldier, one-man guards while the other man searches. First, he kicks the enemy weapon away.
  - -- Second, he rolls the body over (if on the stomach) by lying on top and when given the go ahead by the guard (who is positioned at the enemy's head), the searcher rolls the body over on him. This is done for protection in case the enemy Soldier has a grenade with the pin pulled underneath him.
  - -- The searchers then conduct a systematic search of the dead Soldier from head to toe removing all papers and anything new (different type rank, shoulder boards, different unit patch, pistol, weapon, or NVD). They note if the enemy has a fresh or shabby haircut and the condition of his uniform and boots. They note the radio frequency, and then they secure the SOI, maps, documents, and overlays.
  - -- Once the body has been thoroughly searched, the search team will continue in this manner until all enemy personnel in and near the kill zone have been searched.
- Identify, collect, and prepare all equipment to be carried back or destroyed.
- Evacuate and treat friendly wounded first, then enemy wounded, time permitting.

- The demolition team prepares dual-primed explosives or incendiary grenades and awaits the signal to initiate. This is normally the last action performed before the unit departs the objective and may signal the security elements to return to the ORP.
- Actions on the objective with stationary assault line; all actions are the same with the exception of the search teams. To provide security within the teams to the far side of the kill zone during the search, they work in three-Ranger teams. Before the search begins, the Rangers move all KIAs to the near side of the kill zone.

(p) If enemy reinforcements try to penetrate the kill zone, the flank security will engage to prevent the assault element from being compromised.

(q) The platoon leader directs the unit's withdrawal from the ambush site:

- Elements normally withdraw in the reverse order that they established their positions.
- The elements may return to the RP or directly to the ORP, depending on the distance between elements.
- The security element of the ORP must be alert to assist the platoon's return to the ORP. It maintains security for the ORP while the rest of the platoon prepares to leave.
- If possible, all elements should return to the location at which they separated from the main body. This location should usually be the RP.

(r) The PL and PSG direct actions at the ORP, to include accountability of personnel and equipment and recovery of rucksacks and other equipment left at the ORP during the ambush.

(s) The platoon leader disseminates information, or moves the platoon to a safe location (no less than one kilometer or one terrain feature away from the objective) and disseminates information.

(t) As required, the PL and FO execute indirect fires to cover the platoon's withdrawal.

Figure 5-5. ACTIONS ON THE OBJECTIVE -- DELIBERATE AMBUSH

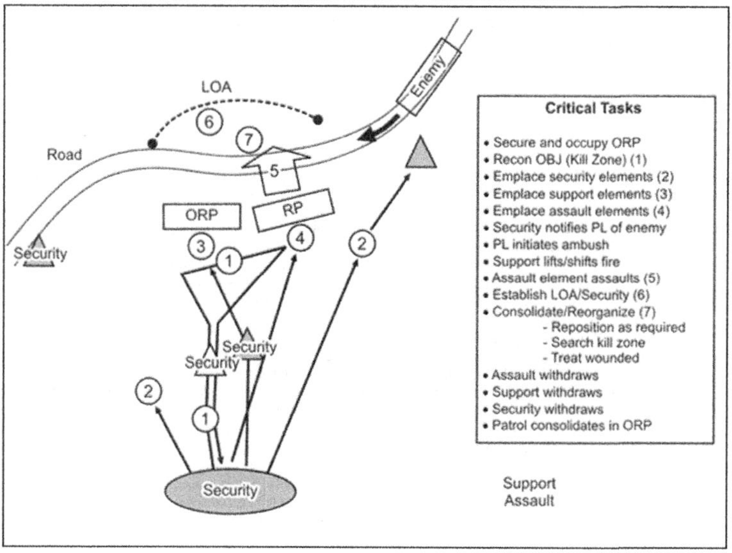

e. **Perform Raid**. The patrol initiates the raid NLT the time specified in the order, surprises the enemy, assaults the objective, and accomplishes its assigned mission within the commander's intent. The patrol does not become decisively engaged en route to the objective. The patrol obtains all available PIR from the raid objective and continues follow-on operations.

(1) ***Planning Considerations***. A Raid is a form of attack, usually small scale, involving a swift entry into hostile territory to secure information, confuse the enemy, or destroy installations followed by a planned withdrawal. Squads do not conduct raids. The sequence of platoon actions for a raid is similar to those for an ambush. Additionally, the assault element of the platoon may have to conduct a breach of an obstacle. It may have additional tasks to perform on the objective such as demolition of fixed facilities. Fundamentals of the raid include--

- Surprise and speed. Infiltrate and surprise the enemy without being detected.
- Coordinated fires. Seal off the objective with well-synchronized direct and indirect fires.
- Violence of action. Overwhelm the enemy with fire and maneuver.
- Planned withdrawal. Withdraw from the objective in an organized manner, maintaining security.

(2) ***Actions on the Objective (Raid)***. See Figure 5-6.

(a) The patrol moves to and occupies the ORP IAW the patrol SOP. The patrol prepares for the leader's recon.

(b) The PL, squad leaders, and selected personnel conduct a leader's recon.

- PL leaves a five point contingency plan with the PSG.
- PL establishes the RP, pinpoints the objective, contacts the PSG to prep men, weapons, and equipment, emplaces the surveillance team to observe the objective, and verifies and updates intelligence information.
- Leader's recon verifies location of and routes to security, support, and assault positions.
- Security teams are brought forward on the leader's reconnaissance and emplaced before the leader's recon leaves the RP
- Leaders conduct the recon without compromising the patrol.
- Leaders normally recon support by fire position first, then the assault position.

(c) The PL confirms, denies, or modifies his plan and issues instructions to his squad leaders.

- Assigns positions and withdrawal routes to all elements.
- Designates control measures on the objective (element objectives, lanes, limits of advance, target reference points, and assault line).
- Allows SLs time to disseminate information, and confirm that their elements are ready.

(d) Security elements occupy designated positions, moving undetected into positions that provide early warning and can seal off the objective from outside support or reinforcement.

(e) The support element leader moves the support element to designated positions. The support element leader ensures his element can place well-aimed fire on the objective.

(f) The PL moves with the assault element into the assault position. The assault position is normally the last covered and concealed position before reaching the objective. As it passes through the assault position the platoon deploys into its assault formation; that is, its squads and fire teams deploy to place the bulk of their firepower to the front as they assault the objective.

- Makes contact with the surveillance team to confirm any enemy activity on the objective.
- Ensures that the assault position is close enough for immediate assault if the assault element is detected early.
- Moves into position undetected, and establish local security and fire control measures.

(g) Element leaders inform the PL when their elements are in position and ready.

(h) The PL directs the support element to fire.

(i) Upon gaining fire superiority, the PL directs the assault element to move towards the objective.

- Assault element holds fire until engaged, or until ready to penetrate the objective.
- PL signals the support element to lift or shift fires. The support element lifts or shifts fires as directed, shifting fire to the flanks of targets or areas as directed in the FRAGO.

(j) The assault element attacks and secures the objective. The assault element may be required to breech a wire obstacle. As the platoon, or its assault element, moves onto the objective, it must increase the volume and accuracy of fires. Squad leaders assign specific targets or objectives for their fire teams. Only when these direct fires keep the enemy suppressed can the rest of the unit maneuver. As the assault element gets closer to the enemy, there is more emphasis on suppression and less on maneuver. Ultimately, all but one fire team may be suppressing to allow that one fire team to break into the enemy position. Throughout the assault, Soldiers use proper individual movement techniques, and fire teams retain their basic shallow wedge formation. The platoon does not get "*on-line*" to sweep across the objective.

- Assault element assaults through the objective to the designated LOA.
- Assault element leaders establish local security along the LOA, and consolidate and reorganize as necessary. They provide ACE reports to the PL and PSG. The platoon establishes security, operates key weapons, provides first aid, and prepares wounded Soldiers for MEDEVAC. They redistribute ammunition and supplies, and they relocate selected weapons to alternate positions if leaders believe that the enemy may have pinpointed them during the attack. They adjust other positions for mutual support. The squad and team leader provide ammunition, casualty, and equipment (ACE) reports to the platoon leader. The PL/PSG reorganizes the patrol based on the contact.
  - -- On order, special teams accomplish all assigned tasks under the supervision of the PL, who positions himself where he can control the patrol.
  - -- Special team leaders report to PL when assigned tasks are complete.

(k) On order or signal of the PL, the assault element withdraws from the objective. Using prearranged signals, the assault line begins an organized withdrawal from the objective site, maintaining control and security throughout the withdrawal. The assault element bounds back near the original assault line, and begins a single file withdrawal through the APL's choke point. All Rangers must move through the choke point for an accurate count. Once the assault element is a safe distance from the objective and the headcount is confirmed, the platoon can withdraw the support element. If the support elements were a part of the assault line, they withdraw together, and security is signaled to withdraw. Once the support is a safe distance off the objective, they notify the platoon leader, who contacts the security element and signals them to withdraw. All security teams link up at the release point and notify the platoon leader before moving to the ORP. Personnel returning to the ORP immediately secure their equipment and establish all-round security. Once the security element returns, the platoon moves out of the objective area as soon as possible, normally in two to three minutes.

- Before withdrawing, the demo team activates demo devices and charges.
- Support element or designated personnel in the assault element maintain local security during the withdrawal.
- Leaders report updated accountability and status (ACE report) to the PL and PSG.

(l) Squads withdraw from the objective in the order designated in the FRAGO to the ORP.

- Account for personnel and equipment.
- Disseminate information.
- Redistribute ammunition and equipment as required.

(m) The PL reports mission accomplishment to higher and continues the mission.

- Reports raid assessment to higher.
- Informs higher of any IR/PIR gathered.

NOTES

Figure 5-6. ACTIONS ON THE OBJECTIVE -- RAID

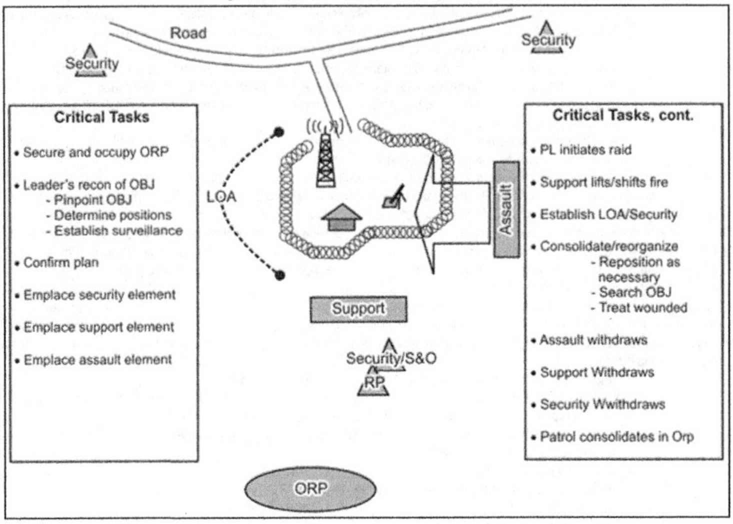

5-5. **SUPPORTING TASKS**. This section covers Linkup, Patrol Debriefing, and Occupation of an ORP.

a. **Linkup**. A linkup is a meeting of friendly ground forces. Linkups depend on control, detailed planning, communications, and stealth.

(1) ***Task Standard***. The units link up at the time and place specified in the order. The enemy does not surprise the main bodies. The linkup units establish a consolidated chain of command.

(2) ***Linkup Site Selection***. The leader identifies a tentative linkup site by map reconnaissance, other imagery, or higher headquarters designates a linkup site. The linkup site should have the following characteristics:

- Ease of recognition.
- Cover and concealment.
- No tactical value to the enemy.
- Location away from natural lines of drift.
- Defendable for a short period of time.
- N=Multiple access and escape routes.

(3) ***Execution***. Linkup procedure begins as the unit moves to the linkup point. The steps of this procedure are:

(a) The stationary unit performs linkup actions.

- Occupies the linkup rally point (LRP) NLT the time specified in the order.
- Establishes all-around security, establishes commo, and prepares to accept the moving unit.
- The security team clears the immediate area around the linkup point. It then marks the linkup point with the coordinated recognition signal. The security team moves to a covered and concealed position and observes the linkup point and immediate area around it.

(b) The moving unit--
- Performs linkup actions.
- The unit reports its location using phase lines, checkpoints, or other control measures.
- Halts at a safe distance from the linkup point in a covered and concealed position (the linkup rally point).

(c) The PL and a contact team--
- Prepare to make physical contact with the stationary unit.
- Issue a contingency plan to the PSG.
- Maintain commo with the platoon; verify near and far recognition signals for linkup (good visibility and limited visibility).
- Exchange far and near recognition signals with the linkup unit; conduct final coordination with the linkup unit.

(d) The stationary unit--
- Guides the patrol from its linkup rally point to the stationary unit linkup rally point.
- Linkup is complete by the time specified in the order.
- The main body of the stationary unit is alerted before the moving unit is brought forward.

(e) The patrol continues its mission IAW the order.

(4) ***Coordination Checklist***. The PL coordinates or obtains the following information from the unit that his patrol will link up with:
- Exchange frequencies, call signs, codes, and other communication information.
- Verify near and far recognition signals.
- Exchange fire coordination measures.
- Determine command relationship with the linkup unit; plan for consolidation of chain of command.
- Plan actions following linkup.
- Exchange control measures such as contact points, phase lines, contact points, as appropriate.

b. **Debrief**. Immediately after the platoon or squad returns, personnel from higher headquarters conduct a thorough debrief. This may include all members of the platoon or the leaders, RTOs, and any attached personnel. Normally the debriefing is oral. Sometimes a written report is required. Information on the written report should include--
- Size and composition of the unit conducting the patrol.
- Mission of the platoon such as type of patrol, location, and purpose.
- Departure and return times.
- Routes. Use checkpoints, grid coordinates for each leg or include an overlay.
- Detailed description of terrain and enemy positions that were identified.
- Results of any contact with the enemy.
- Unit status at the conclusion of the patrol mission, including the disposition of dead or wounded Soldiers.
- Conclusions or recommendations.

c. **Objective Rally Point**. The ORP is a point out of sight, sound, and small arms range of the objective area. It is normally located in the direction that the platoon plans to move after completion of actions on the objective. The ORP is tentative until the objective is pinpointed.

(1) ***Occupation of the ORP***. See Figure 5-7.

(a) The patrol halts beyond sight and sound of the tentative ORP (200-400m in good visibility, 100-200m in limited visibility).

(b) The patrol establishes a security halt IAW the unit SOP.

(c) After issuing a five-point contingency plan to the PSG, the PL moves forward with a recon element to conduct a leader's recon of the ORP.

(d) For a squad-sized patrol, the PL moves forward with a compass man and one member of each fire team to confirm the ORP.
- After physically clearing the ORP location, the PL leaves two Rangers at the 6 o'clock position facing in opposite directions.
- The PL issues a contingency plan and returns with the compass man to guide the patrol forward.
- The PL guides the patrol forward into the ORP, with one team occupying from 3 o'clock through 12 o'clock to 9 o'clock, and the other occupying from 9 o'clock through 6 o'clock to 3 o'clock.

(e) For a platoon-size patrol, the PL, RTO, WSL, three ammo bearers, a team leader, a SAW gunner, and riflemen go on the leaders recon for the ORP and position themselves at 10, 2, and 6 o'clock.

- The first squad in the order of march is the base squad, occupying from 10 to 2 o'clock.
- The trail squads occupy from 2 to 6 o'clock and 6 to 10 o'clock, respectively.
- The patrol headquarters element occupies the center of the triangle.

(2) ***Actions in the ORP***. The unit prepares for the mission in the ORP. Once the leader's recon pinpoints the objective, the PSG generally lines up rucks IAW unit SOP in the center of the ORP.

Figure 5-7. OCCUPATION OF THE ORP

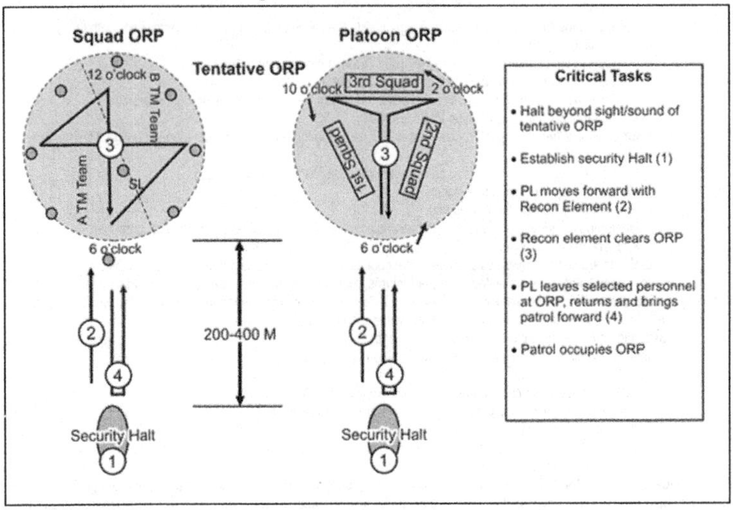

d. **Patrol Base**. A patrol base is a security perimeter that is set up when a squad or platoon conducting a patrol halts for an extended period. Patrol bases should not be occupied for more than a 24-hour period (except in emergency). A patrol never uses the same patrol base twice.

(1) ***Use***. Patrol bases are typically used--

- To avoid detection by eliminating movement.
- To hide a unit during a long detailed reconnaissance.
- To perform maintenance on weapons, equipment, eat and rest.
- To plan and issue orders.
- To reorganize after infiltrating on an enemy area.
- To establish a base from which to execute several consecutive or concurrent operations.

(2) ***Site Selection***. The leader selects the tentative site from a map or by aerial reconnaissance. The site's suitability must be confirmed and secured before the unit moves into it. Plans to establish a patrol base must include selecting an alternate patrol base site. The alternate site is used if the first site is unsuitable or if the patrol must unexpectedly evacuate the first patrol base.

(3) ***Planning Considerations***. Leaders planning for a patrol base must consider the mission and passive and active security measures. A patrol base (PB) must be located so it allows the unit to accomplish its mission.

- Observation posts and communication with observation posts.
- Patrol or platoon fire plan.
- Alert plan.
- Withdrawal plan from the patrol base to include withdrawal routes and a rally point, rendezvous point, or alternate patrol base.
- A security system to make sure that specific Soldiers are awake at all times.
- Enforcement of camouflage, noise, and light discipline.
- The conduct of required activities with minimum movement and noise.
- Priorities of Work.

(4) ***Security Measures***.

- Select terrain the enemy would probably consider of little tactical value.
- Select terrain that is off main lines of drift.
- Select difficult terrain that would impede foot movement, such as an area of dense vegetation, preferably bushes and trees that spread close to the ground.
- Select terrain near a source of water.
- Select terrain that can be defended for a short period and that offers good cover and concealment.
- Avoid known or suspected enemy positions.
- Avoid built-up areas.
- Avoid ridges and hilltops, except as needed for maintaining communications.
- Avoid small valleys.
- Avoid roads and trails.

(5) ***Occupation***. See Figure 5-8.

(a) A PB is reconned and occupied in the same manner as an ORP, with the exception that the platoon will typically plan to enter at a 90 degree turn. The PL leaves a two-man OP at the turn, and the patrol covers any tracks from the turn to the PB.

(b) The platoon moves into the PB. Squad-sized patrols will generally occupy a cigar-shaped perimeter; platoon-sized patrols will generally occupy a triangle-shaped perimeter.

(c) The PL and another designated leader inspect and adjust the entire perimeter as necessary.

(d) After the PL has checked each squad sector, each SL sends a two-man R&S team to the PL at the CP. The PL issues the three R&S teams a contingency plan, reconnaissance method, and detailed guidance on what to look for (enemy, water, built-up areas or human habitat, roads, trails, or possible rally points).

(e) Where each R&S team departs is based on the PLs guidance. The R&S team moves a prescribed distance and direction, and reenters where the PL dictates.

- Squad-sized patrols do not normally send out an R&S team at night.
- R&S teams will prepare a sketch of the area to the squad front if possible.
- The patrol remains at 100 % alert during this recon.
- If the PL feels the patrol was tracked or followed, he may elect to wait in silence at 100 % alert before sending out R&S teams.
- The R&S teams may use methods such as the "*I*," the "*Box*," or the "*T*." Regardless of the method chosen, the R&S team must be able to provide the PL with the same information.
- Upon completion of R&S, the PL confirms or denies the patrol base location, and either moves the patrol or begins priorities of work.

(6) ***Passive (Clandestine) Patrol Base (Squad)***.

- The purpose of a passive patrol base is for the rest of a squad or smaller size element.
- Unit moves as a whole and occupies in force.
- Squad leader ensures that the unit moves in at a 90-degree angle to the order of movement.
- A claymore mine is emplaced on route entering patrol base.
- Alpha and Bravo teams sit back-to-back facing outward, ensuring that at least one individual per team is alert and providing security.

(7) ***Priorities of Work (Platoon and Squad).*** Once the PL is briefed by the R&S teams and determines the area is suitable for a patrol base, the leader establishes or modifies defensive work priorities in order to establish the defense for the patrol base. Priorities of work are not a laundry list of tasks to be completed; to be effective, priorities of work must consist of a task, a given time, and a measurable performance standard. For each priority of work, a clear standard must be issued to guide the element in the successful accomplishment of each task. It must also be designated whether the work will be controlled in a centralized or decentralized manner. Priorities of work are determined IAW METT-TC. Priorities of Work may include, but are not limited to the following tasks:

(a) *Security (Continuous).*

- Prepare to use all passive and active measures to cover all of the perimeter all of the time, regardless of the percentage of weapons used to cover that all of the terrain.
- Readjust after R&S teams return, or based on current priority of work (such as weapons maintenance).
- Employ all elements, weapons, and personnel to meet conditions of the terrain, enemy, or situation.
- Assign sectors of fire to all personnel and weapons. Develop squad sector sketches and platoon fire plan.
- Confirm location of fighting positions for cover, concealment, and observation and fields of fire. SLs supervise placement of aiming stakes and claymores.
- Only use one point of entry and exit, and count personnel in and out. Everyone is challenged IAW the unit SOP.
- Hasty fighting positions are prepared at least 18 inches deep (at the front), and sloping gently from front to rear, with a grenade sump if possible.

(b) *Withdrawal Plan.* The PL designates the signal for withdrawal, order of withdrawal, and the platoon rendezvous point and/or alternate patrol base.

(c) *Communication (Continuous).* Commo must be maintained with higher headquarters, OP's, and within the unit. May be rotated between the patrol's RTOs to allow accomplishment of continuous radio monitoring, radio maintenance, act as runners for PL, or conduct other priorities of work.

(d) *Mission Preparation and Planning.* The PL uses the patrol base to plan, issue orders, rehearse, inspect, and prepare for future missions.

(e) *Weapons and Equipment Maintenance.* The PL ensures that machine guns, weapon systems, commo equipment, and night vision devices (as well as other equipment) are maintained. These items are not disassembled at the same time for maintenance (no more than 33 percent at a time), and weapons are not disassembled at night. If one machine gun is down, then security for all remaining systems is raised.

(f) *Water Resupply.* The PSG organizes watering parties as necessary. The watering party carries canteens in an empty rucksack or duffel bag, and must have commo and a contingency plan prior to departure.

(g) *Mess Plan.* At a minimum, security and weapons maintenance are performed prior to mess. Normally no more than half the platoon eats at one time. Rangers typically eat 1 to 3 meters behind their fighting positions.

- Rest/sleep plan management. The patrol conducts rest as necessary to prepare for future operations.
- Alert Plan and Stand-to. The PL states the alert posture and the stand-to time. He develops the plan to ensure all positions are checked periodically, OP's are relieved periodically, and at least one leader is always alert. The patrol typically conducts stand-to at a time specified by unit SOP (such as 30 minutes before and after BMNT or EENT).
- Resupply. Distribute or cross-load ammunition, meals, equipment, etc.
- Sanitation and personal hygiene. The PSG and medic ensure a slit trench is prepared and marked. All Soldiers will brush teeth, wash face, shave, wash hands, armpits, groin, feet, and darken (brush shine) boots daily. The patrol will not leave trash behind.

Figure 5-8. PATROL BASE

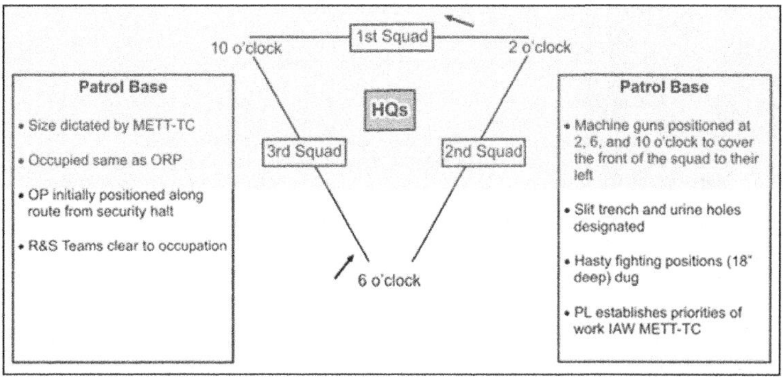

5-6. **MOVEMENT TO CONTACT**. The MTC is one of the five types of offensive operations. An MTC gains or regains contact with the enemy. Once contact is made, the unit develops the situation. Normally a platoon conducts an MTC as part of a larger force.

a. **Techniques**. The two techniques of conducting a movement to contact are search and attack and approach march.

(1) ***Search and Attack (S&A)***. This technique is used when the enemy is dispersed, is expected to avoid contact, disengage or withdraw, or you have to deny his movement in an area. The search and attack technique involves the use of multiple platoons, squads, and fire teams coordinating their actions to make contact with the enemy. Platoons typically try to find the enemy and then fix and finish him. They combine patrolling techniques with the requirement to conduct hasty or deliberate attacks once the enemy has been found.

(a) *Planning Considerations*.

- The factors of METT-TC.
- The requirement for decentralized execution.
- The requirement for mutual support.
- The length of operations.
- Minimize Soldier's load to improve stealth and speed.
- Resupply and MEDEVAC.
- Positioning key leaders and equipment.
- Employment of key weapons.
- Requirement for patrol bases.
- Concept for entering the zone of action.
- The concept for linkups while in contact.

(b) *Critical Performance Measures*.

- The platoon locates the enemy without being detected.
- Once engaged, fixes the enemy in position and maneuvers against the enemy.
- Maintains security throughout actions to avoid being flanked.

(2) ***Approach March***. The concept of the approach march is to make contact with the smallest element, allowing the commander the flexibility of destroying or bypassing the enemy. A platoon uses the approach march method as part of a larger unit. It can be tasked as the advance guard, move as part of the main body, or provide flank or rear security for the company or battalion. They may also receive on-order missions as part of the main body.

(a) *Fundamentals*. These basics are common to all movements to contact.

- Make enemy contact with smallest element possible.
- Rapidly develop combat power upon enemy contact.

- Provide all-round security for the unit.
- Support higher unit's concept.
- Reports all information rapidly and accurately and strives to gain and maintain contact with the enemy.
- Requires decentralized execution.

(b) *Planning Considerations*. The following issues should be considered heavily for MTC operations:

- Factors of METT-TC.
- Reduced Soldier's load.

(c) *Critical Performance Measures*.

- PL selects the appropriate movement formation based on likelihood of enemy contact.
- Maintains contact, once contact is made, until ordered to do otherwise.

b. **Task Standards**. The platoon moves NLT the time specified in the order, the platoon makes contact with the smallest element possible, and the main body is not surprised by the enemy. Once the platoon makes contact, it maintains contact. The platoon destroys squad and smaller-sized elements, and fixes elements larger than a squad. The platoon maintains sufficient fighting force capable of conducting further combat operations. Reports of enemy locations and contact are forwarded. If not detected by the enemy, the PL initiates a hasty attack. The platoon sustains no casualties from friendly fire. The platoon is prepared to initiate further movement within 25 minutes of contact, and all personnel and equipment are accounted for.

NOTES

Chapter 6
BATTLE DRILLS

Infantry battle drills describe how platoons and squads apply fire and maneuver to commonly encountered situations. They require leaders to make decisions rapidly and to issue brief oral orders quickly.

Section I. INTRODUCTION

This section defines and describes the format for battle drills.

6-1. **DEFINITION.** FM 7-1 defines a battle drill as "*a collective action rapidly executed without applying a deliberate decision-making process.*"

a. Characteristics of a battle drill are—

- They require minimal leader orders to accomplish and are standard throughout the Army.
- Sequential actions are vital to success in combat or critical to preserving life.
- They apply to platoon or smaller units.
- They are trained responses to enemy actions or leader's orders.
- They represent mental steps followed for offensive and defensive actions in training and combat.

b. A unit's ability to accomplish its mission often depends on Soldiers and leaders to execute key actions quickly. All Soldiers and their leaders must know their immediate reaction to enemy contact, as well as follow-up actions. Drills are limited to situations requiring instantaneous response; therefore, Soldiers must execute drills instinctively. This results from continual practice. Drills provide small units with standard procedures essential for building strength and aggressiveness.

(1) They identify key actions that leaders and Soldiers must perform quickly.

(2) They provide for a smooth transition from one activity to another; for example, from movement to offensive action to defensive action.

(3) They provide standardized actions that link Soldier and collective tasks at platoon level and below. (Soldiers perform individual tasks to CTT or SDT standard.)

(4) They require the full understanding of each individual and leader, and continual practice.

6-2. **FORMAT.** Drills include a ***title***, a ***situation*** to cue the unit or leader to start the drill, the ***required actions*** in sequence, and supporting illustrations. Where applicable, drills are cross-referenced with material in other chapters, or other drills, or both. Training standards for battle drills are in the mission training plan (MTP).

Section II. DRILLS

This section provides the battle drills for small units.

BATTLE DRILL 1
REACT TO CONTACT

SITUATION: A squad or platoon receives fires from enemy individual or crew-served weapons.

REQUIRED ACTIONS: See Figure 6-1.

1. Soldiers immediately assume the nearest covered positions.

2. Soldiers return fire immediately on reaching the covered positions.

3. Squad/team leaders locate and engage known or suspected enemy positions with well-aimed fire, and they pass information to the platoon/squad leader.

4. Fire team leaders control the fire of their Soldiers by using standard fire commands (initial and supplemental) containing the following elements:

- Alert
- Direction
- Description of target
- Range
- Method of fire (manipulation and rate of fire)
- Command to commence firing

5. Soldiers maintain contact (visual or oral) with the Soldiers on their left and right.

6. Soldiers maintain contact with their team leaders and indicate the location of enemy positions.

7. The leaders (visual or oral) check the status of their personnel.

8. The squad/team leaders maintain visual contact with the platoon/squad leader.

9. The platoon/squad leader moves up to the squad/fire team in contact and links up with its leader.
   a. The platoon leader brings his RTO, platoon FO, the nearest squad's squad leader, and one machine gun team.
   b. The squad leader of the trail squad moves to the front of his lead fire team.
   c. The PSG also moves forward with the second machine gun team and links up with the platoon leader, ready to assume control of the base-of-fire element.

10. The platoon/squad leader determines whether or not his unit must move out of an EA.

11. The platoon/squad leader determines whether his unit can gain and maintain suppressive fires with the element already in contact (based on the volume and accuracy of enemy fires against the element in contact). The platoon/squad leader assesses the situation. He identifies—
   - The location of the enemy position and obstacles.
   - The size of the enemy force engaging the unit in contact. The number of enemy automatic weapons, the presence of any vehicles, and the employment of indirect fires are indicators of the enemy strength.
   - Vulnerable flanks.
   - Covered and concealed flanking routes to the enemy position.

12. The platoon/squad leader determines the next COA, such as fire and movement, assault, breach, knock out bunker, enter and clear a building or trench.

13. The platoon/squad leader reports the situation to the company commander/platoon leader and begins to maneuver his unit.

14. Platoon leader directs platoon FO to call for and adjust indirect fires (mortars or artillery). Squad leaders relay requests through the platoon leader. The platoon/squad leader in conjunction with the platoon FO maintains accurate battle tracking of all friendly elements to facilitate quick clearance of fires.

15. Leaders relay all commands and signals from the platoon chain of command.

16. The PSG positions the base of fire element to observe and to provide supporting fires.

Note: Once the platoon has executed the React to Contact drill, the platoon leader quickly assesses the situation, for example, enemy size and location. He picks a COA. The platoon leader reports the situation to the company commander.

Figure 6-1. REACT TO CONTACT

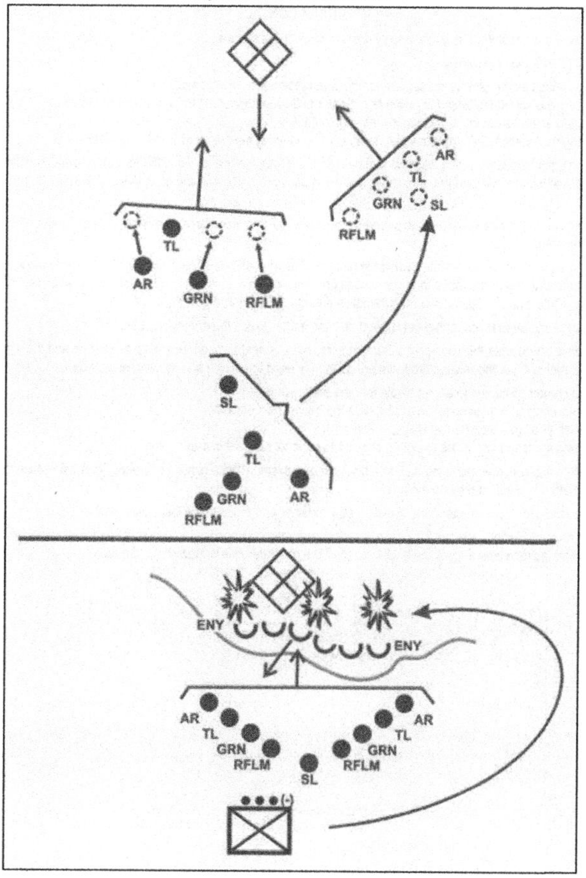

BATTLE DRILL 2
BREAK CONTACT

SITUATION: The squad or platoon is under enemy fire and must break contact.

REQUIRED ACTIONS: See Figure 6-2.

1. The platoon/squad leader directs fire support for the disengagement. He accomplishes this by—
   - Directing one squad/fire team in contact to support the disengagement of the remainder of the unit.
   - Considering the use of indirect fires for breaking contact.
   - Clearing the location, task, purpose, and method of conducting the fire mission with the platoon FO.
2. The platoon/squad leader orders a distance and direction, a terrain feature, or last ORP for the movement of the first squad/fire team. In conjunction with the platoon FO, the leader maintains accurate battle tracking of all friendly elements to facilitate quick clearance of fires.
3. The base of fire squad/team continues to suppress the enemy. The platoon/squad leader directs the platoon FO to execute the fire mission, if needed.
4. The moving squad/team assumes the overwatch position. The squad/team should use M203 grenade launchers, throw fragmentation and concussion grenades, and use smoke grenades to mask movement. The platoon/squad leader directs the platoon FO to execute smoke mission to screen friendly elements movement, if needed.
5. The moving squad/team takes up the designated position and engages the enemy position.
6. The platoon leader directs the base-of-fire element to move to its next location. Based on the terrain and the volume and accuracy of the enemy's fire, the moving fire team or squad may need to use fire and movement techniques.
7. The platoon/squad continues to bound away from the enemy until—
   - It breaks contact. It must continue to suppress the enemy as it does this.
   - It passes through a higher-level SBF position.
   - Its squad/fire teams are in the assigned position to conduct the next mission.
8. The leader should consider changing his unit's direction of movement once contact is broken. This will reduce the ability of the enemy to place effective indirect fires on the unit.
9. If the platoon/squad becomes disrupted, Soldiers stay together and move to the last designated rally point.
10. The platoon/squad leaders account for Soldiers, report, reorganize as necessary, and continue the mission. With the platoon FO, the platoon/squad develops a quick fire plan to support the hasty defense or new route of march.

Figure 6-2. BREAK CONTACT

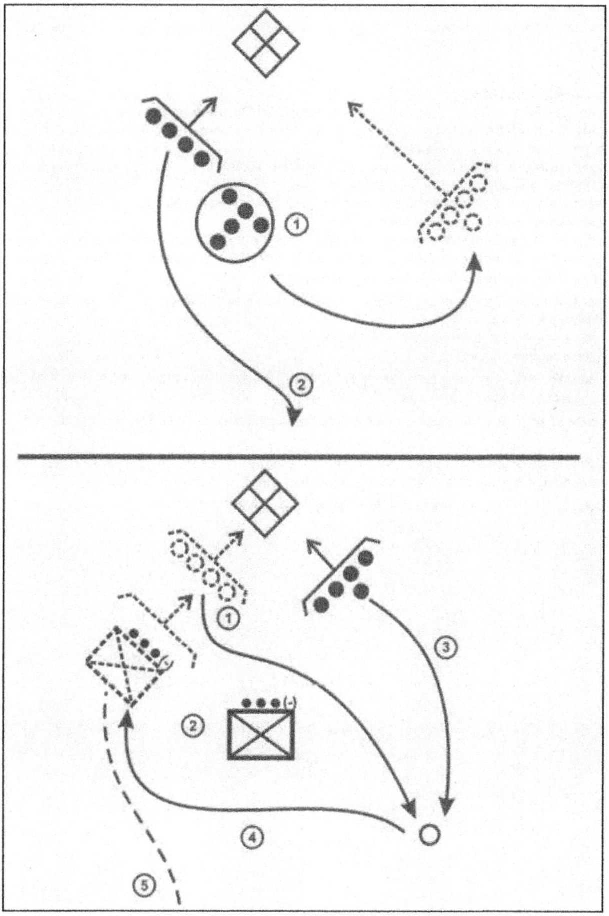

BATTLE DRILL 3
REACT TO AMBUSH

SITUATION: The squad or platoon enters a kill zone and the enemy initiates an ambush with a casualty-producing device and a high volume of fire.

REQUIRED ACTIONS: See Figure 6-3.

1. Near ambush (within hand-grenade range).
   a. Soldiers in the kill zone return fire immediately. How they go about this depends on the terrain. If—
      - Cover is not available, Soldiers immediately, without order or signal, assume prone position and throw concussion or fragmentation and smoke grenades.
      - Cover is available, without order or signal, Soldiers seek the nearest covered position, assume the prone position, and throw fragmentation or concussion and smoke grenades.

   b. Immediately after the explosion of the concussion or fragmentation grenades, Soldiers in the kill zone return fire and assault through the ambush position using fire and movement.
   c. Soldiers not in the kill zone identify the enemy's position. Fire is shifted as the personnel in the kill zone begin to assault.
   d. Soldiers in the kill zone continue the assault to eliminate the ambush or until contact is broken.
   e. The platoon/squad conducts consolidation and reorganization.

2. Far ambush (out of hand-grenade range). Soldiers receiving fire immediately return fire, take up covered positions, and suppress the enemy by—
   - Destroying or suppressing enemy crew-served weapons.
   - Sustaining suppressive fires.

3. Soldiers (squad/teams) not receiving fires move by a covered and concealed route to a vulnerable flank of the enemy position and assault using fire and movement techniques.

4. Soldiers in the kill zone continue suppressive fires and shift fires as the assaulting squad/team fights through the enemy position.

5. The platoon FO calls for and adjusts indirect fires as directed by the platoon leader. On order, he shifts or ceases fires to isolate the enemy position or to attack them with indirect fires as they retreat.

6. The platoon/squad leader reports, reorganizes as necessary, and continues the mission.

Figure 6-3. REACT TO AMBUSH

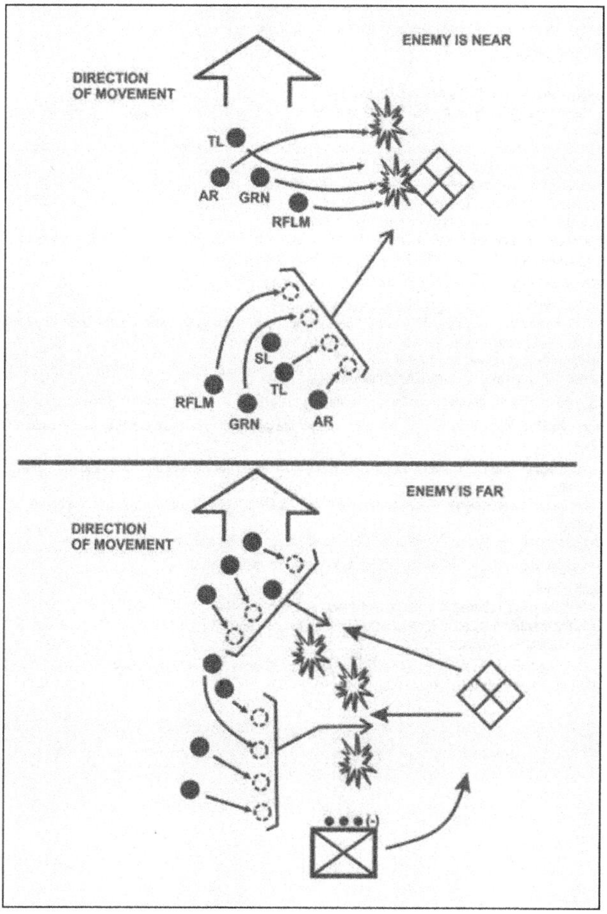

BATTLE DRILL 4
KNOCK OUT BUNKERS

SITUATION: While moving as part of a larger force, the platoon identifies the enemy in bunkers.

REQUIRED ACTIONS: See Figures 6-4 and 6-5.

1. The platoon initiates contact.
   a. The squad in contact establishes a base of fire.
   b. The platoon leader, his RTO, platoon FO, and one machine gun team move forward to link up with the squad leader of the squad in contact.
   c. The PSG moves forward with the second machine gun team and assumes control of the base-of-fire squad.
   d. The base-of-fire squad—
      - Destroys or suppresses enemy crew-served weapons first.
      - Obscures the enemy position with smoke (M203).
      - Sustains suppressive fires at the lowest possible level.
   e. The platoon leader directs platoon FO to call for and adjust indirect fires. The platoon leader in conjunction with the platoon FO maintains accurate battle tracking of all friendly elements to facilitate quick clearance of fires.
2. The platoon leader determines that he can maneuver by identifying—
   - Enemy bunkers, other supporting positions, and any obstacles.
   - Size of the enemy force engaging the platoon. The number of enemy automatic weapons, the presence of any vehicles, and the employment of indirect fires are indicators of enemy strength.
   - A vulnerable flank of at least one bunker.
   - A covered and concealed flanking route to the flank of the bunker.
3. The platoon leader determines which bunker is to be assaulted first and directs one squad (not in contact) to knock it out.
4. If necessary, the PSG repositions a squad, fire team, or machine gun team to isolate the bunker as well as to continue suppressive fires.
5. The assaulting squad, with the platoon leader and his RTO, move along the covered and concealed route and take action to knock out the bunker. The following occurs.
   a. On the platoon leader's signal, the support squad ceases or shifts fires to the opposite side of the bunker from which the squad is assaulting.
   b. At the same time, the platoon FO shifts indirect fires to isolate enemy positions.
6. The assaulting squad leader reports to the platoon leader and reorganizes his squad.
7. The platoon leader—
   - Directs the supporting squad to move up and knock out the next bunker.
   - Directs the assaulting squad to continue and knock out the next bunker.
   - Rotates squads as necessary.
8. The platoon leader reports, reorganizes as necessary, and continues the mission. The company follows up the success of the platoon attack and continues to assault enemy positions.

Figure 6-4. KNOCK OUT A BUNKER (SQUAD)

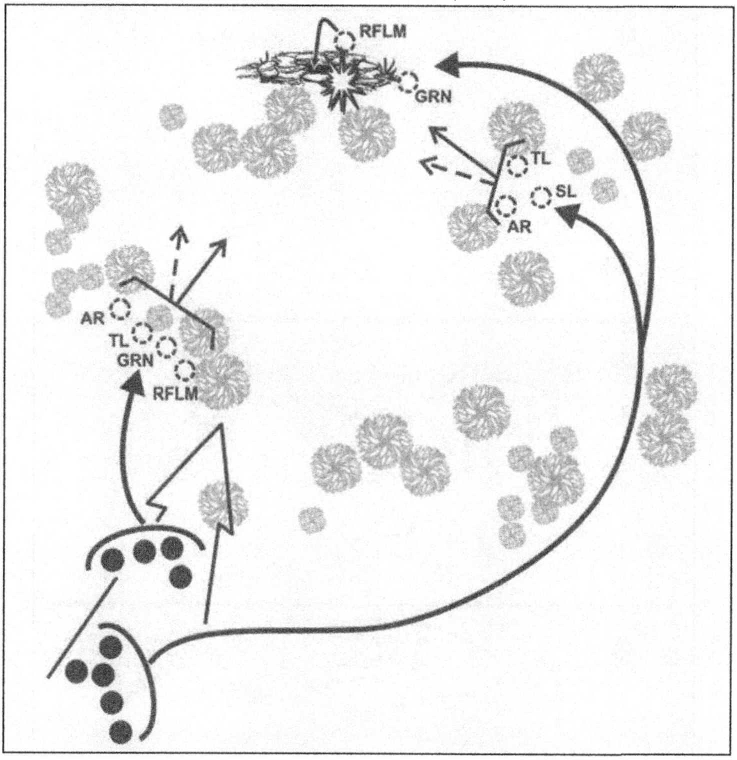

Figure 6-5. KNOCK OUT BUNKERS (PLATOON)

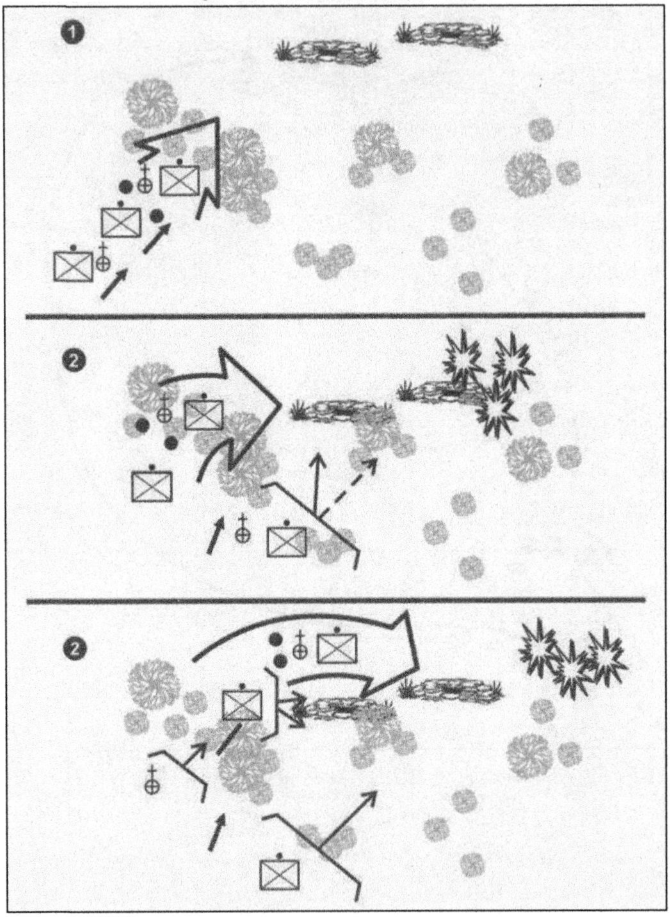

BATTLE DRILL 5
ENTER BUILDING/CLEAR ROOM

SITUATION: Operating as part of a larger force (in daylight or darkness), the squad is tasked to participate in clearing a building. The platoon leader directs the squad to enter the building or to clear a room. An entry point or breach has already been identified or will be created before initiating the entry. For a detailed discussion of urban entry breaching techniques, see FM 3.06-11. Enemy forces and noncombatants may or may not be present in the room and or building to be cleared.

REQUIRED ACTIONS: See Figures 6-6 through 6-14.

1. Platoon and squad leaders must consider the task and purpose they have been given and the method they are to use to achieve the desired results. They must operate IAW the ROE and must be aware of the affects that platoon weapons will have on the type and composition of the buildings.

2. To seize or gain control of a building may not always require committing troops into the structure or closing with the enemy. Before initiating this action and exposing members of the clearing squad to direct enemy contact and risking casualties, the platoon leader should consider/direct employment of all organic, crew-served, and supporting weapon systems onto the objective area to suppress and neutralize the threat, providing the mission, purpose, building composition, and ROE permit.

3. When conducting urban operations, Soldiers must be equipped at all times with a properly-mounted and immediately-useable NVD or light source to illuminate the immediate area.

Note: The following discussion assumes that only the platoon's organic weapons are supporting the infantry squad. Urban situations may require precise application of firepower. This situation is especially true of an urban environment in which the enemy is mixed with noncombatants. Noncombatants may be found in the room, which can restrict the use of fires and reduce the combat power available to a squad leader. His squad may have to operate in no-fire areas. ROE can prohibit the use of certain weapons until a specific hostile action takes place. All Soldiers must be aware of the ROE. Leaders must include the specific use of weapons in their planning for precision operations in urban terrain. Leaders should always consider the use of snipers or designated marksmen to apply precise fires to the objective.

4. Clearing team members must approach the entry point quickly, quietly, and in standard order. The squad leader must ensure he is in a position to control the actions of both fire teams. This approach preserves the element of surprise and allows for quick entry and domination of the room. If a breach is required, the order may be slightly modified based on the breach technique (FM 3-06.11). The members of the fire team are assigned numbers one through four. The rifleman is # 1 and the grenadier is # 3. If one member of the clearing team is armed with the SAW rather than an M16 rifle or carbine, he should be designated # 4. The team leader is normally the # 2 man, because he will have the most immediate decision to make as he enters the room.

5. The entire team enters the room as quickly and smoothly as possible and clears the doorway immediately. If possible, the team moves from a covered or concealed position already in their entry order. Ideally, the team arrives and passes through the entry point without having to stop. If the team must stop to effectively "*stack*" outside the entry point, it must do so only momentarily, and it must provide cover.

6. The door is the focal point of anyone in the room. It is known as the "*fatal funnel*," because it focuses attention at the precise point where the individual team members are the most vulnerable. Moving into the room quickly reduces the chance of anyone being hit by enemy fire directed at the doorway

7. For this battle to be effectively employed, each member of the team must know his sector of fire and how his sector overlaps and links with the sectors of the other team members. Team members do not move to the points of domination and then engage their targets. Rather, they engage targets as they calmly and quickly move to their designated points. Engagements must not slow movement to their points of domination. Team members may shoot from as short a range as 1 to 2 inches. They engage the most immediate threat first and then the less immediate threats in sector. Immediate threats are personnel who—

- Are armed and prepared to return fire immediately.
- Block movement to the position of domination.
- Are within arm's reach of a clearing team member.
- Are within 3 to 5 feet of the breach point.

8. The squad leader designates the assault team and identifies the location of the entry point for them.

9. The squad leader positions the follow-on assault team to provide overwatch and supporting fires for the initial assault team.

10. Assault team members use available cover and concealment, and move as close to the entry point as possible.

a. If an explosive breach or a ballistic breach is to be performed by a supporting element, the assault team remains in a covered position until the breach is made. They may provide overwatch and fire support for the breaching element if necessary.

b. All team members must signal one another that they are ready before the team moves to the entry point.

c If stealth is a consideration, team members avoid the use of verbal signals, which may alert the enemy and remove the element of surprise.

d. Assault team members must move quickly from the covered position to the entry point, minimizing the time they are exposed to enemy fire.

11. The assault team enters through the entry point or breach. Unless a grenade will be thrown prior to entry, the team should avoid stopping outside the point of entry.

a. If required, the # 2 man throws a grenade of some type (fragmentary, concussion, stun) into the room before entry.

b. The use of grenades should be consistent with the ROE and building structure. The grenade should be cooked off before being thrown, if applicable to the type of grenade used.

c. If stealth is not a factor, the thrower should sound off with a verbal indication ("*Frag out*," "*Concussion out*," "*Stun out*") that a grenade of some type is being thrown. If stealth is a factor, only visual signals are given as the grenade is thrown.

CAUTION

If walls and floors are thin, fragments from fragmentation grenades and debris created by concussion grenades can injure Soldiers outside the room. If the structure has been stressed by previous explosive engagements, the use of these grenades could cause it to collapse. Leaders must determine the effectiveness of these types of grenades compared to possibilities of harm to friendly troops.

12. On the signal to go, or after the grenade detonates, the assault team moves through the entry point (Figure 6-6 through Figure 6-9) and quickly takes up positions inside the room that allow it to completely dominate the room and eliminate the threat. Unless restricted or impeded, team members stop movement only after they have cleared the door and reached their designated point of domination. In addition to dominating the room, all team members are responsible for identifying possible loopholes and mouseholes in the ceiling, walls, and floor.

Note: Where enemy forces may be concentrated and the presence of noncombatants is highly unlikely, the assault team can precede their entry by throwing a fragmentation or concussion grenade (structure dependent) into the room, followed by bursts of automatic small-arms fire by the # 1 man as he enters. Carefully consider the ROE and building composition before employing this method.

13. The # 1 and # 2 men are initially concerned with the area directly to their front, then along the wall on either side of the door or entry point. This area is in their path of movement, and it is their primary sector of fire. Their alternate sector of fire is from the wall they are moving toward, back to the opposite far corner.

14. The # 3 and # 4 men start at the center of the wall opposite their point of entry and clear to the left if moving toward the left, or to the right if moving toward the right. They stop short of their respective team member (either the # 1 man or the # 2 man).

15. The team members move toward their points of domination, engaging all threat or hostile targets in sequence in their sector. Team members must exercise fire control and discriminate between hostile and noncombatant room occupants. (The most practical way to do this is to identify whether or not the target has a weapon in his or her hands.) Shooting is done without stopping, using reflexive shooting techniques. Because the Soldiers are moving and shooting at the same time, they must move using the careful hurry. Figure 6-10, shows all four team members at their points of domination for a room with a center door and their overlapping sectors of fire.

16. The first man (rifleman) enters the room and eliminates the immediate threat. He has the option of going left or right, normally moving along the path of least resistance to one of two corners. When using a doorway as the point of entry, the team uses the path of least resistance, which they determine initially based on the way the door opens. If the door opens inward, the first man plans to move away from the hinges. If the door opens outward, he plans to move toward the hinged side. Upon entering, his direction is influenced by the size of the room, the enemy situation, and furniture or other obstacles that hinder or channel his movement.

17. The direction each man moves in should not be preplanned unless the exact room layout is known. Each man should go in a direction opposite the man in front of him (Figure 6-6). Every team member must know the sectors and duties of each position.

18. As the first man goes through the entry point, he can usually see into the far corner of the room. He eliminates any immediate threat and continues to move along the wall if possible and to the first corner, where he assumes a position of domination facing into the room.

Figure 6-6. FIRST MAN ENTERS A ROOM, FOLLOWED BY TEAM LEADER

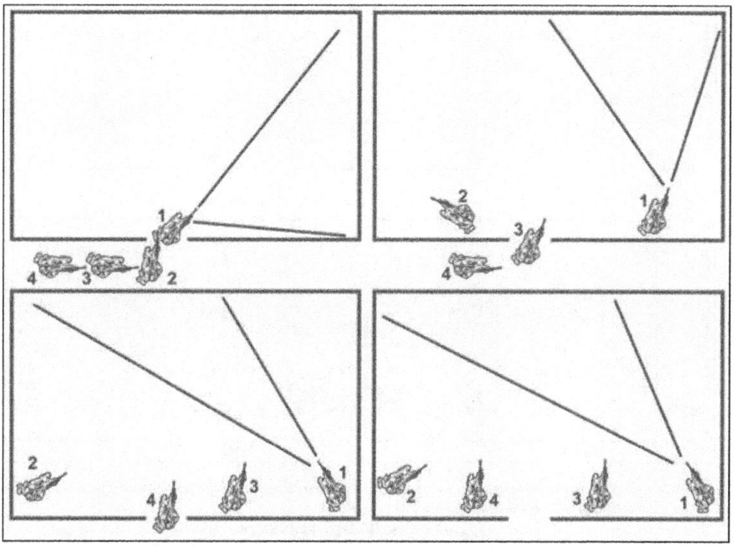

Note: Team members must always stay within 1 meter of the wall. If a team member finds his progress blocked by some object that will force him more than 1 meter from the wall, he must either step over it (if able) or stop where he is and clear the rest of his sector from where he is. If this action creates dead space in the room, the team leader directs which clearing actions to take once other members of the team have reached their points of domination.

19. The second man (normally the team leader), entering almost simultaneously with the first, moves in the opposite direction, following the wall (Figure 6-7). The second man must clear the entry point, clear the immediate threat area, clear his corner, and move to a dominating position on his side of the room. The second man must also immediately determine if he is entering a center door or corner door and act accordingly (Figure 6-7 and Figure 6-10).

Figure 6-7. SECOND MAN (TEAM LEADER) ENTERS A ROOM

20. The third man (normally the grenadier) simply goes opposite of the second man inside the room, moves at least 1 meter from the entry point, and takes a position that dominates his sector (Figure 6-8).

Figure 6-8. THIRD MAN ENTERS A ROOM

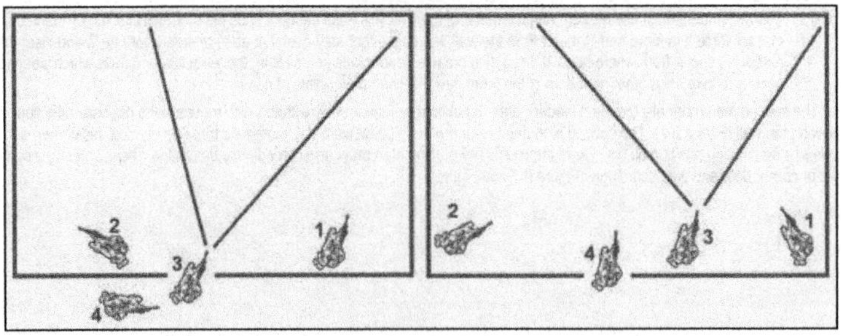

21. The fourth man (normally the SAW gunner) moves opposite of the third man, clears the doorway by at least 1 meter, and moves to a position that dominates his sector (Figure 6-9).

Figure 6-9. FOURTH MAN IN A ROOM

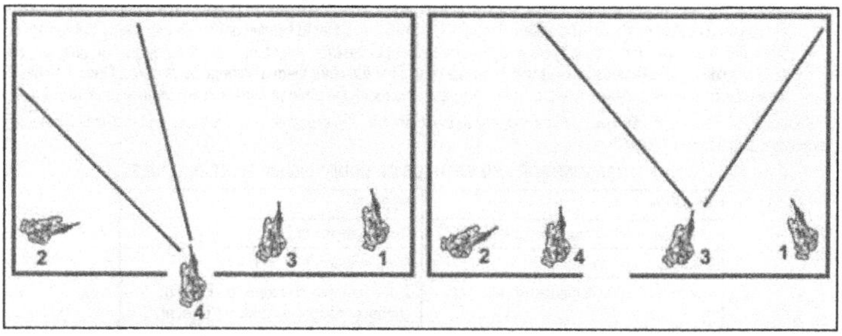

Note: If the path of least resistance takes the first man to the left, then all points of domination are the mirror image of those shown in the diagrams.

22. Points of domination should not be in front of doors or windows so team members are not silhouetted to the outside of the room (Figure 6-10). No movement should mask the fire of any of the other team members.

Figure 6-10. POINTS OF DOMINATION AND SECTORS OF FIRE, CENTER DOOR VERSUS CORNER DOOR

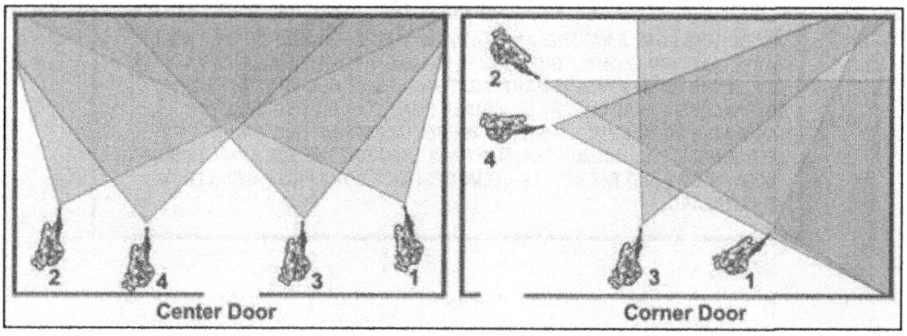

23. On order, any member of the assault team may move deeper into the room overwatched by the other team members. The team leader must control this action.

24. Once the room is cleared, the team leader signals to the squad leader that the room has been cleared. The squad leader marks the room IAW unit SOP (see FM 3.06-11). The squad leader determines whether or not his squad can continue to clear through the building.

25. The squad reorganizes as necessary. Leaders redistribute the ammunition. The squad leader reports to the platoon leader when the room is clear.

Note: If at any point a team member experiences a weapon malfunction in the presence of enemy combatants, he has to make an immediate decision. If the enemy is outside immediate danger distance from the Soldier, then the Soldier should drop to one knee, which indicates that he has experienced a weapon malfunction and prevents fratricide by ensuring that the next Soldier's fields of fire are clear. Once on a knee, the Soldier must remain there until the team leader directs him to stand up. If the kneeling Soldier corrects his weapon malfunction, he can continue to engage targets from his kneeling position. If, however, the enemy Soldier is so close that he presents an immediate threat, and if dropping to a knee would expose the US Soldier to immediate harm, then the US Soldier must try to subdue or disable the enemy Soldier. He can strike the enemy with the muzzle of his weapon or a bayonet to the face, throat, or chest. Or, he could grapple with the enemy soldier and take him to the ground immediately to clear the other team members' fields of fire. Once they have cleared their sectors and eliminated any other threats in the room, the other members of the clearing team assist.

26. Although this battle drill is effective, leaders might have to modify it for the situation. Some reasons and methods for modifying the technique are shown in Table 6-1.

Table 6-1. REASONS FOR AND METHODS OF MODIFYING ENTRY TECHNIQUES

REASON	METHOD
Objective rooms are consistently small	Clear in teams of 2 or 3
Shortage of personnel	Clear in teams of 2 or 3
Enemy poses no immediate threat	2 or 3 men search rooms to ensure no enemy or noncombatants are present
No immediate threat, speed is critical	1 to 3 men visually search each room

27. When full four-man teams are not available for room clearing, two- and three-man teams can be used. If the # 1 or # 2 man discovers that the room is very small, he can yell, "*Short room*" or "*Short*," which tells the # 3 or # 4 man (whoever following the # 1 or # 2 man) to stay outside the room. Figures 6-11 and 6-12 show the points of domination and sectors of fire for a two-man clearing team. Figures 6-13 and 6-14 show the actions for a three-man team.

DANGER

RICOCHETS POSE A HAZARD. ALL SOLDIERS MUST BE AWARE OF THE TYPE OF WALL CONSTRUCTION OF THE ROOM BEING CLEARED. THE WALLS OF AN ENCLOSED ROOM PRESENT MANY RIGHT ANGLES. COMBINED WITH HARD SURFACES SUCH AS CONCRETE, A BULLET MAY CONTINUE TO RICOCHET AROUND A ROOM UNTIL ITS ENERGY IS SPENT. AFTER HITTING THREAT PERSONNEL, BALL AMMUNITION MAY PASS THROUGH THE BODY AND RICOCHET. BODY ARMOR AND THE KEVLAR HELMET PROVIDE SOME PROTECTION FROM THIS HAZARD.

Figure 6-11. POINTS OF DOMINATION AND SECTORS OF FIRE (TWO-MAN TEAM, CENTER DOOR)

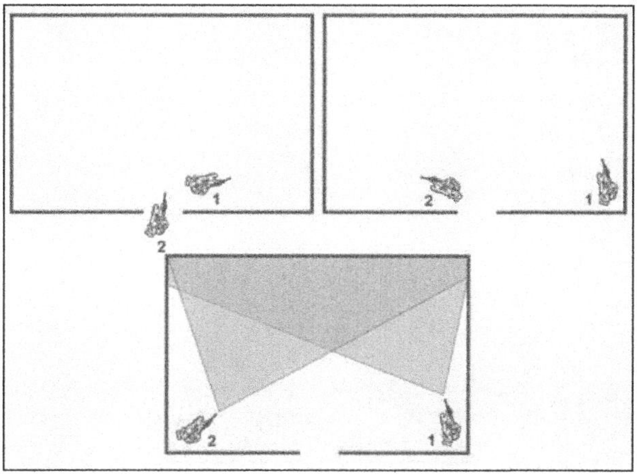

Figure 6-12. POINTS OF DOMINATION AND SECTORS OF FIRE (TWO-MAN TEAM, CORNER DOOR)

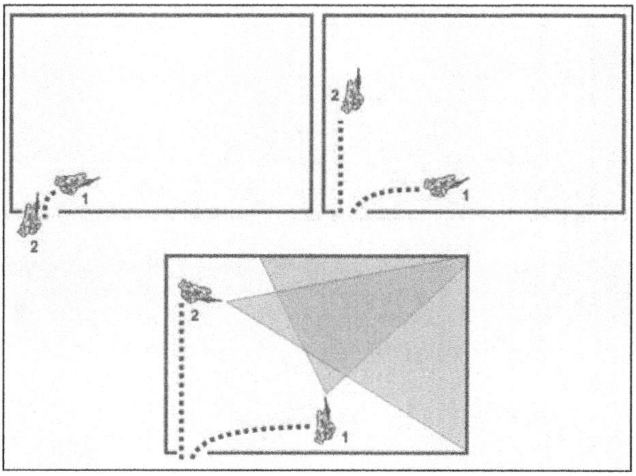

Figure 6-13. POINTS OF DOMINATION AND SECTORS OF FIRE (THREE-MAN TEAM, CENTER DOOR)

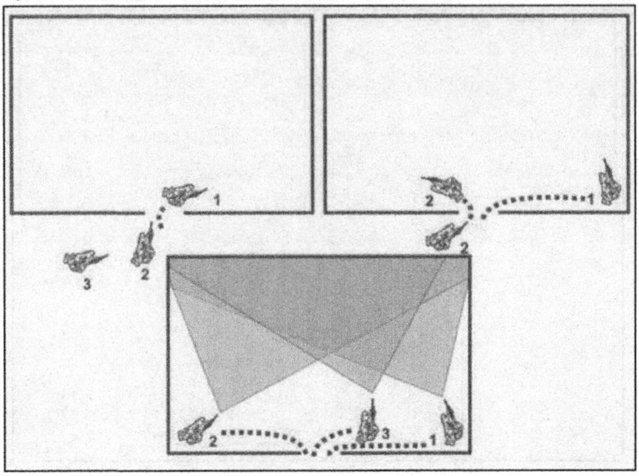

Figure 6-14. POINTS OF DOMINATION AND SECTORS OF FIRE (THREE-MAN TEAM, CORNER DOOR)

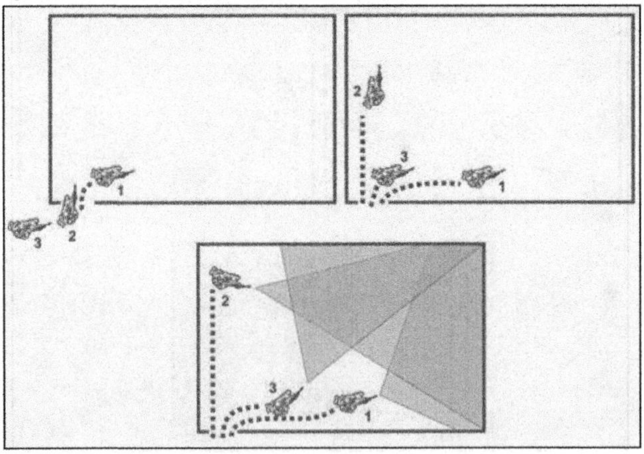

28. While moving through a building Soldiers may encounter the following architectural building features.

a. **Multiple Team/Multiple Rooms**. Figure 6-15 shows this.

STEP 1: First team enters and clears Room 1.

STEP 2: Squad leader determines the direction from which the second clearing team must enter Room 1, based on the location of the Room 2 entry point.

STEP 3: First team collapses inward to allow the second team to move into the room.

STEP 4: Second team 'stacks left' and prepares to enter Room 2.

Figure 6-15. MULTIPLE TEAMS / MULTIPLE ROOMS

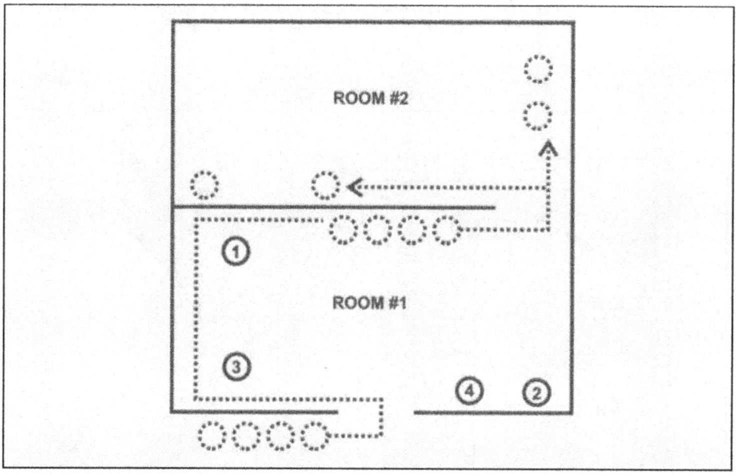

b. **Open Stairwell**. This is a gap between flights of stairs that allows a person to look up and down between the flights. Figure 6-16 shows an open stairwell.

STEP 1: The # 1 man pulls security on the highest point he can see / engage.

STEP 2: The # 2 man moves up the stairs on the inside with the # 3 man to a point that he can see / engage the next landing, where he turns around and continues to move up to the next landing.

STEP 3: The # 3 man moves up the stairs with the # 2 man on the outside and engages the threat on the immediate landing.

STEP 4: The # 4 man moves up the stairs with the # 1 man, on the squeeze, the # 2 man turns around to engage the next landing.

STEP 5: The flow continues with the # 2 man picking up the sector of the # 1 man had. The # 3 man picks up where the # 2 man was. The # 4 man picks up where the # 3 man was. The # 1 man picks up where the # 4 man was.

Note: Most stairwells require a second team.

Figure 6-16. OPEN STAIRWELL

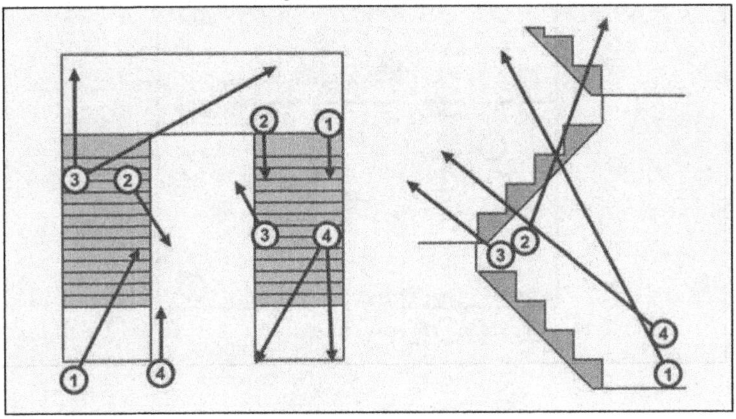

c. **Closed Stairwell**. This is any stairwell with walls separating the flights of stairs. Figure 6-17 shows a closed stairwell.

STEP 1: The # 1 man checks high to ensure there is no opening on the landing or between the stairs.

STEP 2: The # 2 man pulls long security to the next bend or landing.

STEP 3: The # 1 or # 3 man moves up the steps with the # 2 man. As they approach the corner, the # 2 man signals his presence by tapping the # 1 man on the shoulder.

STEP 4: Keying off the # 1 man's movement, the other men simultaneously break around the corner.

STEP 5: If no fire is received, the # 2 man moves to the opposite wall. Both men continue to move up until they reach their objective.

STEP 6: The # 3 and # 4 men will continue to move 3 to 4 steps behind.

Note: Avoid getting locked into a security position such as inside a stairwell. Also, avoid spreading yourselves thin or getting separated by more than one floor of stairs.

Figure 6-17. CLOSED STAIRWELL

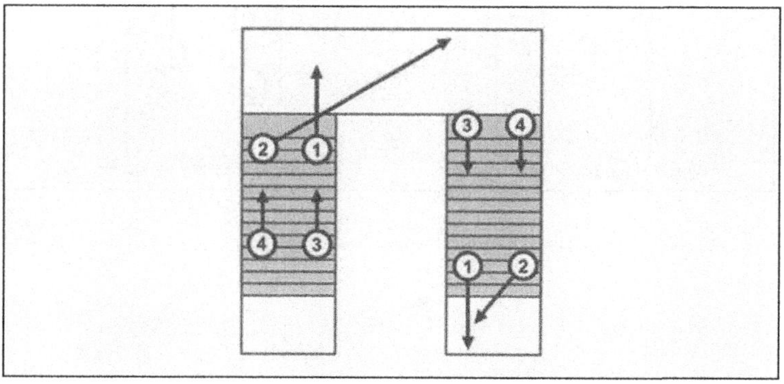

d. **Hallway Movement**. Clearing team(s) move down the hallway using the frontal security (cross cover) technique (Figure 6-18).

Figure 6-18. HALLWAY MOVEMENT

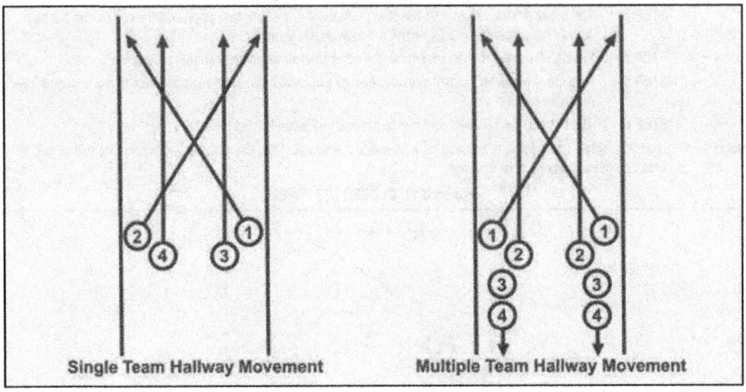

e. **T-Shaped Intersection**. See Figure 6-19. This technique can incorporate the dynamic corner clear (Figure 6-20).

STEP 1: Each # 1 man goes to a knee covering his sector.

STEP 2: On a predetermined signal, each two-man team will break the corner, picking up their sectors of fire.

Figure 6-19. T-SHAPED INTERSECTION

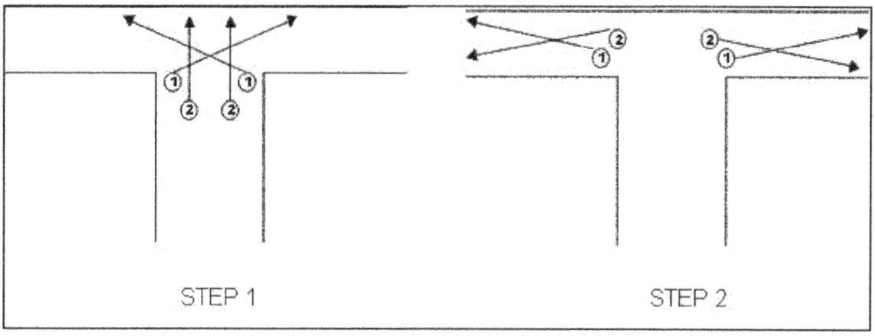

f. **Dynamic Corner**. (See Figure 6-20.)

STEP 1: The #1 and #2 man, as they approach the corner, have to clear. Do not slow down.

STEP 2: The #2 man will tap the #1 man on the shoulder about 2-3 meters away from the corner letting the #1 man know the #2 man is with him.

STEP 3: Keying off of the #1 man's movement, they both break the corner simultaneously.

STEP 4: The #1 man goes low to a knee, the #2 man stays high.

STEP 5: If the Rangers are not receiving fire the #2 man rabbits, or moves, to the far side.

STEP 6: The #1 and #2 man take up sectors of fire.

STEP 7: The #3 and #4 man take long security in the direction of movement.

Figure 6-20. DYNAMIC CORNER

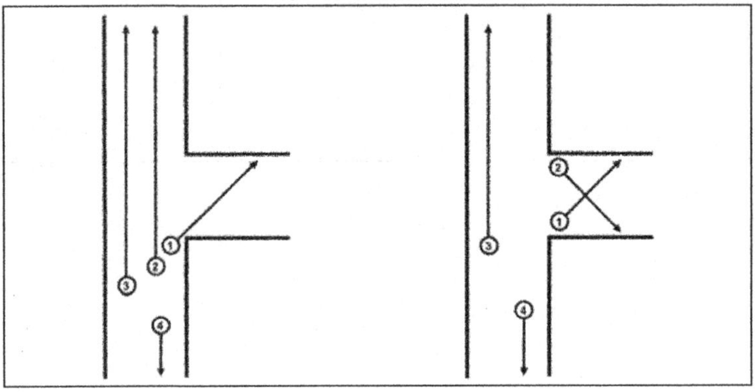

g. **Three-Way Intersection**. (See Figure 6-21.)

STEP 1: The #1 through #4 man will use one of the corner clearing techniques to clear the intersection.

STEP 2: When the intersection is secure, the #5 and #6 man move through the intersection and enter either more hallway or a room. If they are entering more hallway, the #1 through #4 man provide cross coverage as the #5 and #6 man move through the intersection into more hallway. If they are moving into a room, the #5 and #6 man move to one side of the door and signal the #1 and #3 or #2 and #4 man to enter the stack depending on which side of the door they stack on. The remaining members of the team, not in the stack, will continue to provide security down the hallway.

STEP 3: If the #5 and #6 man move into more hallway the #1 through #4 man will enter the stack and proceed down the hallway with the #3 and #4 man providing rear security.

Figure 6-21. THREE WAY INTERSECTION

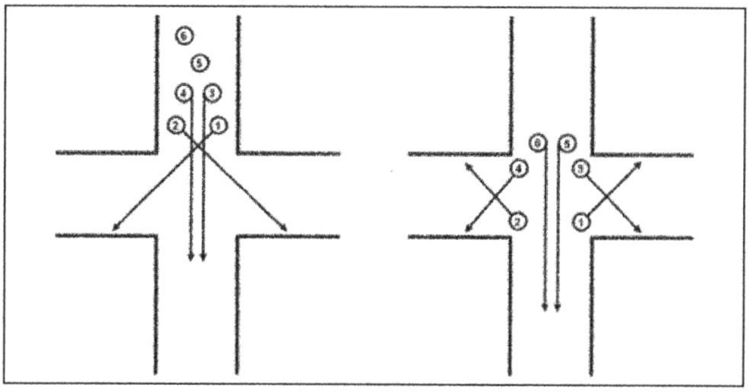

BATTLE DRILL 6
ENTER/CLEAR A TRENCH

SITUATION: The platoon is attacking as part of a larger force and identifies the enemy in a trench line. The platoon deploys and establishes a base of fire. The platoon leader determines that he has sufficient combat power to maneuver and assault the trench line.

REQUIRED ACTIONS: See Figure 6-22 and Figure 6-23.

1. The platoon leader directs one squad to enter the trench and secure a foothold.
2. The platoon leader designates the entry point of the trench line and the direction of movement once the platoon begins clearing.
3. The PSG positions Soldiers and machine guns to suppress the trench and isolate the entry point.
4. Platoon leader directs platoon FO to initiate fire mission, if necessary, in support of assault. They maintain accurate battle tracking of all friendly elements to facilitate quick clearance of fires. The platoon FO ceases or shifts fires to isolate the OBJ as assault team advances.
5. The assaulting squad executes actions to enter the trench and establish a foothold. The squad leader signals to the platoon leader that the foothold is secure, and the follow on elements can move into the trench. The squad leader remains at the entry point and marks it. The platoon follows the success of the seizure of the foothold with the remainder of the platoon as part of the platoon actions to clear a trench line.
6. The platoon leader moves into the trench with the assaulting squad.

7 The platoon leader directs one of the base-of-fire squads to move into the trench and begin clearing it in the direction of movement from the foothold.

8. The base-of-fire element repositions as necessary to continue suppressive fires.

9. The assaulting squad passes the squad that has secured the foothold and executes actions to take the lead and clear the trench.

a. The squad leader designates a lead fire team and a trail fire team.

b. The lead fire team and the squad leader move to the forward most secure corner or intersection. The squad leader tells the team securing that corner or intersection that his squad is ready to continue clearing the trench. The trail fire team follows, maintaining visual contact with the last Soldier of the lead team.

Notes: 1. The fire support element must be able to identify the location of the lead fire team in the trench at all times.

2. Throughout this battle drill, the team leader positions himself at the rear of the fire team to have direct control (physically, if necessary) of his Soldiers. Other Soldiers in the fire team rotate the lead to change magazines and prepare grenades. Rotating the lead provides constant suppressive fires down the trench and maintains the momentum of the attack as the squad clears the trench.

c. The lead fire team passes the element securing the foothold. The following then occurs.

(1) The lead Soldier of the fire team moves abreast of the Soldier securing the corner or intersection, taps him, and announces, "*Taking the lead.*"

(2) The Soldier securing the corner or intersection acknowledges that he is handing over the lead by shouting, "*Okay!*" He allows the fire team to pass him.

d. The lead fire team starts clearing in the direction of movement. They arrive at a corner or intersection. The following then occurs.

(1) Allowing for cook-off (2 seconds maximum) and shouting, "*Frag out*," the second Soldier prepares and throws a grenade around the corner.

(2) Upon detonation of the grenade, the lead Soldier moves around the corner firing three-round bursts and advancing as he fires. The entire fire team follows him to the next corner or intersection.

e. The squad leader—

- Follows behind the team.
- Ensures that the trailing fire team moves up and is ready to pass the lead at his direction.
- Rotates fire teams as necessary to keep his Soldiers fresh and to maintain the momentum of the attack.
- Requests indirect fires, if required, through the platoon leader. The squad leader also directs the employment of the M203 to provide immediate suppression against positions along the trench line.
- Ensures fire teams maintain sufficient interval to prevent them from being engaged by the same enemy fires.

f. At each corner or intersection, the lead fire team performs the same actions previously described.

g. If the lead Soldier finds that he is nearly out of ammunition before reaching a corner or intersection, he announces, "*Ammo*," and then--

(1) The lead Soldier stops and moves against one side of the trench, ready to let the rest of the team pass. He continues to aim his weapon down the trench in the direction of movement.

(2) The next Soldier ensures that he has a full magazine, moves abreast of the lead Soldier, taps him, and announces, "*Taking the lead.*"

(3) The lead Soldier acknowledges that he is handing over the lead by shouting, "*Okay.*" Positions rotate and the squad continues forward.

h. The trailing fire team secures intersections and marks the route within the trench as the squad moves forward. The trailing fire team leader ensures that follow-on squads relieve his buddy teams to maintain security.

i. The squad leader reports the progress of the clearing operation. The base-of-fire element must be able to identify the location of the lead fire team in the trench at all times.

10. The platoon leader rotates squads to keep the Soldiers fresh and to maintain the momentum of the assault.

11. The PSG calls forward ammunition resupply and organizes teams to move it forward into the trench.

12. The base-of-fire element ensures that all friendly forces move into the trench only through the designated entry point to avoid fratricide.

13. The platoon leader reports to the company commander that the trench line is secured, or he is no longer able to continue clearing. If trench line is secured, then platoon leader directs platoon FO to develop a fire plan to support the defense of the platoon position.

DANGER

FIRE TEAMS: MAINTAIN SUFFICIENT INTERVALS TO AVOID ENGAGEMENT BY THE SAME ENEMY FIRES.

Figure 6-22. ENTER A TRENCH (SQUAD)

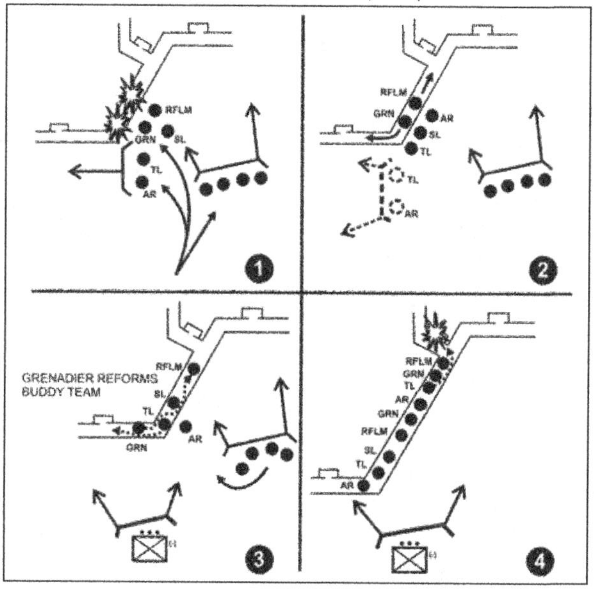

Figure 6-23. CLEAR A TRENCH LINE (PLATOON)

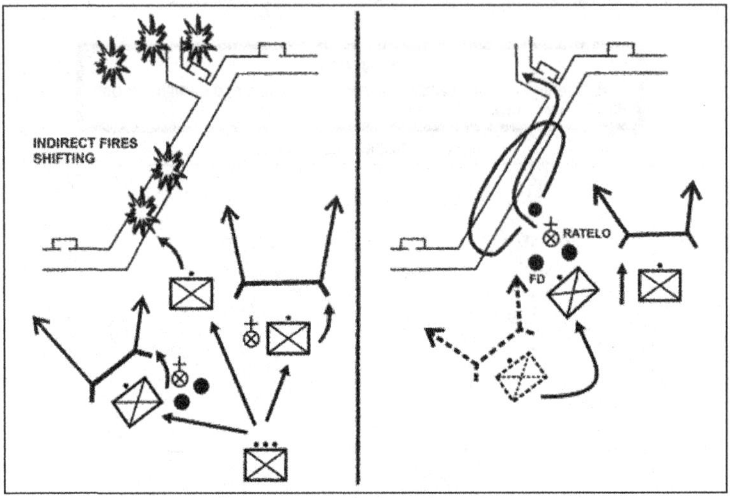

BATTLE DRILL 7
CONDUCT INITIAL BREACH OF A MINED WIRE OBSTACLE (PLATOON)

SITUATION: The platoon is operating as part of a larger force. The lead squad identifies a wire obstacle reinforced with mines that cannot be bypassed and enemy positions on the far side of the obstacle.

REQUIRED ACTIONS: See Figure 6-24 and Figure 6-25.

1. The squad in contact reacts to contact.
2. The platoon gains suppressive fires. The following then occurs.
   a. The squad in contact establishes a base-of-fire position.
   b. The platoon leader, his RTO, platoon FO, and the squad leader of the next squad with one machine gun team move forward to link up with the squad leader of the squad in contact.
3. The platoon leader determines that he can maneuver by identifying—
   - The obstacle and enemy positions.
   - The size of the enemy force engaging the squad, for example, the number of enemy automatic weapons, the presence of any vehicles, and the employment of indirect fires are indicators of enemy strength.
   - A breach point.
   - A covered and concealed route to the breach point.
4. The platoon leader directs the squad in contact to support the movement of another squad to the breach point. He indicates the base-of-fire position and the route to it, the enemy position to be suppressed, and the breach point and route the rest of the platoon will take to it. The platoon leader also clears the location, task, purpose, and method of conducting the fire mission with the platoon FO.

5. On the platoon leader's signal, the base-of-fire squad—
- Destroys or suppresses enemy weapons that are firing effectively against the platoon.
- Obscures the enemy position with smoke.
- Continues suppressive fires at the lowest possible level.

6. The platoon leader designates one squad as the breach squad and the remaining squad as the assault squad once the breach has been made. The assault squad may add its fires to the base-of-fire squad. Normally, it follows the covered and concealed route of the breach squad and assaults through immediately after the breach is made.

7. The base-of-fire squad moves to the breach point and establishes a base of fire.

8. The PSG moves forward to the base-of-fire squad with the second machine gun and assumes control of the squad.

9. The platoon leader leads the breach and assault squads along the covered and concealed route.

10. The platoon FO calls for and adjusts indirect fires as directed by the platoon leader. The platoon leader in conjunction with the platoon FO maintains accurate battle tracking of all friendly elements to facilitate quick clearance of fires.

11. The breach squad executes actions to breach the obstacle.

a. The squad leader directs one fire team to support the movement of the other fire team to the breach point.

b. The squad leader identifies the breach point.

c. The base-of-fire element continues to provide suppressive fires and isolates the breach point.

d. The breaching fire team, with the squad leader, moves to the breach point using the covered and concealed route. From there, the following takes place:

(1) The squad leader and breaching fire team leader employ smoke grenades to obscure the breach point. The platoon base-of-fire element shifts direct fires away from the breach point and continues to suppress key enemy positions. The platoon FO ceases indirect fires or shifts them beyond the obstacle.

(2) The breaching fire team leader positions himself and the automatic rifleman on one flank of the breach point to provide close-in security.

(3) The grenadier and rifleman of the breaching fire team probe for mines and cut the wire obstacle, marking their path as they proceed. If available, a bangalore is preferred.

(4) Once the obstacle has been breached, the breaching fire team leader and the automatic rifleman move to the far side of the obstacle and take up covered and concealed positions with the rifleman and grenadier. The team leader signals to the squad leader when they are in position and ready to support.

(5) The squad leader signals the base-of-fire team leader to move his fire team up and through the breach. He then moves through the obstacle and joins the breaching fire team, leaving the grenadier (or anti-armor specialist) and rifleman of the supporting fire team on the near side of the breach to guide the rest of the platoon through.

(6) Using the same covered and concealed route as the breaching fire team, the base-of-fire team moves through the breach and takes up covered and concealed positions on the far side.

12. The squad leader reports the situation to the platoon leader and posts guides at the breach point.

13. The platoon leader leads the assault squad through the breach in the obstacle and positions them beyond the breach to support the movement of the remainder of the platoon or assaults the enemy position covering the obstacle. Platoon leader directs the platoon FO to shift indirect fires off the enemy position.

14. The breaching squad continues to widen the breach to allow vehicles to pass through.

15. The platoon leader reports the situation to the company commander, and then he directs his base-of-fire squad to move up and through the obstacle. The platoon leader appoints Soldiers to guide the company through the breach point.

16. The PSG brings the remaining elements forward and through the breach on the platoon leader's command.

17. The company follows up the success of the platoon as it conducts the breach and continues the assault against the enemy positions.

Figure 6-24. CONDUCT INITIAL BREACH OF A MINED WIRE OBSTACLE (SQUAD)

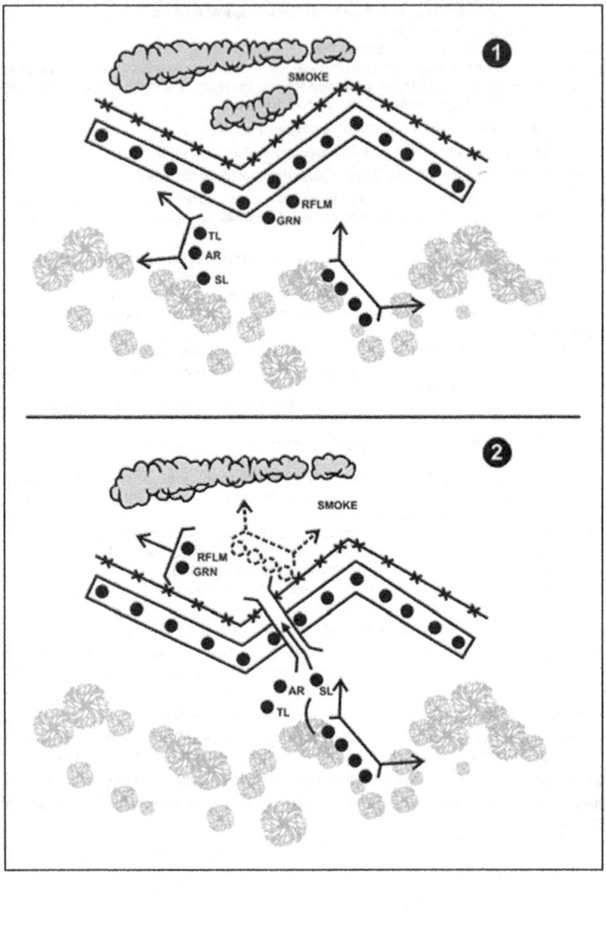

Figure 6-25. CONDUCT INITIAL BREACH OF A MINED WIRE OBSTACLE (PLATOON)

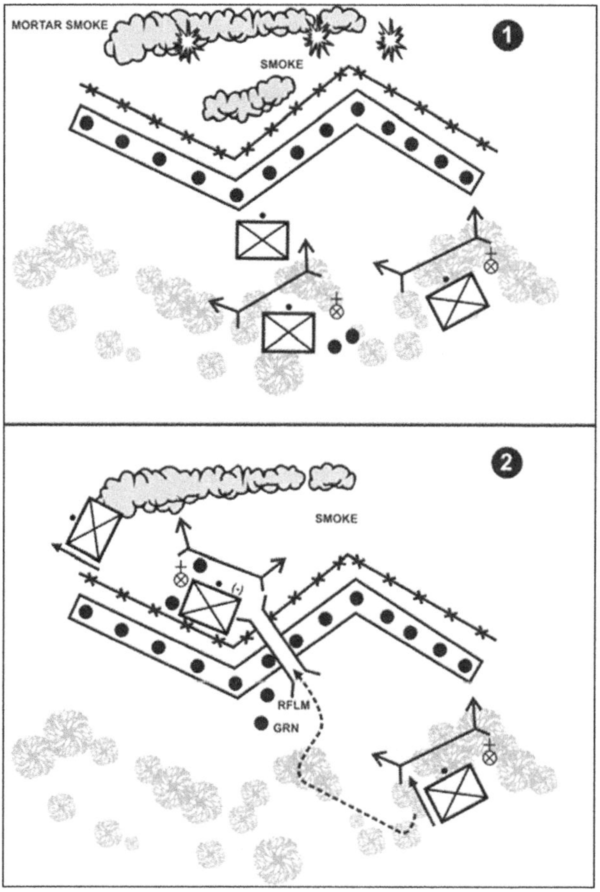

BATTLE DRILL 8
REACT TO INDIRECT FIRE

SITUATION: A platoon or squad while moving or at the halt (not dug in) receives indirect fire.

REQUIRED ACTIONS: NA

Any Soldier announces, "*Incoming*." Then, if the platoon/squad is moving, the following takes place:

1. The platoon/squad leader gives direction and distance for the platoon/squad to move to a rally point by ordering direction and distance, for example, "*Four o'clock, three hundred meters*."

2. Platoon/squad members move rapidly along the direction and distance to the rally point.

3. At the rally point, the leader immediately accounts for personnel and equipment, and forms the platoon/squad for a move to an alternate position.

4. The senior leader submits a SITREP to higher headquarters.

5. If the platoon/squad is halted (not dug in) or is preparing to move because they hear incoming artillery, the following takes place:

   a. The platoon/squad leader gives direction and distance for the platoon/squad to move to a rally point by ordering direction and distance, for example, "*Four o'clock, three hundred meters*."

   b. Platoon/squad members secure all mission-essential equipment and ammunition and move rapidly along the direction and distance to the rally point.

   c. At the rally point, the leader immediately accounts for personnel and equipment, and forms the platoon/squad for a move to an alternate position.

   d. The senior leader submits a SITREP to higher headquarters.

Note: If platoon/squad members are in defensive (dug in) positions, then members will remain in those positions if appropriate. Senior leader submits SITREP.

NOTES

Chapter 7
COMMUNICATIONS

The basic requirement of combat communications is to provide rapid, reliable, and secure interchange of information. Communications are a vital aspect in successful mission accomplishment. The information in this chapter helps the Ranger squad/platoon maintain effective communications and correct any radio antenna problems.

Section I. INTRODUCTION

This section discusses military frequency modulated radios and automated net control devices (ANCDs).

7-1. **MILITARY FREQUENCY MODULATED RADIOS**. Military frequency modulated (FM) radios are composed of a receiver and a transmitter. Together, the receiver-transmitter has many capabilities and features that enable you to perform your mission more effectively. The radio can operate in Single Channel mode or frequency hopping mode. The radio has about 2,320 SC Channels and includes voice and digital communication. The operating voltage for the manpack radio is 13.5 volts from the primary battery. The range of the manpack radio is 5-10 KM on Hi power. This range is based upon line of sight and is derived from averages achieved under normal atmospheric and weather conditions. Ranges depend upon location, sighting, weather, and surrounding noise level, among other factors. the automated net control device (ANCD) is an essential tool. It replaces the signal operating instructions booklet. The ANCD has many capabilities. The main purpose of the ANCD for your mission is to obtain signal operating instructions electronically. The following discusses assembly, operation, and troubleshooting for the AN/PRC-119; provides step-by-step instructions to obtain SOI from an ANCD; and explains how to repair broken antennas:

a. **Manpack Radio Assembly**. To assemble a manpack radio you must first check and install a battery.
  (1) Inspect the battery box for dirt or damage.
  (2) Stand RT on front panel guards.
  (3) Check battery life condition (you will be using the rechargeable BB-390 batteries).
  (4) Place battery in box.
  (5) Close battery cover, and secure using latches.
  (6) Return radio to upright position.
  (7) If used battery was installed, enter the battery life condition into radio by performing the following:
    (a) Set **FCTN** to **LD**.
    (b) Press **BAT**, then **CLR**.
    (c) Enter number recorded on side of battery.
    (d) Press **STO**.
    (e) Set **FCTN** to **SQ ON**.

b. **Antenna**.
  (1) Inspect whip antenna connector on antenna and on radio for damage.
  (2) Screw whip antenna into base.
  (3) Hand tighten.
  (4) Carefully mate antenna base with RT ANT connector.
  (5) Hand tighten.
  (6) Position antenna as needed by bending goose neck.

Note: Keep antenna straight, if possible. If the antenna is bent to a horizontal position, it may be necessary to turn the radio in order to receive and transmit messages.

c. **Handset**.
  (1) Inspect the handset for damage.
  (2) Push handset on **AUD/DATA** and twist clockwise to lock in place.

d. **Field Pack**.
  (1) Place radio transmitter (RT) in field pack with antenna on the left shoulder.
  (2) Fold top flap of field over RT and secure flap to field pack using straps and buckles.

e. **Presets**.
  (1) Set **CHAN** to **1**.
  (2) Set **MODE** to **SC**.
  (3) Set **RF PWR** to **HI**.

(4) Set **VOL** to mid range.
(5) Set **DIM** full clockwise.
(6) Set **FCTN** to **LD**.
(7) Set **DATA RATE** to **OFF**.

f. **Single Channel Loading Frequencies**.
(1) Obtain Ranger SOI.
(2) Set **FCTN** to **LD**.
(3) Set **MODE** to **SC**.
(4) Set **CHAN** to **MAN**, **Cue**, or desired channel (**1** to **6**) where frequency is to be stored.
(5) Press **FREQ** (display will show "**00000**," or the frequency the RT is currently set to).
(6) Press **CLR** (display will show five lines).
(7) Enter the number of the new frequency.
(8) If you make a mistake with a number, press **CLR**.
(9) Press **STO** (display will blink).
(10) Set **FCTN** to **SQ ON**.

g. **Clearing of Frequencies**.
(1) Set **MODE** to **SC**.
(2) Set **CHAN** to **MAN**, **Cue** or the channel whose frequency is to be cleared.
(3) Press **FREQ**.
(4) Press **CLR**.
(5) Press **Load**; then press **STO**.
(6) Set **FCTN** to **SQ ON**.

h. **Scanning of Multiple Frequencies**.
(1) Load all desired frequencies using "Single Channel Loading Frequencies" instructions (see 7-2h).
(2) Set **CHAN** to **CUE**.
(3) Set **SC** to **FH**.
(4) Set **FCTN** to **SQ ON**.
(5) Press **STO** (display will say **SCAN**).
(6) Press **8**. You will now be able to scan more than one frequency.

7-2. **AUTOMATED NET CONTROL DEVICE**. The RTO retrieves all necessary COMSEC and information from the ANCD. To retrieve the SOI from an ANCD:

a. Press the **ON** button on the ANCD keypad.
b. Press the letter lock button to unlock the keys on the ANCD.
c. Press **Main Menu** key.
d. Once you press the main menu key, this is what should show on the screen.

Appl Date Time Setup
Util Bit.

e. Using the arrow keys, scroll over to **Appl** (as shown above) and press the enter button on the keypad.

f. Once you have entered into the **Appl** (applications) menu using the arrow keys scroll over to **SOI** and press the enter button on the key pad. For example--

RDS SOI Radio.
(Appl) <--*(Indicates that you are in the applications menu).*

SOI menu will look like this:
Qref Group Net Sufx Pyro
Tmpd Set C/S Find Memo

g. Using the arrow keys scroll over to **Set** and press the enter button on the keypad.
h. Press enter on choose once in the **Set** menu.

> *EXAMPLE:*
> SELECT:
> **Choose Send Receive**.

i. Once in the **Choose** menu press the enter key on set **1-5** or **6-10**.

j. After you choose your selection out of the **Set** menu, the ANCD will automatically return back to the **SOI** menu.

k. Using the keypad scroll over to **Tmpd** and enter the number out of the set you need for that day. After this operation, the ANCD will automatically return back to the **SOI** menu.

l. Using the arrow keys on the keypad scroll over to **Net** and press the enter button on the keypad. The Net menu should look like the following example:

> *EXAMPLE:*
> T05 <--(time period) 1 <--(platoon) / A <--(company) /A3Q <--(prefix).
> C 60.000.

Note: When in **Net** menu, this is where you will obtain your frequency and the first part of your call sign.

m. After you obtain the information you need out of **Net**, press the **Abort** button on the keypad. This will return you to the **SOI** menu.

n. Using the keypad, scroll over to **Sufx** and press the **Enter** button.

o. Once you are in the **Sufx** menu, use the up and down arrow keys on the ANCD to locate your two-digit designator. These two digits will go at the end of your prefix; your prefix plus the two digits equal your call sign.

p. Once you are done, press the **OFF** button on the ANCD to end operation.

Section II. ANTENNAS

This section discusses repair techniques, construction and adjustment, field-expedient antennas, antenna length and orientation, and improvement of marginal communications.

7-3. **REPAIR**. Antennas are sometimes broken or damaged, causing either a communications failure or poor communications. If a spare antenna is available, the damaged antenna is replaced. When there is no spare, the Ranger squad/platoon may have to construct an emergency antenna. The following paragraphs contain suggestions for repairing antennas and antenna supports and the construction and adjustment of emergency antennas.

> **DANGER**
>
> **SERIOUS INJURY OR DEATH CAN RESULT FROM CONTACT WITH THE RADIATING ANTENNA OF A MEDIUM-POWER OR HIGH-POWER TRANSMITTER. TURN THE TRANSMITTER OFF WHILE ADJUSTING THE ANTENNA.**

a. **Whip Antennas**. When a whip antenna is broken into two sections, the part of the antenna that is broken off can be connected to the part attached to the base by joining the sections.

(1) Use the method shown in A, Figure 7-1 when both parts of the broken whip are available and usable.

(2) Use the method in B, Figure 7-1, when the part of the whip that was broken off is lost, or when the whip is so badly damaged that it cannot be used.

(3) To restore the antenna to its original length, a piece of wire is added that is nearly the same length as the missing part of the whip. The pole support is then lashed securely to both sections of the antenna. The two antenna sections are cleaned thoroughly to ensure good contact before connecting them to the pole support. If possible, the connections are soldered.

b. **Wire Antennas**. Emergency repair of a wire antenna may involve the repair or replacement of the wire used as the antenna or transmission line; or the repair or replacement of the assembly used to support the antenna.

(1) When one or more wires of an antenna are broken, the antenna can be repaired by reconnecting the broken wires. To do this, lower the antenna to the ground, clean the ends of the wires, and twist the wires together. Whenever possible, solder the connection.

(2) If the antenna is damaged beyond repair, construct a new one. Make sure that the lengths of the wires on the substitute antenna match the length of the original.

(3) Antenna supports may also require repair or replacement. A substitute item may be used in place of a damaged support and, if properly insulated, can be of any material of adequate strength. If the radiating element is not properly insulated, field antennas may be shorted to ground and be ineffective. Many commonly found items can be used as field-expedient insulators.

(4) The best of these items are plastic or glass to include plastic spoons, buttons, bottle necks, and plastic bags. Though less effective than plastic or glass but still better than no insulator at all are wood and rope. The radiating element--the actual antenna wire-should touch only the antenna terminal and should be physically separated from all other objects, other than the supporting insulator. (Figure 7-2 shows how to make emergency insulators.)

Figure 7-1. USE OF BROKEN ANTENNA

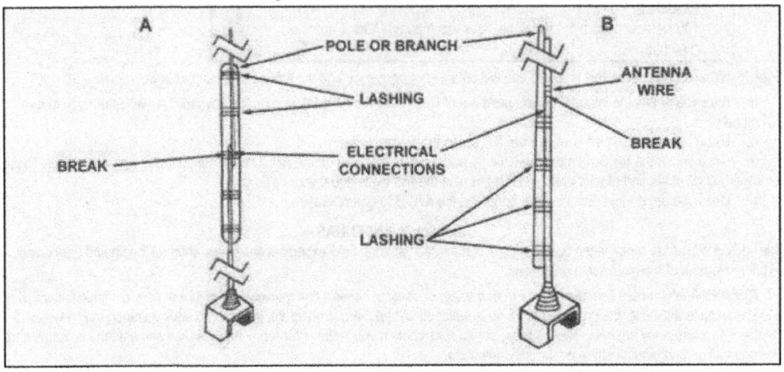

Figure 7-2. METHODS OF CONSTRUCTING EMERGENCY INSULATORS

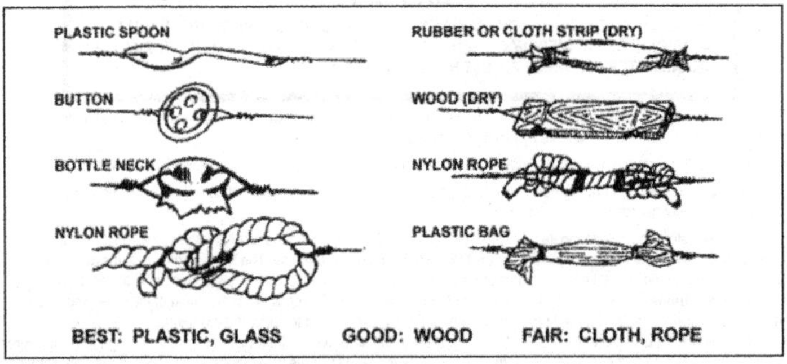

7-4. **CONSTRUCTION AND ADJUSTMENT**. Ranger squad/platoons may use the following methods to construct and adjust antennas:

a. **Construction**. The best kinds of wire for antennas are copper and aluminum. In an emergency, however, Rangers use any type of wire that is available.

(1) The exact length of most antennas is critical. The emergency antenna should be the same length as the antenna it replaces.

(2) Antennas supported by trees can usually survive heavy wind storms if the trunk of a tree or a strong branch is used as a support. To keep the antenna taut and to prevent it from breaking or stretching as the trees sway, the Ranger attaches a spring or old inner tube to one end of the antenna. Another technique is to pass a rope through a pulley or eyehook. The rope is attached to the end of the antenna and loaded with a heavyweight to keep the antenna tightly drawn.

(3) Guidelines used to hold antenna supports are made of rope or wire. To ensure the guidelines will not affect the operation of the antenna, the Ranger cuts the wire into several short lengths and connects the pieces with insulators.

b. **Adjustment**. An improvised antenna may change the performance of a radio set. The following methods can be used to determine if the antenna is operating properly:

(1) A distant station may be used to test the antenna. If the signal received from this station is strong, the antenna is operating satisfactorily. If the signal is weak, the Ranger adjusts the height and length of the antenna and the transmission line to receive the strongest signal at a given setting on the volume control of the receiver. This is the best method of tuning an antenna when transmission is dangerous or forbidden.

(2) In some radio sets, the Ranger uses the transmitter to adjust the antenna. First, he sets the controls of the transmitter to normal; then, he tunes the system by adjusting the antenna height, the antenna length, and the transmission line length to obtain the best transmission output.

7-5. **FIELD-EXPEDIENT OMNIDIRECTIONAL ANTENNAS**. Vertical antennas are omnidirectional. The omnidirectional antenna transmits and receives equally well in all directions.

a. **Vertical Antennas**. Most tactical antennas are vertical. For example, the man-pack portable radio uses a vertical whip and so do the vehicular radios in tactical vehicles. A vertical antenna can be made by using a metal pipe or rod of the correct length, held erect by means of guidelines. The lower end of the antenna should be insulated from the ground by placing it on a large block of wood or other insulating material. A vertical antenna may also be a wire supported by a tree or a wooden pole (Figure 7-3).

Figure 7-3. VERTICAL ANTENNA MADE FROM A WIRE SUPPORTED BY A TREE OR WOODEN POLE

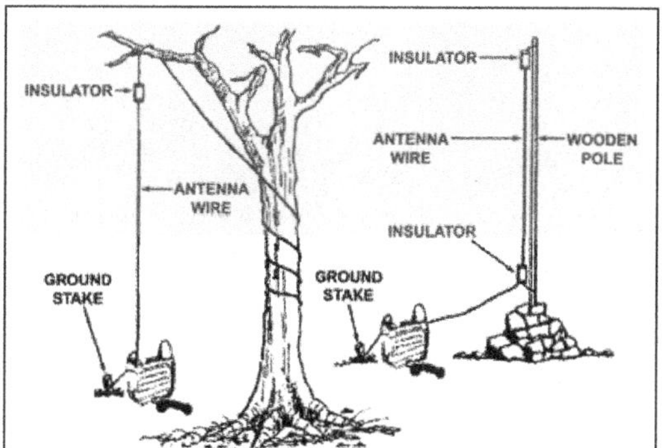

b. **Modification of Connection on Shorter Vertical Antennas**. For short vertical antennas, the pole may be used without guidelines (if properly supported at the base). If the length of the vertical mast is not long enough to support the wire upright, it may be necessary to modify the connection at the top of the antenna (Figure 7-4; see FM 24-18).

Figure 7-4. MODIFIED CONNECTION AT TOP OF ANTENNA

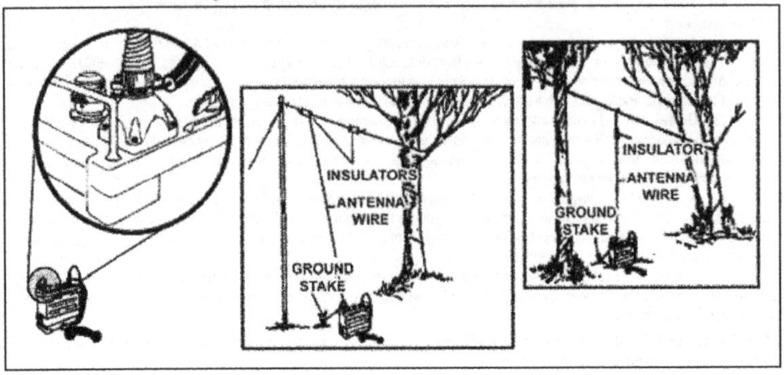

c. **End-Fed, Half-Wave Antenna**. An emergency, end-fed half-wave antenna (Figure 7-5) can be constructed from available materials such as field wire, rope, and wooden insulators. The electrical length of this antenna is measured from the antenna terminal on the radio set to the far end of the antenna. The best performance can be obtained by constructing the antenna longer than necessary and then shortening it, as required, until the best results are obtained. The ground terminal of the radio set should be connected to a good earth ground for this antenna to function efficiently.

Figure 7-5. END-FED, HALF-WAVE ANTENNA

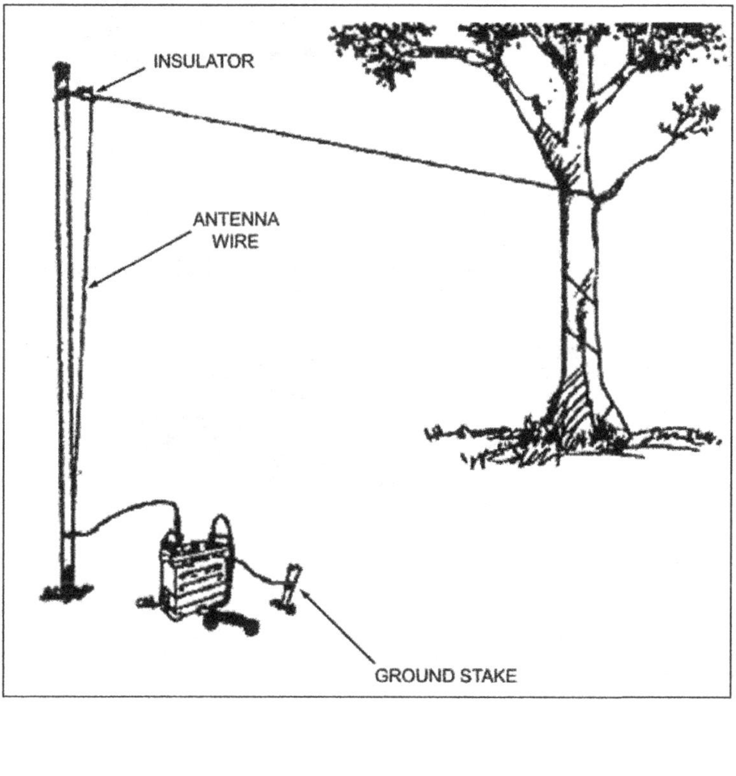

d. **Center-Fed, Doublet Antenna**. The center-fed doublet is a half-wave antenna consisting of two quarter wavelength sections on each side of the center (Figure 7-6). Doublet antennas are directional broadside to their length, which makes the vertical doublet antenna omnidirectional. This is because the radiation pattern is doughnut-shaped and bidirectional.

Figure 7-6. CENTER-FED, DOUBLET ANTENNA

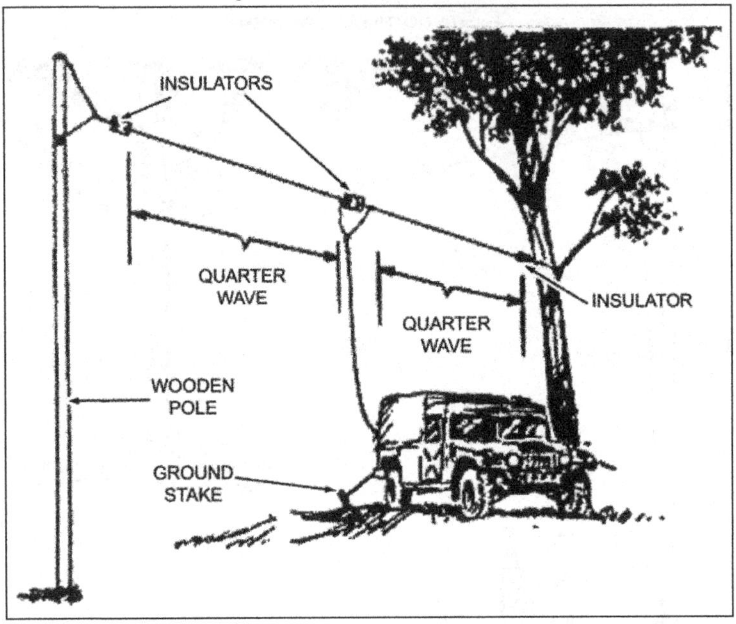

(1) Compute the length of a half-wave antenna by using the formula previously discussed. Cut the wires as close as possible to the correct length; this is very important.

(2) Uses transmission line for conducting electrical energy from one point to another and for transferring the output of a transmitter to an antenna. Although it is possible to connect an antenna directly to a transmitter, the antenna is usually located some distance away.

(3) Support center-fed half-wave FM antennas entirely with pieces of wood. (A horizontal antenna of this type is shown in A, Figure 7-7 and a vertical antenna in B, Figure 7-7.) Rotate these antennas to any position to obtain the best performance.

(a) If the antenna is erected vertically, bring out the transmission line horizontally from the antenna for a distance equal to at least one-half of the antenna's length before it is dropped down to the radio set.

Figure 7-7. CENTER-FED ANTENNAS SUPPORTED WITH WOOD

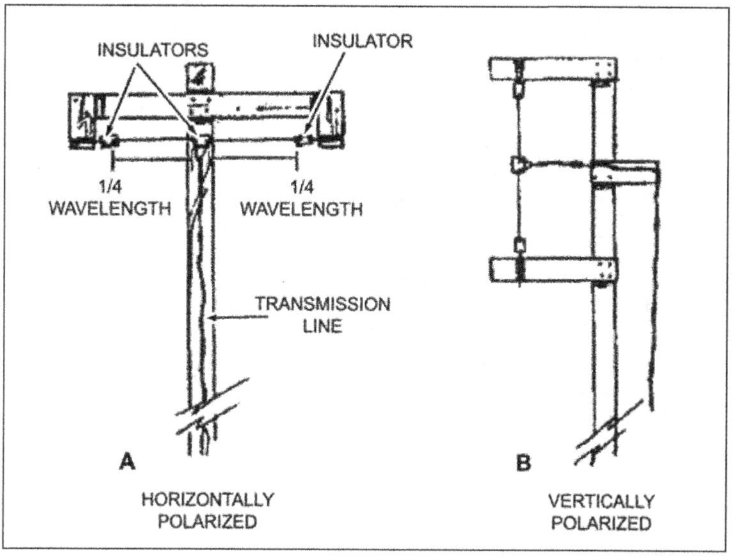

(b) The half-wave antenna is used with FM radios (Figure 7-8). It is effective in heavily wooded areas to increase the range of portable radios. Connect the top guidelines to a limb or pass it over the limb and connect it to the tree trunk or a stake.

Figure 7-8. HALF-WAVE ANTENNA USED WITH FM RADIO

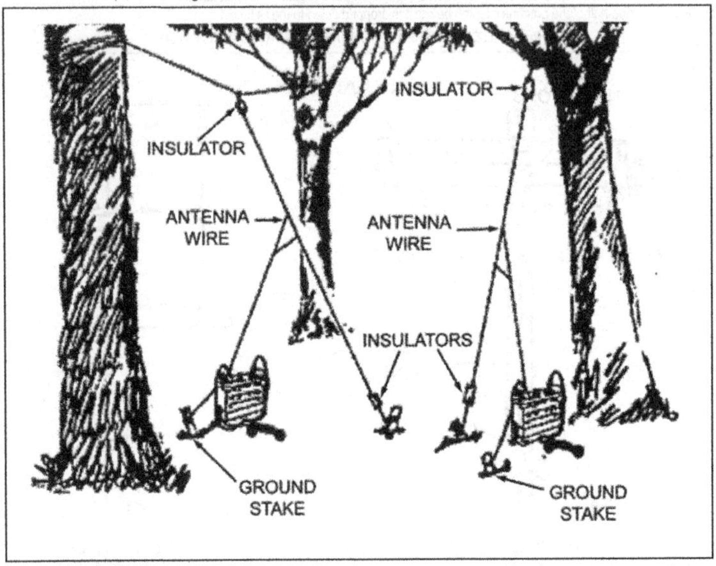

c. **Expedient 292-Type Antenna**. The expedient 292-type antenna was developed for use in the jungle and, if used properly, can increase the team's ability to communicate. In its entirety, the antenna is bulky, heavy, and not acceptable for sniper team operations. The team can, however, carry only the mast head and antenna sections, mounting these on wood poles or hanging them from trees; or, the team can make a complete expedient 292-type antenna (Figure 7-9), using WD-1, wire, and other readily available material. The team can also use almost any plastic, glass, or rubber objects for insulators. Dry wood is acceptable when nothing else is available. (See Figure 7-2 for types of insulators that may be used.) The following describes how to make this antenna:

(1) Use the quick-reference table (Table 7-2) to determine the length of the elements (one radiating and three ground planes) for the frequency that will be used. Cut these elements (A, Figure 7-9) from WD-1 field wire (or similar wire). Cut spacing sticks (B, Figure 7-9) the same length. Place the ends of the sticks together to form a triangle and tie the ends with wire, tape, or rope. Attach an insulator to each corner. Attach a ground-plane wire to each insulator. Bring the other ends of the ground-plane wires together, attach them to an insulator (C, Figure 7-9), and tie securely. Strip about 3 inches of insulation from each wire and twist them together (D, Figure 7-9).

(2) Tie one end of the radiating element wire to the other side of insulator C and the other end to another insulator (D, Figure 7-9). Strip about 3 inches of insulation from the radiating element at insulator C.

(3) Cut enough WD-1 field wire to reach from the proposed location of the antenna to the radio set. Keep this line as short as possible, because excess length reduces the efficiency of the system. Tie a knot at each end to identify it as the *hot* lead. Remove insulation from the *hot* wire and tie it to the radiating element wire at insulator C. Remove insulation from the other wire and attach it to the bare ground-plane element wires at insulator C. Tape all connections and do not allow the radiating element wire to touch the ground-plane wires.

(4) Attach a rope to the insulator on the free end of the radiating element and toss the rope over the branches of a tree. Pull the antenna as high as possible, keeping the lead-in routed down through the triangle. Secure the rope to hold the antenna in place.

(5) At the radio set, remove about 1 inch of insulation from the *hot* lead and about 3 inches of insulation from the other wire. Attach the *hot* line to the antenna terminal (doublet connector, if so labeled). Attach the other wire to the metal case-the handle, for example. Be sure both connections are tight or secure.

(6) Set up correct frequency, turn on the set, and proceed with communications.

Figure 7-9. COMPLETE EXPEDIENT 292-TYPE ANTENNA

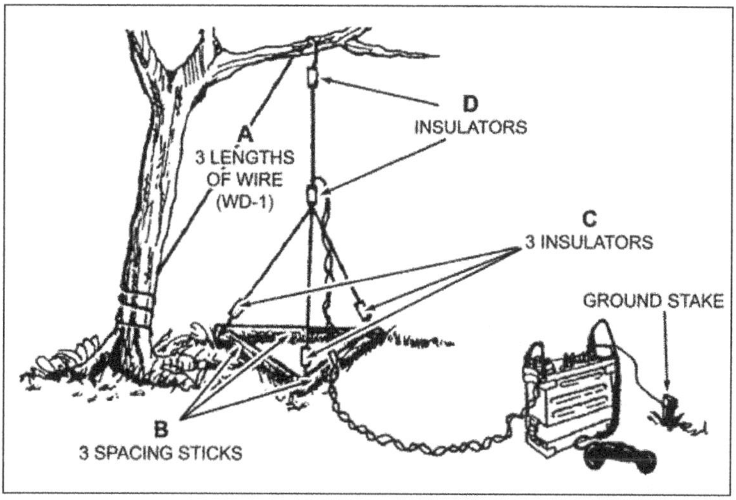

Table 7-2. QUICK REFERENCE TABLE

OPERATING FREQUENCY IN MHz	ELEMENT LENGTH (RADIATING ELEMENTS AND GROUND-PLANE ELEMENTS)	
30	2.38 m	(7 ft 10 in)
32	2.23 m	(7 ft 4 in)
34	2.1 m	(6 ft 11 in)
36	1.98 m	(6 ft 6 in)
38	1.87 m	(6 ft 2 in)
40	1.78 m	(5 ft 10 in)
43	1.66 m	(5 ft 5 in)
46	1.55 m	(5 ft 1 in)
49	1.46 m	(4 ft 9 in)
52	1.37 m	(4 ft 6 in)
55	1.3 m	(4 ft 3 in)
58	1.23 m	(4 ft 0 in)
61	1.17 m	(3 ft 10 in)
64	1.12 m	(3 ft 8 in)
68	1.05 m	(3 ft 5 in)
72	0.99 m	(3 ft 3 in)
76	0.94 m	(3 ft 1 in)

7-6. **FIELD-EXPEDIENT DIRECTIONAL ANTENNAS**. The vertical half-rhombic antenna (Figure 7-10) and the long-wire antenna (Figure 7-11) are two field-expedient directional antennas. These antennas consist of a single wire, preferably two or more wavelengths long, supported on poles at a height of 3 to 7 meters (10 to 20 feet) above the ground. The antennas will, however, operate satisfactorily as low as 1 meter (about 3 feet) above the ground–the radiation pattern is directional. The antennas are used mainly for either transmitting or receiving high-frequency signals.

Figure 7-10. VERTICAL HALF-RHOMBIC ANTENNA

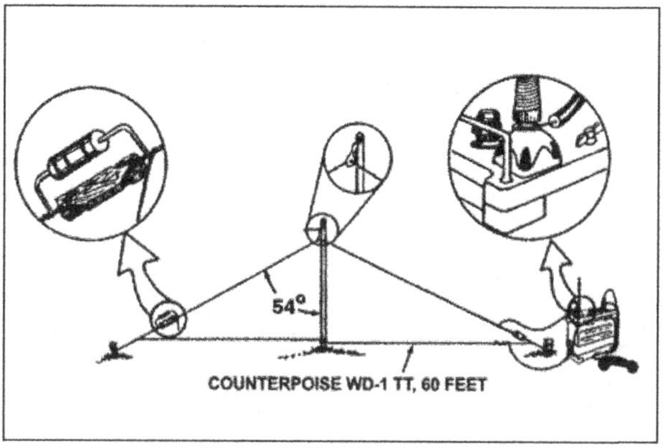

Figure 7-11. LONG-WIRE ANTENNA

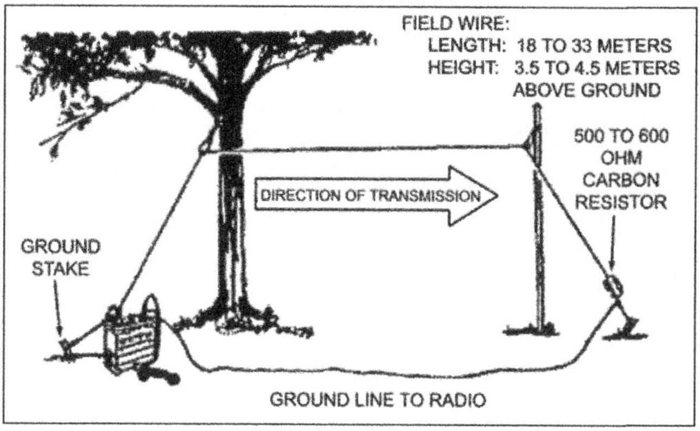

a. The vee antenna (Figure 7-12) is another field-expedient directional antenna. It consists of two wires forming a V with the open area of the "V" pointing toward the desired direction of transmission or reception. To make construction easier, the legs should slope downward from the apex of the "V." This is called a sloping vee antenna (Figure 7-13). The angle between the legs varies with the length of the legs to achieve maximum performance. (Use Table 7-1 to determine the angle and the length of the legs.)

Figure 7-12. VEE ANTENNA

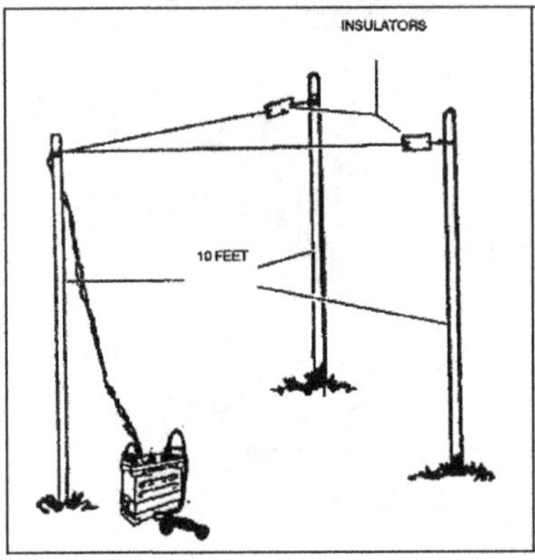

Figure 7-13. SLOPING VEE ANTENNA

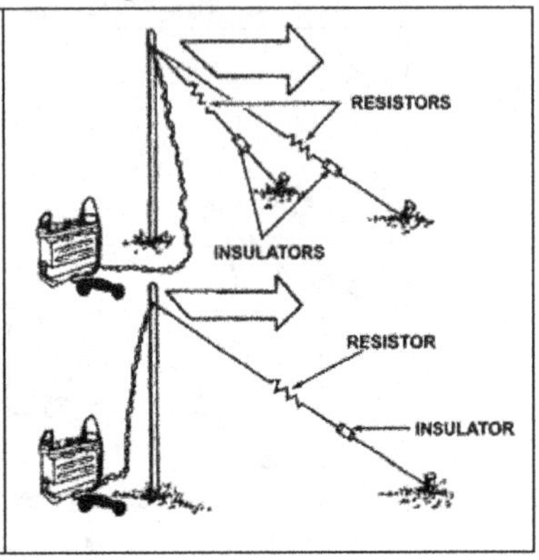

b. When the antenna is used with more than one frequency or wavelength, use an apex angle that is midway between the extreme angles determined by the chart. To make the antenna radiate in only one direction, add noninductive terminating resistors from the end of each leg (not at the apex) to ground. (See TM 11-666.)

7-7. **ANTENNA LENGTH**. The length of an antenna must be considered in two ways: both a physical and an electrical length. These two lengths are never the same. The reduced velocity of the wave on the antenna and a capacitive effect (known as end *effect*) make the antenna seem longer electrically than it is physically. The contributing factors are the ratio of the diameter of the antenna to its length and the capacitive effect of terminal equipment, such as insulators and clamps, used to support the antenna.

a. To calculate the physical length of an antenna, use a correction of 0.95 for frequencies between 3.0 and 50.0 MHz The figures given below are for a half-wave antenna.

b. Use the following formula to calculate the length of a long-wire antenna (one wavelength or longer) for harmonic operation:

EXAMPLE:

If N = the number of half-wavelengths in the total length of the antenna = 3,
and if the frequency in MHz = 7, then--

$$\text{Length (meters)} = \frac{150(N-0.05)}{\text{Frequency in MHz}} = \frac{150(3-.05)}{7} =$$

$$\frac{150 \times 2.95}{7} = \frac{442.50}{7} = 63.2 \text{ meters}$$

7-8. **ANTENNA ORIENTATION**. If the azimuth of the radio path is not provided, the azimuth should be determined by the best available means. The accuracy required in determining the azimuth of the path depends on the radiation pattern of the directional antenna. In transportable operation, the rhombic and V antennas may have such a narrow beam as to require great accuracy in azimuth determination. The antenna should be erected for the correct azimuth. Great accuracy is not required in erecting broad-beam antennas. Unless a line of known azimuth is available at the site, the direction of the path is best determined by a magnetic compass.

7-9. **IMPROVEMENT OF MARGINAL COMMUNICATIONS**. Under certain situations, it may not be feasible to orient directional antennas to the correct azimuth of the desired radio path. As a result, marginal communications may suffer. To improve marginal communications, the following procedure can be used:

a. Check, tighten, and tape cable couplings and connections.
b. Return all transmitters and receivers in the circuit.
c. Ensure antennas are adjusted for the proper operating frequency.
d. Change the heights of antennas.
e. Move the antenna a short distance away and in different locations from its original location.

NOTES

Chapter 8
ARMY AVIATION

Army aviation and infantry units can be fully integrated with other members of the combined arms team to form powerful and flexible air assault task forces that can project combat power throughout the entire depth, width, and breadth of the modern battlefield with little regard for terrain barriers. These combat operations are deliberate, precisely planned, and vigorously executed. They are assigned to strike the enemy when and where he is most vulnerable (See page 2-28, *Air Movement Annex* and page 2-37, *Army Aviation Coordination Checklist*).

8-1. **REVERSE PLANNING SEQUENCE**. Successful air assault execution is based on a careful analysis of METT-TC and detailed, precise reverse planning. Five basic plans that comprise the reverse planning sequence are developed for each air assault operation. The battalion is the lowest level that has sufficient personnel to plan, coordinate, and control air assault operations. When company size or lower operations are conducted, the bulk of the planning takes place at battalion or higher headquarters. They are:

a. **Ground Tactical Plan**. The foundation of a successful air assault operation is the commander's ground tactical plan. All additional plans must support this plan. The plan specifies actions in the objective area to ultimately accomplish the mission and address subsequent operations.

b. **Landing Plan**. The landing plan must support the ground tactical plan. This plan outlines a sequence of events that allows elements to move into the area of operations, and ensures that units arrive at designated locations at prescribed times prepared to execute the ground tactical plan.

c. **Air Movement Plan**. The air movement plan is based on the ground tactical and landing plans. It specifies the schedule and provides instructions for air movement of troops, equipment, and supplies from PZs to LZs.

d. **Loading Plan**. The loading plan is based on the air movement plan. It ensures that troops, equipment, and supplies are loaded on the correct aircraft. Unit integrity is maintained when aircraft loads are planned. Cross-loading may be necessary to ensure survivability of command and control assets, and that the mix of weapons arriving at the LZ is ready to fight.

e. **Staging Plan**. The staging plan is based on the loading plan and prescribes the arrival time of ground units (troops, equipment and supplies) at the PZ in the order of movement

8-2. **SELECTION AND MARKING OF PICKUP AND LANDING ZONES**

a. Small unit leaders should consider the following when selecting a PZ/LZ:

(1) ***Size***. Minimal circular landing point separation from other aircraft and obstacles is needed:

- OH-58D – 25 meters.
- UH-1, AH-1 – 35 meters.
- UH-60, AH-64 – 50 meters.
- Cargo helicopters – 80 meters.

(2) ***Surface Conditions***. Avoid potential hazards such as sand, blowing dust, snow, tree stumps, or large rocks.

(3) ***Ground Slope***.

- 0% - 6 % -- land upslope.
- 7% - 15% -- land sideslope.
- over 15% -- no touchdown (aircraft may hover).

(4) ***Obstacles***. An obstacle clearance ratio of 10 to 1 is used in planning approach and departure of the PZ and LZ, for example, a 10-foot tall tree requires 100 feet of horizontal distance for approach or departure. Obstacles will be marked with a red chemlight at night, or red panels in daytime. Avoid using markings if the enemy would see them.

(5) ***Approach/Departure***. Approach and departure are made into the wind and along the long axis of the PZ/LZ.

(6) ***Loads***. The greater the load, the larger the PZ/LZ must be to accommodate the insertion or extraction.

b. **Marking of PZs and LZs**.

(1) ***Day***. A ground guide will mark the PZ or LZ for the lead aircraft by holding an M16/M4 rifle over his head, by displaying a folded VS-17 panel chest high, or by other coordinated and identifiable means.

(2) ***Night***. The code letter "Y" (inverted "Y") is used to mark the landing point of the lead aircraft at night (Figure 8-1). Chemical lights or "beanbag" lights are used to maintain light discipline. A swinging chemlight may also be used to mark the landing point.

Figure 8-1. INVERTED "Y"

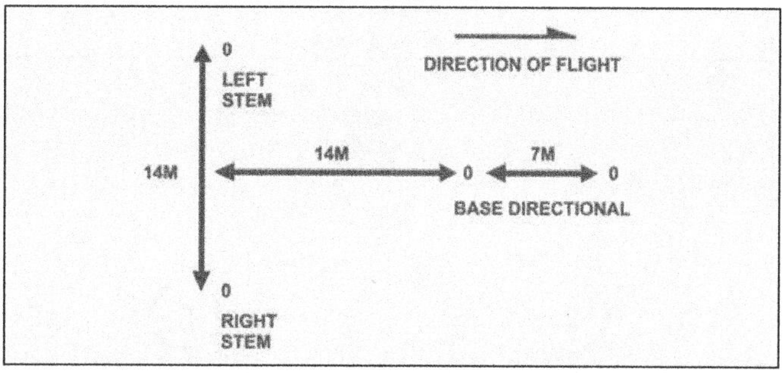

8-3. **AIR ASSAULT FORMATIONS**. Aircraft supporting an operation may use any of the following PZ/LZ configurations which are prescribed by the air assault task force (AATF) commander working with the air mission commander (AMC):

a. ***Heavy Left or Right***. Requires a relatively long, wide landing area; presents difficulty in pre-positioning loads; restricts suppressive fire by inboard gunners; provides firepower to front and flank (Figure 8-2).

Figure 8-2. HEAVY LEFT/HEAVY RIGHT

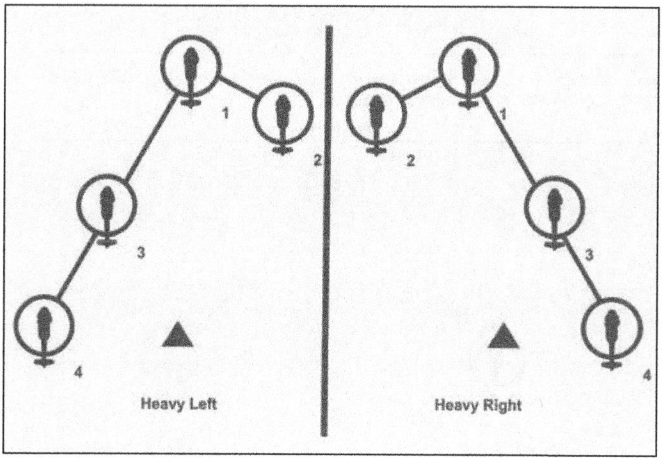

b. **Diamond**. Allows rapid deployment for all-round security; requires small landing area; presents some difficulty in pre-positioning loads; restricts suppressive fire of inboard gunners (Figure 8-3).

Figure 8-3. DIAMOND

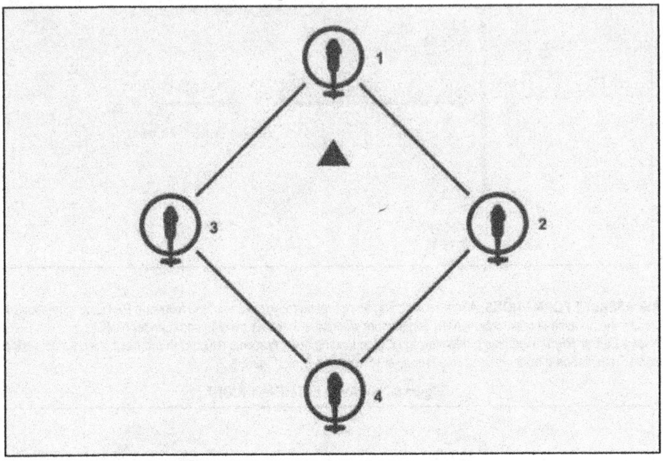

c. **Vee**. Requires a relatively small landing area; allows rapid deployment of forces to the front; restricts suppressive fire of inboard gunners; presents some difficulty in pre-positioning loads (Figure 8-4).

Figure 8-4. VEE

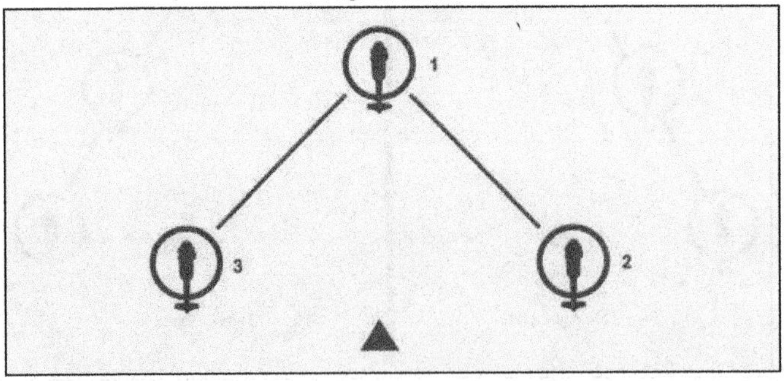

d. **Echelon Left or Right**. Requires a relatively long, wide landing area; presents some difficulty in pre-positioning loads; allows rapid deployment of forces to the flank; allows unrestricted suppressive fire by gunners (Figure 8-5).

Figure 8-5. ECHELON LEFT/ECHELON RIGHT

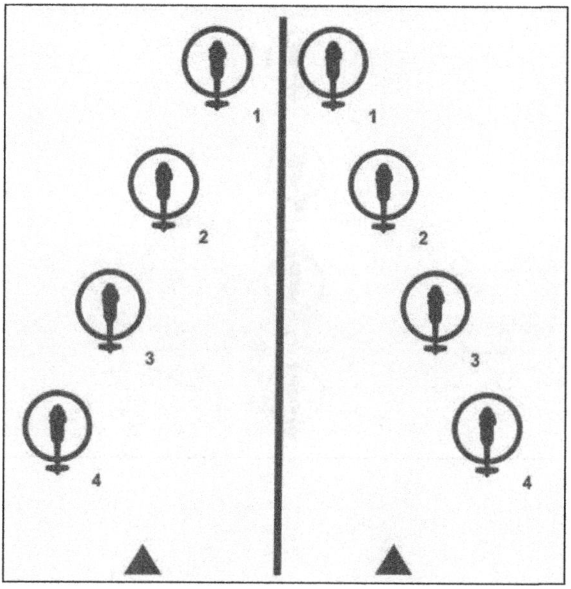

e. **Trail**. Requires a relatively small landing area; allows rapid deployment of forces to the flank; simplifies pre-positioning loads; allows unrestricted suppressive fire by gunners (Figure 8-6).

Figure 8-6. TRAIL

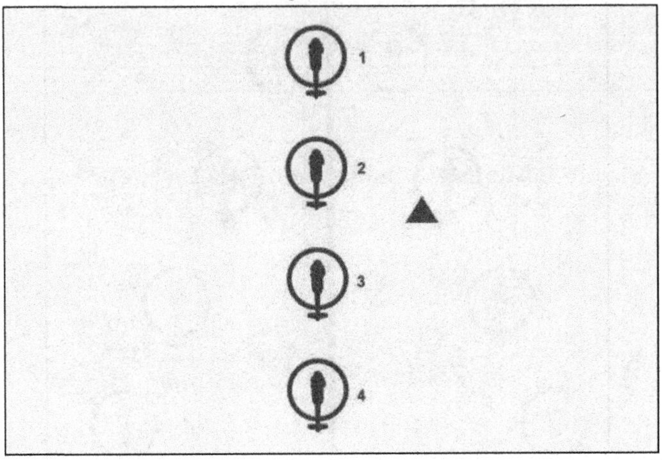

f. **Staggered Trail Left or Right**. Requires a relatively long, wide landing area; simplifies pre-positioning loads; allows rapid deployment for all round security; gunners' suppressive fire restricted somewhat (Figure 8-7).

Figure 8-7. STAGGERED TRAIL LEFT/STAGGERED TRAIL RIGHT

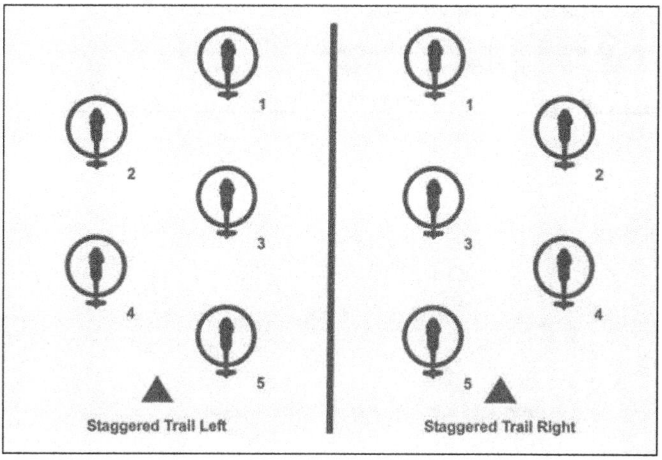

8-4. **PICKUP ZONE OPERATIONS**. Prior to arrival of aircraft, the PZ is secured, PZ control party is positioned, and the troops and equipment are positioned in platoon/squad assembly areas. In occupying a patrol/squad assembly area, the patrol/squad leader does the following. Figure 8-8 shows an example of a large, one-sided PZ. Figures 8-8 through 8-12 show loading/unloading procedures and techniques:

- Maintain all-around security of the assembly area.
- Maintain communications.
- Organize personnel and equipment into chalks and loads.
- Conduct safety briefing and equipment check of troops.

Figure 8-8. LARGE, ONE SIDED PZ

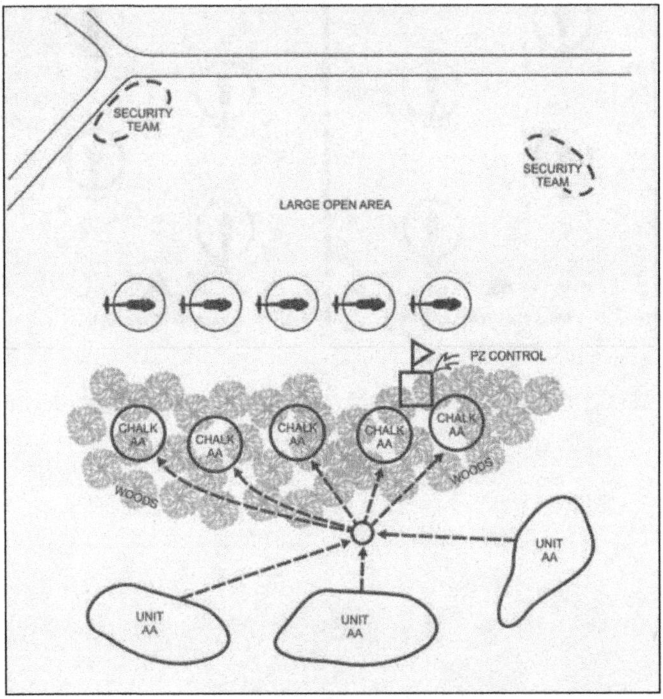

Figure 8-9. UH-60 LOADING SEQUENCE

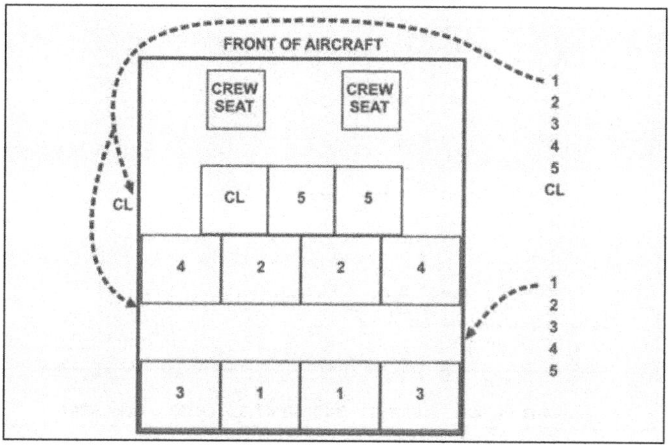

Figure 8-10. UNLOADING SEQUENCE

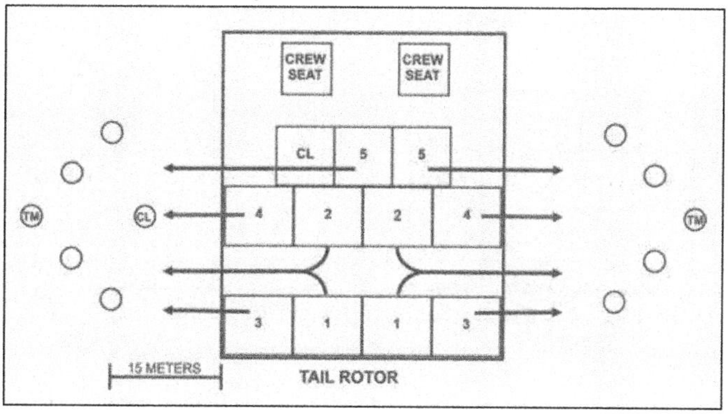

Figure 8-11. TACTICAL LOADING SEQUENCE

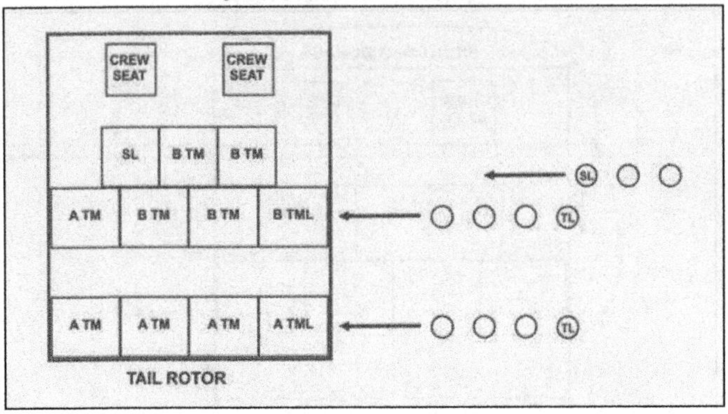

Figure 8-12. TACTICAL UNLOADING USING DOOR NEAREST COVER, CONCEALMENT

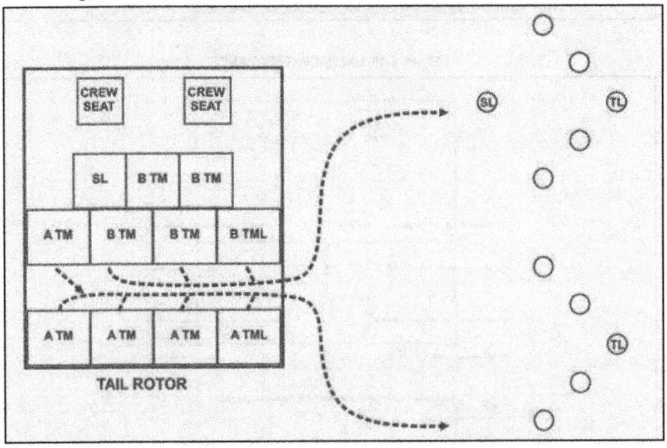

8-5. **SAFETY**. Safety is the primary concern of all leaders when operating in/around aircraft. The inclusion of aircraft into Ranger operations carries an inherent high-risk factor. The following guidelines are to be considered.

a. Approach the aircraft from 90 degrees to 45 degrees off the nose.

b. Weapons with blank firing adapters are pointed muzzles up.

c. Weapons loaded with live ammunition are muzzles down.

d. Wear the ballistic helmet.

e. When possible, conduct an air crew safety brief with all personnel. At a minimum, cover loading/off-loading, emergencies, and egress procedures.

f. Leaders need to carry a manifest and turn a copy into higher.

8-6. **CLOSE COMBAT ATTACK AVIATION**. Close combat attack is defined as a hasty or deliberate attack in support of units engaged in close combat. During CCA, armed helicopters engage enemy units with direct fire that impacts near friendly forces. Targets may range from a few hundred meters to a few thousand meters. Close combat attack is coordinated and directed by a team, platoon or company level ground unit soldiers using standardized CCA procedures in unit SOP's.

a. **Aircraft Capabilities and Limitations**.

(1) AH-64D (Apache):

- Air speed (knots): 164 max, 120 Cruise combat radius: 200 KM
- 30 mm chain gun: 1200 rds, 3,500 m range
- 2.75" rockets: 19 per pod (76), 3-5 km range
- Hellfire Missile: 8 per side (16), 5-8 km range

(2) Capabilities.

- MOBILITY - AO can cover corps or division area
- SPEED - 100-120 knots day/80-100 Knots night
- VERSATILITY - scout-weapons teams vs. pure attack
- LETHALITY - Attack bn can engage 288 targets
- VIDEO RECONNAISSANCE - Provide near real time intelligence

(3) ***Limitations***.

- Threat ID with FLIR
- Low ceilings (Clouds) <500' AGL degrades hellfire capability
- Combat service support consume large amounts of class III, V, and IX

(4) **OH-58D (Kiowa)**.

- Air speed (knots): 125 max, 100 cruise combat radius: 120 KM
- .50 cal MG: 500 rds, 2,000 m range
- 2.75" rockets: 7 per pod (14), 3-5 km range
- Hellfire: 2 per side (4), 5-8 km range
- Stinger: 2 per side (4), 4 km range

(5) ***Capabilities***.

- Mast-mounted sight
- Ability to designate targets while remaining masked
- Thermal imaging system (day/night)
- Laser designator / Aim1 laser
- Video image crosslink (VIXL)
- Moving map display
- Video

(6) ***Limitations***.

- Power limited
- Infrared crossover
- Battlefield obscurants
- Low ceilings - Hellfire
- Remote designation constraints
- Instrument meteorological conditions

c. **CCA Call-for-Fire Format**.

(1) IP/BP/ABF or friendly location__________

- Grid/Lat Long

(2) Known point

- Terrain feature

(3) HDG to TGT __________ (magnetic)

- Specify from IP/BP/ABF or friendly location

(4) DST to TGT ___________ (m) (Specify from IP/BP/ABF or friendly location.)

(5) TGT elevation__________ (feet, mean sea level)

(6) TGT Description ___________________

(7) TGT location ______________

- Grid
- Known point
- Terrain Feature

(8) Type TGT Mark ___________ (day/night)

(9) location of friendly ___________

- Omit if previously given
- Grid/Lat Long
- Known point
- Terrain feature

(10) Egress direction____________

- Cardinal direction not to over fly threats

d. **Marking Method Chart**. Table 8-1 shows this chart.

Table 8-1. MARKING METHOD CHART

Method	D	N	NVG	NVS	Friendly	Target
Smoke	X				X	X
Smoke IR	X		X		X	X
Signal Mirror	X				X	
IR Laser			X		X	X
Tracer	X	X	X			X
Glint Tape				X	X	
CID	X				X	
Strobe		X	X		X	
IR Strobe			X		X	
IR Panel				X	X	
Chem Light		X	X		X	
AN/PAQ-4			X			X
VS-17 Panel	X				X	
Spot Light		X	X		X	
MRE Heater				X	X	
AN/PEQ-2			X			X
Hydra 70 Illum	X	X	X	X		X
Ref FM 3-04.111 Appendix Q for more detail						

e. **Risk-Estimate Distances**. REDs for CCA munitions are shown in Table 8-2.

Table 8-2. RISK ESTIMATE DISTANCES FOR CCA MUNITIONS

WEAPON	*DANGER CLOSE*	*MAX EFF RANGE)*
50 cal	150 meters	2,000 meters
30 mm	150 meters	3,500 meters
2.75 rocket	175 meters	8,000 meters
Hellfire Missile	500 meters	8,000 meters
Danger Close = Risk-Estimate Distances for 0.1% (1/10th of a percent) probability of incapacitation (PI) per FM 3-09.32.		

Chapter 9
WATERBORNE OPERATIONS

This chapter discusses rope bridges, poncho rafts, and watercraft.

9-1. **ROPE BRIDGE TEAM.** The Ranger patrol seldom has ready-made bridges, so they must know how to employ field-expedient stream crossing techniques.

a. **Personnel.**

(1) # 1 man................Lead safety swimmer and far side lifeguard.
(2) # 2 man................Rope puller, swims water obstacle pulling 150-foot rope, ties off rope on far side anchor point.
(3) # 3 man................Near side lifeguard is the last man to cross water obstacle.
(4) # 4 man................Bridge team commander (BTC; most knowledgeable person on team).
(5) # 5/6 men............Rope tighteners.

b. **One-Rope Bridge (Wet Crossing).** Special equipment follows:

- Two snaplinks per piece of heavy equipment.
- Two snaplinks for every 120 feet of rope.
- One 14-foot utility rope per person (Swimmer safety line).
- Two snaplinks per person.
- One waterproof bag per RTO.
- Two B-7 life preservers.
- Three flotation work vests.
- Two 150-foot nylon ropes.

c. **Planning**. A stream crossing annex is prepared in conjunction with the unit's operation order. Special organization is accomplished at this time. For a platoon size patrol, a squad is normally given the task of providing the bridge team. The squad leader designates the most technically proficient soldier in the squad as the bridge team commander (BTC).

d. **Rehearsals and Inspections.**

(1) Rehearse the entire stream crossing, emphasizing--

- Security and actions on enemy contact.
- Actual construction of the rope bridge within the 8-minute time standard on dry land.
- Individual preparation.
- Order of crossing.
- All signals and control measures.
- Reorganization.

(2) Conduct rehearsals as realistically as possible.
(3) Ensure personnel are proficient in the mechanics of a stream crossing operation.
(4) Inspect for equipment completeness, correct rigging and preparation, personnel knowledge and understanding of the operation.
(5) Actions of the # 4 man (BTC) during the preparation phase.

(a) Rehearse the bridge team.
(b) Account for all equipment in the bridge kit.
(c) Ensure 150-foot rope is coiled.

e. **Execution Phase.**

(1) Steps for the establishment and conduct of bridge stream crossing:

(a) Leader halts short of the river, local security is established, and a recon is conducted of the area for the presence of the enemy and for crossing site suitability/ necessity. He directs the BTC to construct the bridge.

(b) Security is established up and downstream while unit leader briefs BTC on anchor points. The unit leader counts individuals across.

(c) The bridge team begins to establish the rope bridge while unit members begin individual preparation.

(d) The BTC is responsible for construction of one-rope bridge and selection of the near side anchor point as well as the far side anchor point if visibility permits. He will tie a swimmer's safety line to anchor himself into the bridge. Tying an end of the line bowline around the waist secured with an overhand knot, and on the free running end, an end of line bowline with an overhand knot. A snaplink will be placed on the loop portion of the end of line bowline, which will extend no further than arms length away from the soldier (standard waterborne uniform). This will ensure that the soldier is never more than one arms length away from the rope bridge should he lose his grip.

(e) Noise and light discipline are enforced and security is maintained.

(f) Individual soldiers put a snaplink in their end of the line bowline and the front sight assembly of every M-4/M16 or M203. M240 gunners put a snaplink through the front sight assembly and rear swivel of their M240 MG. RTOs (and others with heavy rucksacks) place an additional snaplink on their rucksack frame, top center.

(2) The responsibilities of the bridge team while constructing the rope bridge are as follows:

(a) # 1 man (Lead safety swimmer and far side lifeguard). He grounds his rucksack (with snaplink through top of frame) to the rear of the near side anchor point. He wears equipment in the following order (body out). He carries a safety line to assume duties of far side lifeguard:

- Waterborne uniform (top button buttoned, pants unbloused)
- B7 life preserver or engineer work vest
- LBV
- Weapon (across the back)

(b) # 1 man enters the water upstream from the # 2 man and stays an arms length away from the # 2 man on the upstream side. He identifies the far side anchor point upon exiting the water and once the # 2 man has exited the water moves to his far side lifeguard position downstream of the rope bridge with knotted safety line on wrist, LBV LBV/weapon grounded, and work vest held in throwing hand. He continues to wear the B7.

(c) # 2 man (rope puller) in waterborne uniform (same as # 1 man) wears his equipment in the following order. He grounds his rucksack (with snaplink through top of frame) to the rear of the near side anchor point. His duties are to swim across the water obstacle pulling the rope. He ties off the rope on the anchor point identified by the # 1 man with a round turn and two half hitches with a quick release. The direction of the round turn is the same direction as the flow of water (current) to facilitate exit off the rope bridge:

- Work vest.
- LBV.
- Weapon (across the back).
- Swimmer safety line.

(d) # 3 man (near side lifeguard) in the same waterborne uniform as the far side lifeguard. The # 3 man positions himself on the downstream side of the bridge before the # 1/2 men enter the water. He (# 1 man) grounds his rucksack (with snaplink through top of frame) on rear of near-side anchor point. After the PSG crosses and verifies the headcount, the # 1 man unties the quick release at the near side anchor point. The # 3 man reties his safety line into an Australian rappel seat, hooks the end of line bowline into his snaplink, and connects his snaplink to the one on the end-of-line bowline . # 3 man is the last pulled across the water obstacle. Before crossing the water obstacle, he dons his equipment in the following order:

- B7.
- Work vest.
- LBV.
- Weapon.

(e) # 4 man (BTC). He is in the standard waterborne uniform with LBV and sling rope tied in safety line (round the waist bowline with end of line bowline no more than one arms length). He is responsible for construction of rope bridge and organization of bridge team. He is also responsible for back feeding the rope and tying end of line bowlines. He designates the near side anchor point, ties the Figure Eight slip of the transport tightening system, and hooking all personnel to the rope bridge. He ensures that the transport tightening knot is on the upstream side of the rope bridge. He ensures that all individuals are in the waterborne uniform, hooked into the rope facing the current with the safety line routed through the trailing shoulder of the individual's LBV and rucksack. He ensures that the weapon is hooked onto the rope. He controls the flow of traffic on the bridge. He is responsible for crossing with the # 1 man's rucksack. He is generally the next to the last man to cross (follows PSG who is keeping head count).

(f) # 5 and # 6 men (rope tighteners) wear the waterborne uniform with LBV and safety line. They tighten the transport tightening knot. They also take the rucksacks of the # 2 and # 3 men across. Once they reach the far side, # 5 and # 6 men pull the last man (# 3 man) across.

(g) The rucksacks of # 1/2/3 men are transported across by # 4/5/6 men. The rucksacks of the # 1/2/3 men are hooked into the rope by the snaplink through the top of the frames and the # 4/5/6 men pull them across. The weapons of # 4/5/6 men are attached between the # 4/5/6 men and the rucksack that they are pulling across the bridge.

(h) BTC rehearses the bridge team during the planning sessions and directs construction and emplacement. The unit leader selects the crossing site which complements the tactical plan.

- # 3 man positions himself downstream of crossing site.
- # 1 man enters water upstream of # 2. He stays one arm's length from the # 2 man and is prepared to render any assistance to the # 2 man. Both swim in conjunction upstream to compensate for the current. BTC feeds rope out of rucksack positioned on the downstream side of near side anchor point.

(i) The # 1 man exits and identifies the far side anchor point (if BTC cannot identify it for the # 2 man). # 2 man exits on the upstream side of the far side anchor point. The rope is now routed to facilitate movement onto and off the bridge.

(j) Radios and heavy equipment are waterproofed and rigged. All individuals don waterborne uniform and tie safety lines. PSG moves to anchor point and maintains accountability through headcount.

(k) The # 2 man signals the BTC that the rope is temporarily attached to the far side anchor point, and the BTC pulls out excess slack and ties the transport tightening system using a Figure Eight slip. The BTC signals the # 2 man to pull the knot twelve to fifteen feet from the near side anchor point. After this is done the # 2 two man ties round turns 18 to 24 inches off the water with the remaining rope and secures the rope, to itself, with a snaplink. The # 2 man signals the BTC and the pulling team (# 4/5/6) tightens the bridge, pulling the transport tightening system as close as possible to the near side anchor point.

(l) The # 1 man moves downstream and assumes his duties as the far side lifeguard. The bridge team commander will tie off the rope with a round turn and two half hitches around the near side anchor point. The BTC will place himself on the upstream side of the bridge (facing downstream). He begins to hook individuals into the rope inspecting them for safety.

Note: Any Ranger identified as a weak swimmer will cross with no other personnel on the rope bridge. The weak swimmer crosses individually to allow the near and far side lifeguards to focus their attention exclusively on the weak swimmer and not be distracted by other personnel crossing the bridge.

(m) The # 2 man moves to the upstream side of the rope bridge, assists personnel off the rope on the far side, and maintains head count. The # 5 and # 6 men cross with the # 1 and # 2 men's rucksacks..

(n) The BTC maintains the flow of traffic ensuring that no more than three men are on the bridge at any one time (one hooking up, one near the center, and one being unhooked). Once the PSG has accounted for everyone on the near side, he withdraws left and right (L/R) security and sends them across. PSG follows security across. The # 3 man hooks the BTC (with # 3's ruck) onto the rope. Once the BTC has crossed, the # 3 man unhooks near side anchor point and the BTC unties far side anchor point. The # 3 man ties an Australian rappel seat with snaplink to the front, hooks onto the snaplink that is in the end of the line bowline on the 120-foot rope, and then signals the # 4/5/6 men to take in slack. The # 3 man extends arms in front of his head, slightly upstream to fend off debris and is pulled across by the # 4/5/6. All individuals (except # 1/2/3 and RTOs) wear rucksacks across. The # 4/5/6 men hook the rucksacks of the # 1/2/3 men onto the bridge by the snaplink. All the men cross facing upstream.

(o) Once the far side headcount, weapons and equipment are verified (between PSG and # 5 man), personnel reorganize, and then they continue the mission.

(p) Personnel with heavy equipment:

M240 All major groups are tied together with 1/4-inch cord. An anchor line bowline runs through the rear swivel, down the left side of gun. Tie a round turn through the trigger guard. Route the cord down the right side and tie off two half hitches around the forearm assembly with a round turn and two half hitches through the front sight posts. The remainder of the working end is tied off with an end of the rope bowline approximately one foot from the front sight post large enough to place leading hand through. The M240 is secured to the bridge by snaplinks on the front sight post and rear swivel. The M240 is pulled across by the trailing arm of the M240 gunner.

AN-PRC/II9s These are waterproofed before crossing a one-rope bridge. Once far side FM commo is established, the near side RTO breaks down and waterproofs his radio, and prepares to cross the bridge. He puts a snaplink in the top center of the rucksack frame (same as for # 1/2/3 men). The BTC will hook the rucksack to the rope.

Note: Using two snaplinks binds the load on the rope. Adjust arm straps all the way out. RTO pulls radio across the rope bridge.

9-2. **PONCHO RAFT.** Normally a poncho raft is constructed to cross rivers and streams when the current is not swift. A poncho raft is especially useful when the unit is still dry and the platoon leader desires to keep the individuals equipment dry.

a. **Equipment Requirements**.

(1) Two serviceable ponchos.

(2) Two weapons (poles can be used in lieu of weapons).

(3) Two rucksacks per team.

(4) 10 feet of utility cord per team.

(5) One sling rope per team.

b. **Conditions**. Poncho rafts are used to cross water obstacles when any or all of the following conditions are found:

(1) The water obstacle is too wide for 150-foot rope.

(2) No sufficient near or far shore anchor points are available to allow rope bridge construction.

(3) Under no circumstances will poncho rafts be used as a means to cross a water obstacle if an unusually swift current is present.

c. **Choosing a Crossing Site**. Before a crossing site is used, a thorough reconnaissance of the immediate area is made. Analyzing the situation using METT-TC, the patrol leader chooses a crossing site that offers as much cover and concealment as possible and has entrance and exit points that are as shallow as possible. For speed of movement, it is best to choose a crossing site that has near and far shore banks that are easily traversed by an individual Ranger.

d. **Execution Phase**. Steps for the construction of a poncho raft:

(1) Pair off the unit/patrol in order to have the necessary equipment.

(2) Tie off the hood of one poncho and lay out on the ground with the hood up.

(3) Weapons are then placed in the center of the poncho, approximately 18 inches apart, muzzle to butt.

(4) Next, rucksacks and LBV are placed between the weapons with the two individuals placing their rucksacks as far apart as possible.

(5) The two will then start to undress (bottom to top), first with their boots, taking the laces completely out for subsequent use as tie downs if necessary).

(6) The boots are then placed over muzzle/butt of weapon toe in.

(7) Members continue to undress, folding each item neatly and placing it on top of their boots.

(8) Once all of the equipment is placed between the two weapons or poles, the poncho is snapped together. The snapped portion of the poncho is then lifted into the air and tightly rolled down to the equipment. Start at the center and work out to the end of the raft creating pigtails at the end. This is accomplished much easier if done by both soldiers together. The pigtailed ends are then folded in toward the center top of the raft and tied off with a single boot lace.

(9) The other poncho is then laid out on the ground with the hood up and the first poncho with equipment is placed in its center. The second poncho is then snapped, rolled and tied in the same manner as the first poncho. The third and fourth boot laces (or utility cord) are then tied around the raft approximately one foot from each end for added security. The poncho raft is now complete.

Note: The patrol leader must analyze the situation using METT-TC and make a decision on the uniform for crossing the water obstacle such as weapons inside the poncho raft or slung across the back, remaining dressed or stripped down with clothes inside raft.

9-3. **WATERCRAFT.** Use of inland and coastal waterways may add flexibility, surprise, and speed to tactical operations. Use of these waterways will also increase the load carrying capacity of normal dismounted units. Watercraft are employed in reconnaissance and assault operations.

a. **Inflation Method**. Inflatable with foot pumps using four separate valves located on the inside of the buoyancy tubes. Each of the four valves are used to section off the assault boat into eight separate airtight compartments. To pump air into the boat, turn all valves into the "orange" or "inflate" section of the valve. Once the assault boat is filled with air, turn all valves onto the "green" or "navigation" section. This will section the assault boat into eight separate compartments.

b. **Characteristics**.

- Overall length: 15 feet, 5 inches.
- Overall width: 6 feet, 3 inches.
- Weight: 265 pounds.
- Maximum payload: 2,756 pounds.
- Crew : 1 coxswain + either 10 paddlers or a 65 HP short-shaft outboard motor.

c. **Preparation**.

(1) ***Rubber Boat***.

- All rubber boats will have a twelve-foot bowline secured to the front starboard D-Ring. This rope will be tied with an anchor-line bowline, and the knot will be covered with 100-MPH tape.
- All rubber boats will have a fifteen foot center line tied to the rear floor D-Ring. The same procedure for securing the bowline will be used for the centerline.
- All rubber boats will be filled to 240 millibars of air, and checked to ensure that all valve caps are tight, and set in the NAVIGATE position.
- All rubber boats will be equipped with one foot pumps. These will be placed in the front pouches of the rubber boat. If no pouches are present, the foot pumps will be placed on the floor.
- All rubber boats will be inspected with the maintenance chart.

(2) ***Personnel and Equipment***.

- All personnel will wear work vest or kapoks (or another suitable positive flotation device).
- LBV is worn over the work vest, unbuckled at the waist.
- Individual weapon is slung across the back, muzzle pointed down and facing toward the inside of the boat.
- Crew served weapons, radios, ammunition and other bulky equipment is lashed securely to the boat to prevent loss if the boat should overturn. Machine guns with hot barrels are cooled prior to being lashed inside the boats.
- Radios and batteries are waterproofed.
- Pointed objects are padded to prevent puncturing the boat.

NOTES

d. **Positions.** Assign each individual a specific boat position (Figure 9-1).

e. **Duties**.

(1) Designate a commander for each boat, (normally coxswain)

(2) Designate a navigator (normally a leader within the platoon) - observer team as necessary.

(3) Crew is positioned as shown in Figure 9-2.

(4) Duties of the Coxswain.

(a) Responsible for control of the boat and actions of the crew.

(b) Supervises the loading, lashing and distribution of equipment.

(c) Maintains the course and speed of the boat.

(d) Gives all commands.

(5) # 2 paddler (long count) is responsible for setting the pace.

(6) # 1 paddler is the observer, stowing and using the bowline unless another observer is assigned.

f. **Embarkation and Debarkation Procedures**.

(1) When launching, the crew will maintain a firm grip on the boat until they are inside it: similarly, when beaching or debarking, they hold on to the boat until it is completely out of the water. Loading and unloading is done using the bow as the entrance and exit point.

(2) Keep a low center of mass when entering and existing the boat to avoid capsizing. Maintain 3 points of contact at all times.

(3) The long count is a method of loading and unloading by which the boat crew embarks or debarks individually over the bow of the boat. It is used at river banks, on loading ramps, and when deep water prohibits the use of the short count method.

(4) The short count is a method of loading or unloading by which the boat crew embarks or debarks in pairs over the sides of boat while the boat is in the water. It is used in shallow water allowing the boat to be quickly carried out of the water. The short count method of organization is primarily used during surf operations.

(5) Beaching the boat is a method of debarking the entire crew at once into shallow water and quickly carrying the boat out of the water.

g. **Commands**. Commands are issued by the coxswain to ensure the boat is transported over land and controlled in the water. All crew members learn and react immediately to all commands issued by the coxswain. T he various commands are as follows:

(1) "Short Count...count off," Crew counts off their position by pairs, such as 1,2,3,4,5 (passenger # 1, # 2, if applicable) coxswain.

(2) "Long Count...count off," Crew counts off the position by individual, such as 1,2,3,4,5,6,7,8,9,10, (Passenger # 1, # 2, if applicable), coxswain.

(3) "Boat Stations," Crew takes positions along side the boat.

(4) "High Carry...Move," (used for long distance move overland).

(a) On the preparatory command of "High Carry," the crew faces the rear of the boat and squats down grasping carrying handles with the inboard hand.

(b) On the command "Move," the crew swivels around, lifting the boat to the shoulders so that the crew is standing and facing to the front with the boat on their inboard shoulders.

(c) Coxswain guides the crew during movement.

(5) "Low Carry...Move," (used for short distance moves overland).

(a) On preparatory command of "Low Carry," the crew faces the front of the boat, bent at the waist, and grasps the carrying handles with their inboard hands.

(b) On the command of "Move," the crew stands up straight raising the boat approximately six to eight inches off the ground.

(c) Coxswain guides the crew during movement.

(6) "Lower the Boat----Move," Crew lowers the boat gently to the ground using the carrying handles.

(7) "Give Way Together," Crew paddles to front with # 2 setting the pace.

(8) "Hold," Entire crew keeps paddles straight downward motionless in the water stopping the boat.

(9) "Left side hold (Right)," Left crew holds, right continues with previous command.

(10) "Back paddle," Entire crew paddles backward propelling the boat to the rear.

Figure 9-1. BOAT POSITIONS

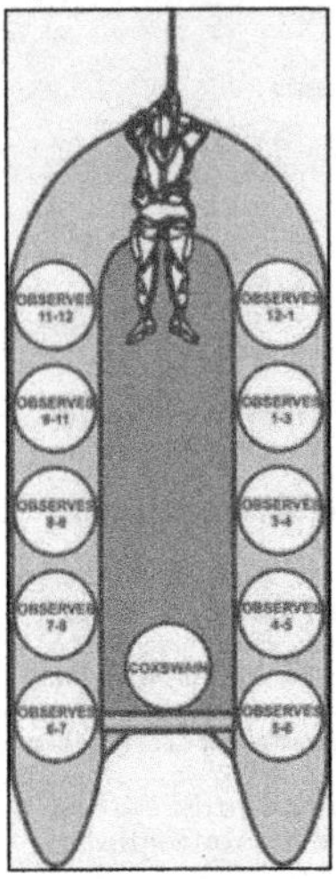

Figure 9-2. CREW POSITIONS, LONG COUNT AND SHORT COUNT

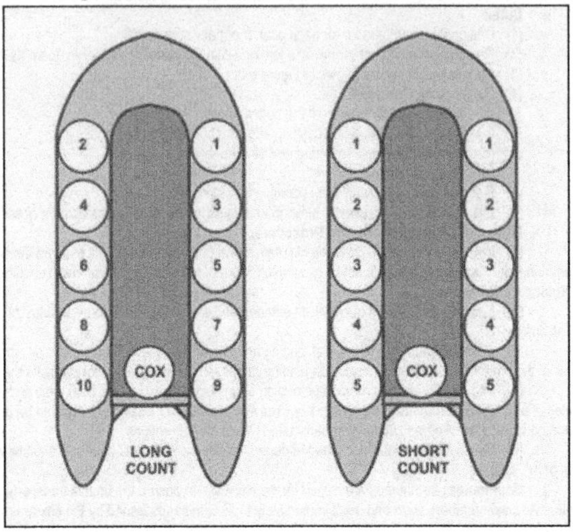

(11) "Back Paddle Left" (Right), Left crew back paddles causing the boat to turn left, right crew continues with previous command.

(12) "Rest Paddles," Crew members place paddles on their laps with blades outboard. This command may be given in pairs such as "# 1s, rest paddles").

h. **Securing of Landing Site.**

(1) If the landing site cannot be secured prior to the waterborne force landing, some form of early warning, such as scout swimmers, is considered. These personnel swim to shore from the assault boats and signal the boats to land. All signals and actions are rehearsed prior to the actual operation.

(2) If the patrol is going into an unsecured landing site, it can provide security by having a security boat land, reconnoiter the landing site and then signal to the remaining boats to land. This is the preferred technique.

(3) The landing site can be secured by force with all the assault boats landing simultaneously in a line formation. While this is the least desirable method of securing a landing site, it is rehearsed in the event the tactical situation requires its use.

(4) Arrival at the debarkation point.

(a) Unit members disembark according to leaders order (Figure 9-3).

(b) Local security is established.

(c) Leaders account for personnel and equipment.

(d) Unit continues movement.

- Soldiers pull security initially with work vest on.
- Coxswains and two men unlash and de-rig rucksacks.
- Soldiers return in buddy teams to secure rucksack and drop off work vest.
- Boats are camouflaged/cached if necessary prior to movement.

Figure 9-3. DEBARKATION

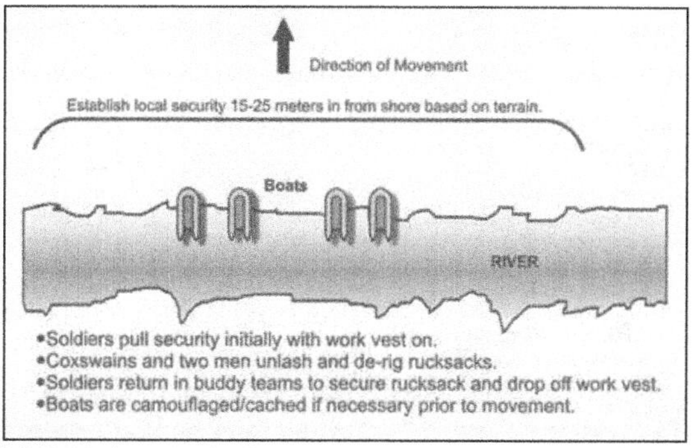

i. **Capsize Drill.** The following commands and procedures are used for capsize drills or to right an over-turned boat,

(1) ***PREPARE TO CAPSIZE***. This command alerts the crew and they raise paddles above their heads, with the blade pointed outward. Before capsizing, the coxswain will conduct a long count.

(2) ***PASS PADDLES***. All paddles are passed back and collected by the # 9/10 men.

(3) ***CAPSIZE THE BOAT***. All personnel slide into the water except the # 3/5/7 men. The # 1 man secures the bowline. The three men in the water grasp the capsize lines (ensuring the lines are routed under the safety lines) and stand on the buoyancy tubes opposite the capsize lines anchor points. The boat is then turned over by the # 3/5/7 men, who lean back and straighten their legs while pulling back on the capsize lines. As the boat lifts off the water, the # 4 man grasps the center carrying handle and rides the boat over. Once the boat is over, the # 4 man helps the # 3/7 men back onto the boat. The # 5 man holds onto the center carrying handle and turns the boat over the same way. The # 5 man rides the boat back over and helps the rest of the crew into the boat. As soon as the boat is capsized, the coxswain commands a long count to ensure that no one sank or was trapped under the boat. Every time the boat is turned over, he conducts another long count.

j. **River Movement.**

(1) ***Characteristics of River***.

(a) Know local conditions prior to embarking on river movement.

(b) A bend is a turn in the river course.

(c) A reach is a straight portion of river between two curves.

(d) A slough is a dead end branch from a river. They are normally quite deep and can be distinguished from the true river by their lack of current

(e) Dead water is a part of the river, due to erosion and changes in the river course that has no current. Dead water is characterized by excessive snags and debris.

(f) An island is usually a pear-shaped mass of land in the main current of the river. Upstream portions of islands usually catch debris and are avoided.

(g) The current in a narrow part of a reach is normally greater than in the wide portion.

(h) The current is greatest on the outside of a curve; sandbars and shallow water are found on the inside of the curve.

(i) Sandbars are located at those points where a tributary feeds into the main body of a river or stream.

(j) Because the # 1 and # 2 men are sitting on the front left and right sides of the boat, they observe for obstacles as the boat moves downriver. If either notices an obstacle on either side of the boat, he notifies the coxswain. The coxswain then adjusts steering to avoid the obstacle.

(2) ***Navigation***. The patrol leader is responsible for navigation. The three acceptable methods of river navigation are--

(a) Checkpoint and general route. These methods are used when the drop site is marked by a well-defined checkpoint and the waterway is not confused by a lot of branches and tributaries. They are best used during daylight hours and for short distances.

(b) Navigator-observer method. This method is the most accurate means of river navigation and is used effectively in all light conditions.

(c) *Equipment Needed.*

- Compass
- GPS
- Photo map (1st choice)
- Topographical map (2nd choice)
- Poncho (for night use)
- Pencil/Grease pencil
- Flashlight (for night use)

(d) *Procedure*. Navigator is positioned in center of boat and does not paddle. During hours of darkness, he uses his flashlight under the poncho to check his map. The observer (or # 1 man) is at the front of the boat.

(a) The navigator keeps his map and compass oriented at all time.

(b) The navigator keeps the observer informed of the configuration of the river by announcing bends, sloughs, reaches, and stream junctions as shown on his map.

(c) The observer compares this information with the bends, sloughs, reaches, and stream junctions he actually sees. When these are confirmed, the navigator confirms the boat's location on his map.

(d) The navigator also keeps the observer informed of the general azimuths of reaches as shown on his map and the observer confirms these with actual compass readings of the river.

(e) The navigator announces only one configuration at a time to the observer and does not announce another until it is confirmed and completed.

(f) A strip-map drawn on clear acetate backed by luminous tape may be used. The drawing is to scale or a schematic. It should show all curves and the azimuth and distance of all reaches. It may also show terrain features, stream junctions, and sloughs.

k. **Formations.** Various boat formations are used (day and night) for control, speed, and security (Figure 9-4). The choice of which is used depends on the tactical situation and the discretion of the patrol leader. He should use hand and arm signals to control his assault boats. The formations are:

- Wedge
- Line
- File
- Echelon
- Vee

Figure 9-4. FORMATIONS

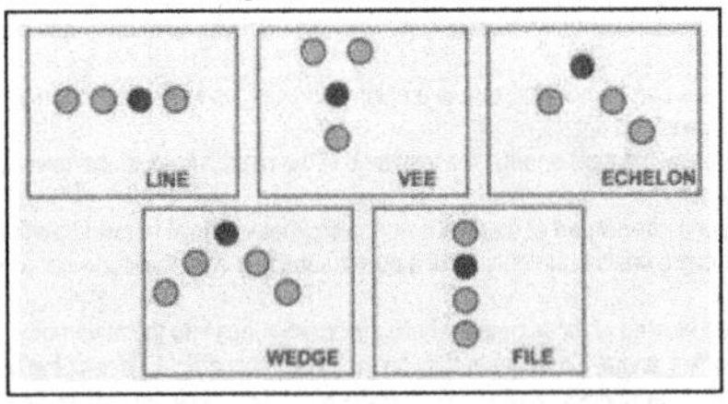

Chapter 10
MILITARY MOUNTAINEERING

In the mountains, commanders face the challenge of maintaining their units' combat effectiveness and efficiency. To meet this challenge, commanders conduct training that provides Rangers with the mountaineering skills necessary to apply combat power in a rugged mountain environment, and they develop leaders capable of applying doctrine to the distinct characteristics of mountain warfare.

10-1. **TRAINING**. Military mountaineering training provides tactical and fundamental mobility skills needed to move units safely and efficiently in mountainous terrain. Rangers should receive extensive training prior to executing combat operations in mountainous environments. Some of the areas to train are as follows:

- Characteristics of the mountain environment.
- Care and use of basic mountaineering equipment.
- Mountain bivouac techniques.
- Mountain communications.
- Mountain travel and walking techniques.
- Mountain navigation, hazard recognition and route selection.
- Rope management and knots.
- Natural and artificial anchors.
- Belay and rappel techniques.
- Installation construction and use such as rope bridges.
- Rock climbing fundamentals.
- Rope bridges and lowering systems.
- Individual movement on snow and ice.
- Mountain stream crossings (to include water survival techniques).

10-2. **DISMOUNTED MOBILITY**. Movement in more technical terrain demands specialized skills and equipment. Before Rangers can move in such terrain, a technical mountaineering team might have to secure the high ground.

10-3. **MOUNTAINEERING EQUIPMENT**. Mountaineering equipment is everything that allows the trained Ranger to accomplish many tasks in the mountains. The importance of this gear to the mountaineer is no less than that of the rifle to the Infantry Soldier.

a. **Ropes and Cords**. Ropes and cords are the most important pieces of mountaineering equipment. They secure climbers and equipment during steep ascents and descents. They are also used to install other ropes and to haul equipment. From WWII until the 1980's, US armed forces mainly used 7/16 *nylon laid rope*, often referred to as *green line* for all mountaineering operations. Since kernmantle ropes were introduced, other, more specialized ropes have in many cases replaced the all-purpose green line. Kernmantle ropes are made much like parachute cord, with a smooth sheath around a braided or woven core. Laid ropes are still used, but should be avoided whenever rope failure could cause injury or loss of equipment.

(1) ***Types***. Kernmantle rope is classified as *dynamic* or *static*:

(a) *Dynamic Ropes*. Ropes used for climbing are classified as *dynamic ropes*. These rope stretch or elongate 8 to 12 percent once subjected to weight or impact. This stretching is critical in reducing the impact force on the climber, anchors, and belayer during a fall by softening the catch. An 11-mm by 150-meter rope is standard for military use. Other sizes are available.

(b) *Static Ropes*. Static ropes are used in situations where rope stretch is undesired, and when the rope is subjected to heavy static weight. Static ropes should never be used while climbing, since even a fall of a few feet could generate enough impact force to injure climber and belayer, and/or cause anchor failure. Static Ropes are usually used when constructing rope bridges, fixed rope installations, vertical haul lines, and so on.

(2) ***Sling Ropes and Cordelettes***. A short section of static rope or static cord is referred to as a sling rope or cordelette. These are a critical piece of personal equipment during mountaineering operations. They are usually 7 to 8mm diameter and up to 21 feet long. The minimum Ranger standard is 8mm by 15 feet.

(3) ***Rope Care***. Ropes used daily should be used no longer that one year. An occasionally used rope can be used generally up to 4 to 5 years if properly cared for.

- Inspect ropes thoroughly before, during, and after use for cuts, frays, abrasion, mildew, and soft or worn spots.
- Never step on a rope or drag it on the ground unnecessarily.
- Avoid running rope over sharp or rough edges (pad if necessary).

- Keep ropes away from oil, acids, and other corrosive substances.
- Avoid running ropes across one another under tension (nylon to nylon contact will damage ropes).
- Do not leave ropes knotted or under tension longer than necessary.
- Clean in cool water, loosely coil, and hang to dry out of direct sunlight. Ultraviolet light rays harm synthetic fibers. When wet, hang rope to drip-dry on a rounded wooden peg, at room temperature (do not apply heat).

(4) ***Webbing and Slings***. Loops of tubular webbing or cord, called slings or runners, are the simplest pieces of equipment and some of the most useful. The uses for these simple pieces are endless, and they are a critical link between the climber, the rope, carabiners, and anchors. Runners are usually made from either 9/16- or 1-inch tubular webbing and are either tied or sewn by a manufacturer.

b. **Carabiners**. The carabiner is one of the most versatile pieces of equipment available in the mountains. This simple piece of gear is the critical connection between the climber, his rope, and the protection attaching him to the mountain. New metal alloys are stronger to hold hard falls; and, they are lighter than steel, which allows the Ranger to carry many at once. Consequently, steel carabiners are used less and less often. Basic carabiners come in several different shapes.

c. **Protection**. This is the generic term used to describe a piece of equipment, natural or artificial, that is used to construct an anchor. Protection, along with a climber, belayer, and climbing rope, forms the lifeline of the climbing team. The rope connects two climbers, and the protection connects them to the rock or ice. The two types of artificial protection are traditional (removable) and fixed (usually permanent) (Figures 10-1 and 10-2).

Figure 10-1. EXAMPLES OF TRADITIONAL (REMOVABLE) PROTECTION USED ON ROCK

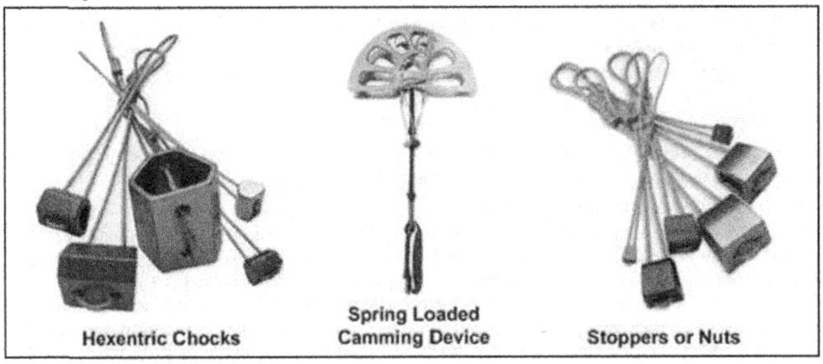

Figure 10-2. EXAMPLES OF FIXED (PERMANENT OR SEMI) PROTECTION USED ON ROCK

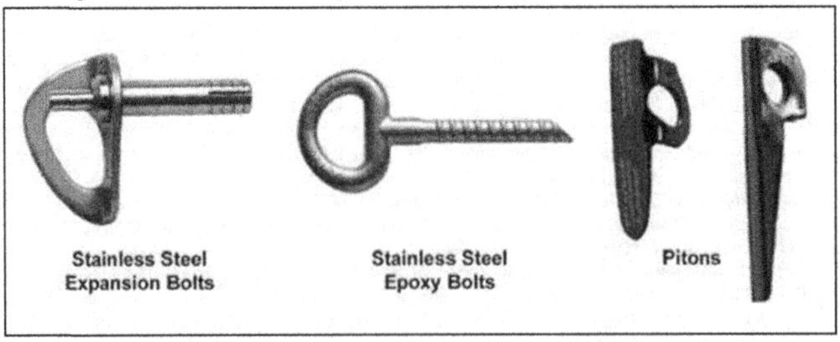

10-4. **ANCHORS**. Anchors form the base for all installations and roped mountaineering techniques. Anchors must be strong enough to support the entire weight of the load or of the impact placed upon them. Several pieces of artificial or natural protection may be incorporated together to make one multipoint anchor. Anchors are classified as artificial or natural:

a. **Artificial Anchors**. Artificial anchors are made from man-made materials. The most common ones use traditional or fixed protection (Figure 10-3).

Figure 10-3. CONSTRUCTION OF THREE-POINT, PRE-EQUALIZED ANCHOR WITH FIXED ARTIFICIAL PROTECTION

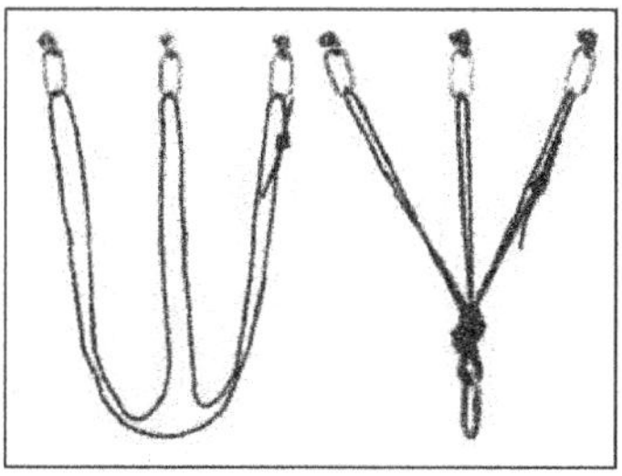

b. **Natural Anchors**. Natural anchors are usually very strong and often simple to construct with little equipment. Trees, shrubs, and boulders are the most common anchors. All natural anchors require is a method of attaching a rope. Regardless of the type of natural anchor used, it must support the weight of the load.

(1) ***Trees***. These are probably the most widely used of all anchors. In rocky terrain, trees usually have a very shallow root system. Check this by pushing or tugging on the tree to see how well it is rooted. Anchor as low as possible to keep the pull on the base of the tree, where it is nearer its own anchor and therefore stronger. Use padding on soft, sap-producing trees to keep sap off ropes and slings.

(2) ***Rock Projections and Boulders***. You can use these, but they must be heavy enough, and they must have a stable enough base to support the load.

(3) ***Bushes and Shrubs***. If no other suitable anchor is available, route a rope around the bases of several bushes. As with trees, place the anchoring rope as low as possible. Make sure all vegetation is healthy and well rooted to the ground.

(4) ***Tensionless Anchor***. Use this to anchor rope on high-load installations such as bridging. The wraps of the rope around the anchor (Figure 10-4) absorb the tension, keeping it off the knot and carabiner. Tie this anchor with at least four wraps around the anchor. A smooth anchor, such as a pipe or rail, might require several more wraps. Wrap the rope from top to bottom. Place a fixed loop into the end of the rope, and attach it loosely back onto the rope with a carabiner.

Figure 10-4. TENSIONLESS NATURAL ANCHOR

10-5. **KNOTS.**

a. **Square Knot**. This joins two ropes of equal diameter (Figure 10-5). It has two interlocking bights; the running ends exit on the same side of the standing portion of the rope. Each tail is secured with an overhand knot on the standing end. When you dress the knot, leave at least a 4-inch tail on the working end.

Figure 10-5. SQUARE KNOT

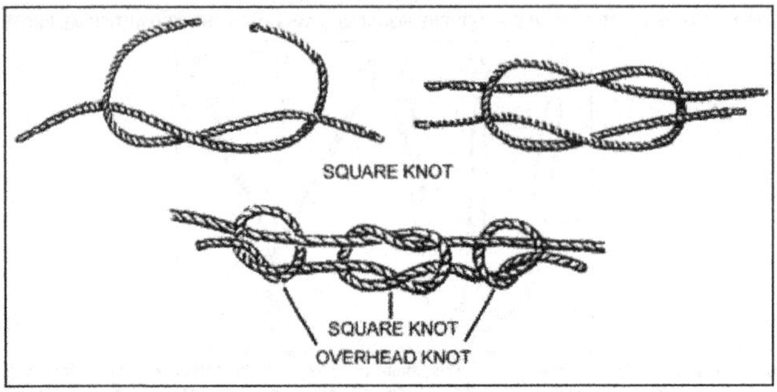

b. **Round-Turn with Two Half Hitches**. This is a constant tension anchor knot (Figure 10-6). The rope forms a complete turn around the anchor point (thus the name round turn), with both ropes parallel and touching, but not crossing. Both half hitches are tightly dressed against the round turn, with the locking bar on top. When you dress the knot, leave at least a 4-inch tail on the working end.

Figure 10-6. ROUND-TURN WITH TWO HALF-HITCHES

c. **End-of-the-Rope Clove Hitch**. The end-of-the-rope clove hitch (Figure 10-7) is an intermediate anchor knot that requires constant tension. Make two turns around the anchor. A locking bar runs diagonally from one side to the other. Leave no more than one rope width between turns of rope. Locking bar is opposite direction of pull. When you dress the knot, leave at least a 4-inch tail on the working end.

Figure 10-7. END-OF-THE-ROPE CLOVE HITCH

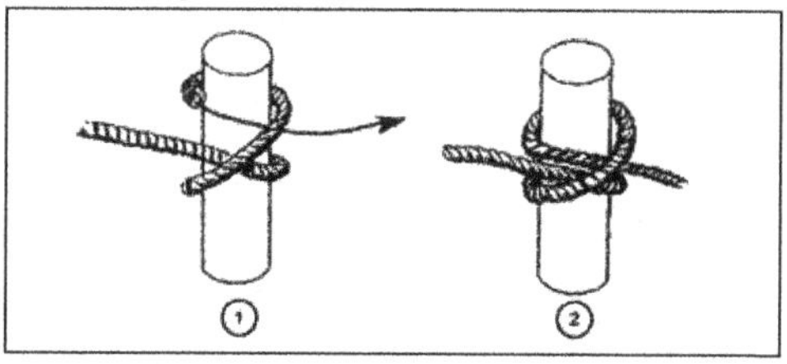

d. **Middle of the Rope Clove Hitch**. This knot (Figure 10-8) secures the middle of a rope to an anchor. The knot forms two turns around the anchor. A locking bar runs diagonally from one side to the other. Leave no more than one rope width between turns. Ensure the locking bar is opposite the direction of pull.

Figure 10-8. MIDDLE-OF-THE-ROPE CLOVE HITCH

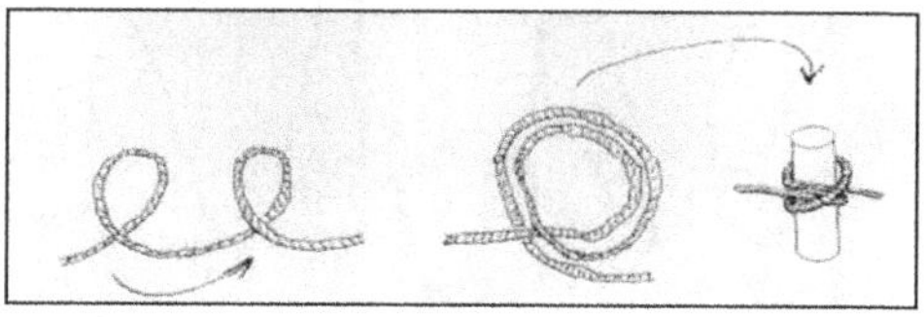

e. **Rappel Seat**. The rappel seat (Figure 10-9) is a rope harness used in rappelling and climbing. It can be tied for use with the left or right hand. Leg straps do not cross, and are centered on buttocks and tight. Leg straps form locking half hitches on rope around waist. A square knot is tied on the right hip and finished with two overhand knots. Tails must be even, no longer than 6 inches. A carabiner is inserted around all ropes, with the gate opening up and away. The carabiner must not contact the square or overhand knot. The rappel seat must be tight enough to prevent a fist from fitting between the rappeller's body and the harness.

Figure 10-9. RAPPEL SEAT

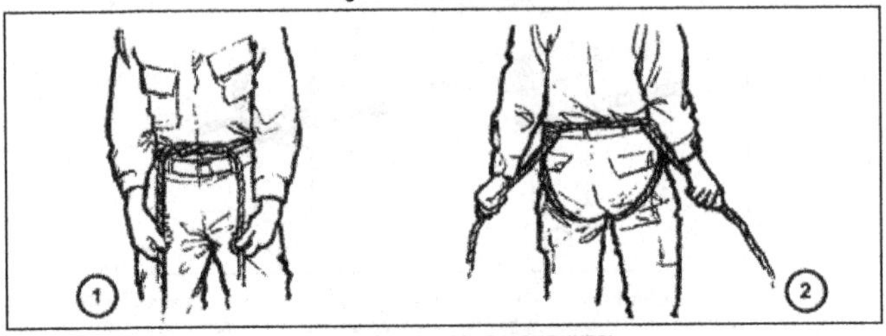

Figure 10-9. RAPPEL SEAT (CONTINUED)

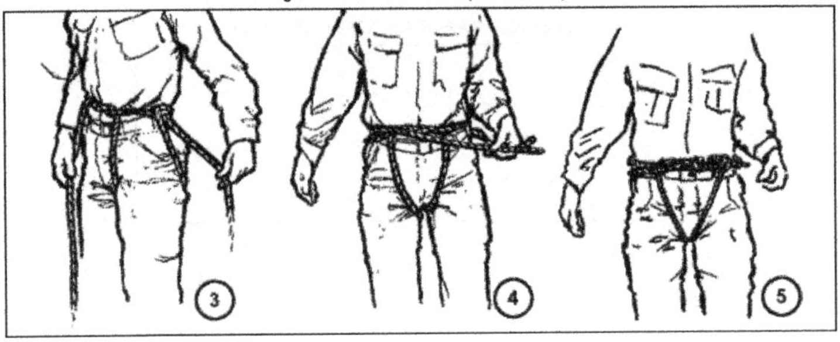

f. **Double Figure Eight**. Use a Figure Eight loop knot (Figure 10-10) to form a fixed loop in the end of the rope. It can be tied at the end of the rope or anywhere along the length of the rope. Figure Eight loop knots are formed by two ropes parallel to each other in the shape of a Figure Eight, no twists are in the Figure Eight. Fixed loops are large enough to insert a carabiner. When you dress the knot, leave at least a 4-inch tail on the working end.

Figure 10-10. DOUBLE FIGURE EIGHT-LOOP KNOT

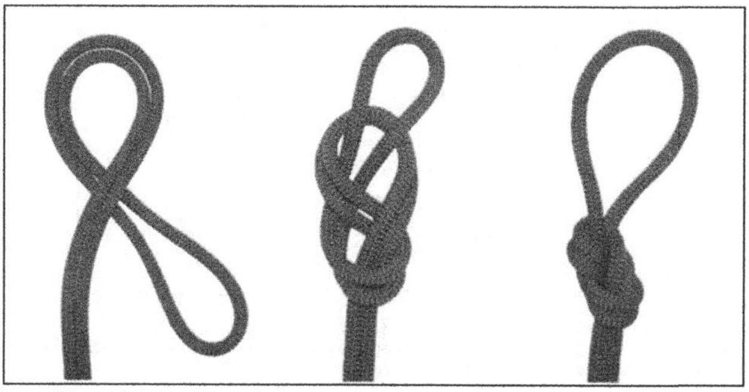

g. **Rerouted Figure Eight Knot**. This anchor knot also attaches a climber to a climbing rope. Form a Figure Eight in the rope, and pass the working end around an anchor. Reroute the end back through to form a double Figure Eight (Figure 10-11). Tie the knot with no twists. When you dress the knot, leave at least a 4-inch tail on the working end.

Figure 10-11. REROUTED FIGURE EIGHT KNOT

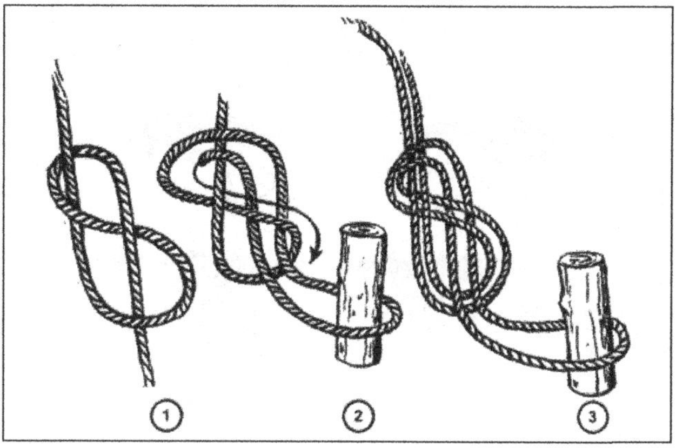

h. **Figure Eight Slip Knot**. The Figure Eight slip knot is used to form an adjustable bight in the middle of a rope. Knot is in the shape of a Figure Eight. Both ropes of the bight pass through the same loop of the Figure Eight. The bight is adjustable by means of a sliding section (Figure 10-12).

Figure 10-12. FIGURE EIGHT SLIP KNOT

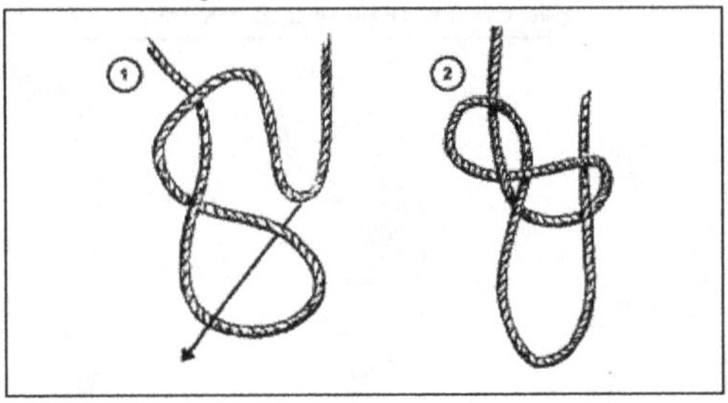

i. **End-of-the-Rope Prusik**. The End-of-the-Rope Prusik (Figure 10-13) attaches a movable rope to a fixed rope. The knot has two round turns, with a locking bar perpendicular to the standing end of the rope. Tie a bowline within 6 inches of the locking bar. When you dress the knot, leave at least a 4-inch tail on the working end.

Figure 10-13. END-OF-THE-ROPE PRUSIK

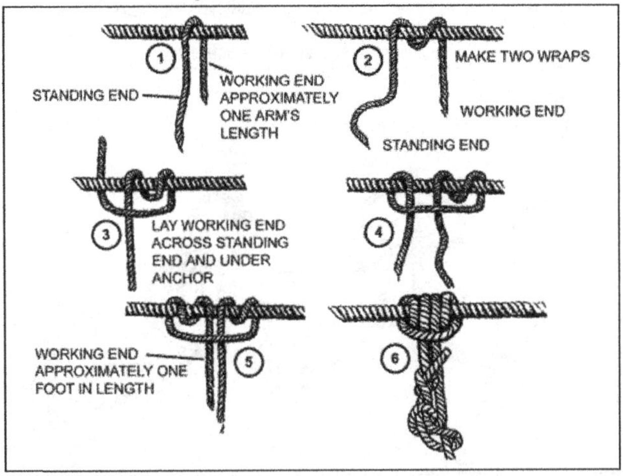

j. **Middle-of-the-Rope Prusik**. The middle-of-the-rope Prusik (Figure 10-14) attaches a movable rope to a fixed rope, anywhere along the length of the fixed rope. Make two round turns with a locking bar perpendicular to the standing end. Ensure the wraps do not cross and that the overhand knot is within 6 inches from the horizontal locking bar. Ensure the knot does not move freely on fixed rope.

Figure 10-14. MIDDLE-OF-THE-ROPE PRUSIK

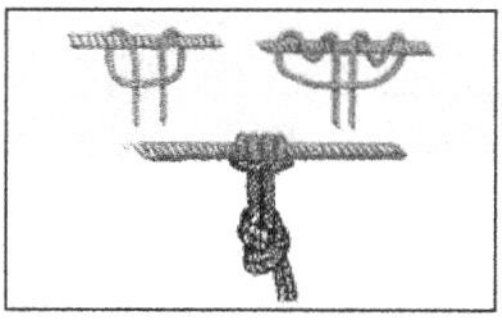

k. **Bowline on a Coil**. This knot (Figure 10-15) secures a climber to the end of a climbing rope. Make at least four parallel wraps around the body, between the hipbone and lower set of ribs. Keep all coils free of clothing, touch and are tight enough to ensure that a fist cannot be inserted between the wraps and the body. Three distinct coils show through the bight of the bowline. The rope coming off the bottom of the coils is on the right side, forward of the hip and forms the bight and the overhand knot. The rope coming off the top of the coils is on the left side, forward of the hips and forms the third and final coil showing through the bight of the bowline. Bowline is centered on the gig line and secured with an overhand knot with a minimum of 4 inch tail remaining after the knot is dressed.

Figure 10-15. BOWLINE ON A COIL

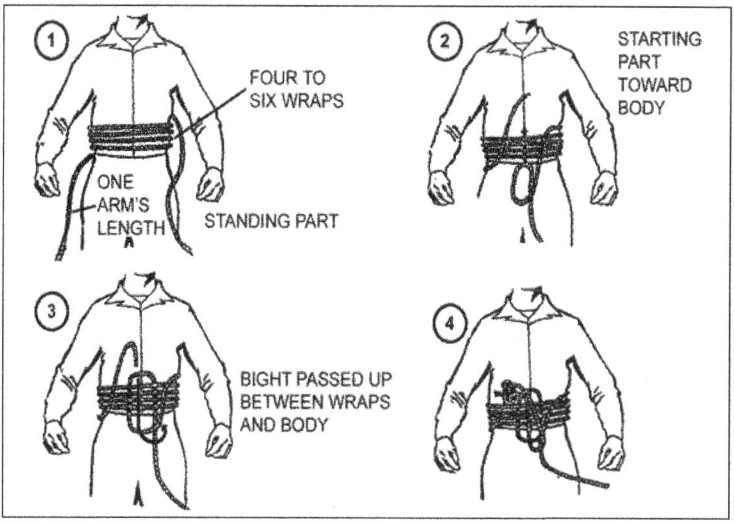

10-6. **BELAYS**. Belaying is any action taken to stop a climber's fall or to control the rate a load descends. The belayer also helps manage the rope for a climber by controlling the amount of rope that is taken out or in, or by controlling a rappeller's rate of descent. The belayer must be stably anchored to prevent him from being pulled out of his position, and losing control of the rope. The two types of belays are body and mechanical belays.

a. **Body Belay**. This belay (Figure 10-16) uses the belayer's body to apply friction by routing the rope around the belayer's body. Caution must be used when belaying from the body since the entire weight of the load may be placed on the belayer's body.

Figure 10-16. BODY BELAY

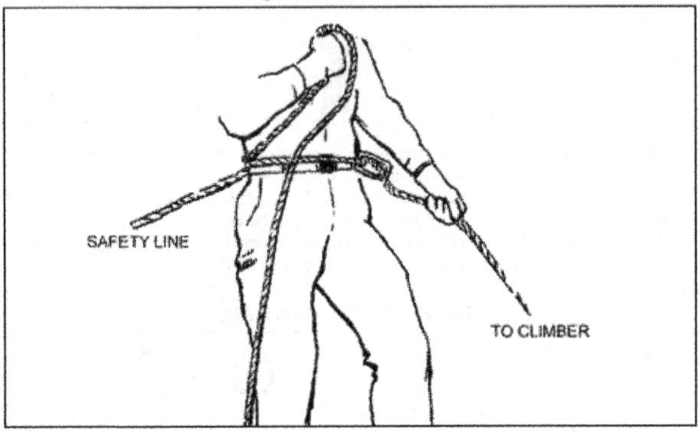

b. **Mechanical Belay**. This belay (Figure 10-17) uses mechanical devices to help the belayer control the rope, as in rappelling. A variety of devices are used to construct a mechanical belay for mountaineering.

Figure 10-17. MECHANICAL BELAY

(1) ***Munter Hitch***. One of the most often used belays, the munter hitch (Figure 10-18), requires very little equipment. The rope is routed through a locking, pear-shaped carabiner, then back on itself. The belayer controls the rate of descent by manipulating the working end back on itself with the brake hand.

Figure 10-18. MUNTER HITCH

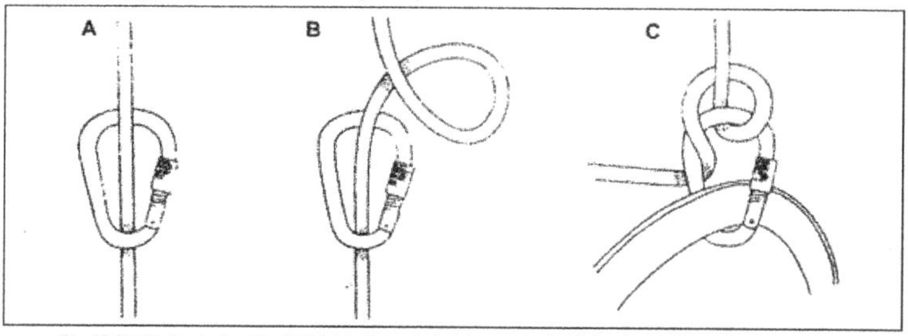

(2) ***Air Traffic Controller***. The ATC (air traffic controller) is a mechanical belay device (Figure 10-19) that locks down on itself when tension is applied in opposite directions. The belayer need apply very little force with his brake hand to control descent or arrest a fall.

Figure 10-19. AIR TRAFFIC CONTROLLER

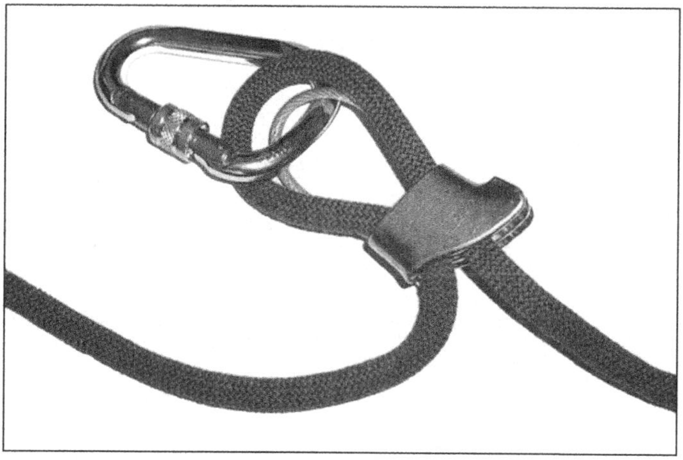

10-7. **CLIMBING COMMANDS**. Table 10-1 shows the sequence of commands used by climber and belayer.

Table 10-1. SEQUENCE OF CLIMBING COMMANDS

SEQUENCE	*COMMAND*	*GIVEN BY*	*MEANING*
1.	BELAY ON, CLIMB	Belayer	Alerts climber that belay is on and climber may climb.
2.	CLIMBING	Climber	Alerts belayer that climber is climbing.
3.	UP-ROPE	Climber	Belayer, remove excess slack in the rope.
4.	BRAKE	Climber	Alerts belayer to immediately apply brake.
5.	FALLING	Climber	Alerts belayer that climber is falling and that the belayer should immediately apply the brake and prepare to arrest the fall.
6.	TENSION	Climber	Alerts belayer to remove all slack from climbing rope until rope is tight, then apply brake and hold position.
7.	SLACK	Climber	Alerts belayer to pull slack into the climbing rope (Belayer may have to assist)
8.	ROCK	Anyone	Alerts everyone that an object is about to fall near them. Belayer immediately applies brake.
9.	POINT	Climber	Alerts belayer that the direction of pull on the climbing rope has changed in the event of a fall.
10.	STAND-BY	Climber or Belayer	Alerts the other to hold position, stand by, I am not ready.
11.	DO YOU HAVE ME?	Climber	Alerts belayer to prepare for a fall or to prepare to lower the climber.
12.	I HAVE YOU	Belay	Alerts climber that the brake is on and the belayer is prepared for the climber to fall, or to lower him.
13.	OFF-BELAY	Climber	Alerts belayer that climber is safetied in or it is safe to come off belay.
14.	3-METERS	Belayer	Alerts climber to the amount of rope between climber and belayer (May be given in feet or meters)
15.	BELAY-OFF	Belayer	Alerts climber that belayer is off belay

10-8. **ROPE INSTALLATIONS**. Rope installations may be constructed by teams to assist units in negotiating natural and manmade obstacles. installation teams consist of a squad sized element, with 2 to 4 trained mountaineers. Installation teams deploy early and prepare the AO for safe, rapid movement by constructing various types of mountaineering installations. Following construction of an installation, the squad, or part of it, remains on site to secure and monitor the system, assist with the control of forces across it, and make adjustments or repairs during its use. After passage of the unit, the installation team may then disassemble the system and deploy to another area as needed.

a. **Fixed Rope Installations**. A fixed rope is anchored in place to help Rangers move over difficult terrain. Its simplest form is a rope tied off on the top of steep terrain. As terrain becomes steeper or more difficult, fixed rope systems may require intermediate anchors along the route. Planning considerations are as follows:

- Does the installation allow you to bypass the obstacle?
- (Tactical considerations) Can obstacle be secured from construction through negotiation, to disassembly?
- Is it in a safe and suitable location? Is it easy to negotiate? Does it avoid obstacles?
- Are natural and artificial anchors available?
- Is the area safe from falling rock and ice?

b. **Vertical Hauling Line**. This installation (Figure 10-20) is used to haul men and equipment up vertical or near vertical slopes. It is often used in conjunction with the fixed rope.

Figure 10-20. VERTICAL HAULING LINE

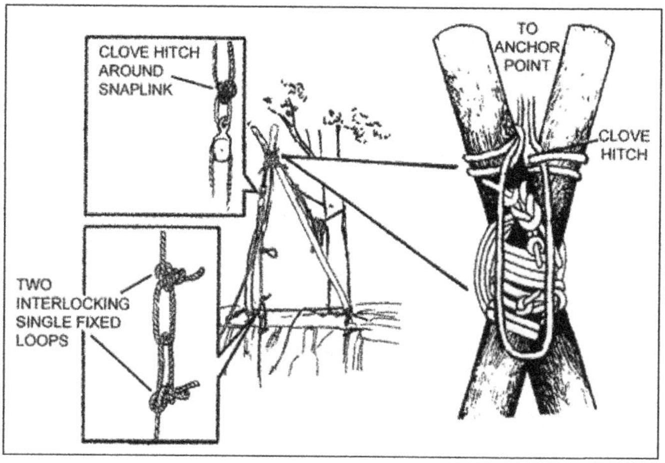

(1) ***Planning Considerations***.
- Does the installation allow you to bypass the obstacle?
- (Tactical considerations) Can you secure the installation from construction through negotiation, to disassembly?
- Does it have good loading and off-loading platforms? Are the platforms natural and easily accessible? Do they provide a safe working area?
- Does it allow sufficient clearance for load? Is there enough space between the slope and the apex of the A-frame to allow easy on-loading and off-loading of troops and equipment?
- Does it have an A-Frame for artificial height?
- Does it allow you to haul line in order to move personnel and equipment up and down slope?
- Does the A frame have a pulley or locking carabiner to ease friction on hauling line?
- Does it have a knotted hand line to help Rangers up the installation?
- Does it allow for Rangers top and bottom to monitor safe operation?

(2) ***Equipment***.
- Three 120-foot (50m) static ropes.
- Three 15-foot sling ropes for constructing A-frame.
- Two A-frame poles, 7 to 9 feet long, 4 to 6 inches in diameter (load dependent).
- Nine carabiners.
- One pulley with steel locking carabiner.

c. **Bridges**. Rope bridges are employed in mountainous terrain to bridge linear obstacles such as streams or rivers where the force of flowing water may be too great or temperatures are too cold to conduct a wet crossing.

(1) ***Construction***. The rope bridge is constructed using a static ropes. The max span that can be bridged is 50 percent of the ropes entire length for a dry crossing, 75 percent for a wet crossing. The ropes are anchored with an anchor knot on the far side

of the obstacle, and are tied off at the near end with a transport tightening system. Rope Bridge Planning Considerations are as follows:

- Does the installation allow you to bypass the obstacle?
- (Tactical considerations) Can you secure the installation from construction through negotiation, to disassembly?
- Is it in the most suitable location such as a bend in the river? Is it easily secured?
- Does it have near and far-side anchors?
- Does it have good loading and off-loading platforms?

(2) ***Equipment (1-Rope Bridge)***.

- One sling rope per-man.
- One steel locking carabineer.
- Two steel ovals.
- Two 120-foot static ropes.

(3) ***Construction Steps***. The first Ranger swims the rope to the far side and ties a tensionless anchor (Fig 10-4) between knee and chest level, with at least 6 to 8 wraps. The BTC ties a transport tightening system (Fig 10-21) to the near side anchor point. Then, he ties a Figure Eight slipknot and incorporates a locking half hitch around the adjustable bight. Insert 2 steel oval carabineers into the bight so the gates are opposite and opposed. The rope is then routed around the near side anchor point at waist level and dropped into the steel oval carabineers.

- A three-Ranger pulling team moves forward from the PLT. No more than three men should tighten the rope. Using more can overtighten the rope, bringing it near failure.
- Once the rope bridge is tight enough, the bridge team secures the transport-tightening system using two half hitches, without losing more than 4 inches of tension.
- Personnel cross using the commando crawl, rappel seat, or monkey crawl method (Figures 10-22 thru 24).

Figure 10-21. TRANSPORT TIGHTENING SYSTEM

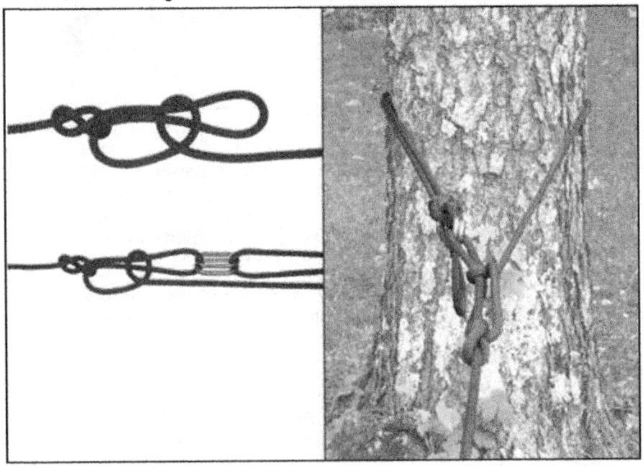

Figure 10-22. COMMANDO CRAWL METHOD

Figure 10-23. RAPPEL SEAT (TYROLEAN TRAVERSE) METHOD

Figure 10-24. MONKEY CRAWL METHOD

(4) ***Bridge Recovery***. Once all but two Rangers have crossed the rope bridge, the bridge team commander (BTC) chooses either the wet or dry method to dismantle it. If he chooses the dry method, he should he should first anchor his tightening system with the transport knot.

- The BTC back-stacks all of the slack coming out of the transport knot, then ties a fixed loop and places a carabiner into the fixed loop.
- The next to last Ranger to cross attaches the carabiner to his rappel seat or harness, and then moves across the bridge using the Tyrolean traverse method.
- The BTC then removes all knots from the system. The far side remains anchored. The rope should now only pass around the near side anchor.
- A three-man pull team, assembled on the far side, takes the end brought across by the next to last Ranger and pulls the rope tight again and holds it.
- The BTC then attaches himself to the rope bridge and moves across.
- Once across, the BTC breaks down the far side anchor, removes the knots, and then pulls the rope across. If the BTC chooses a wet crossing, any method can be used to anchor the tightening system.
- All personnel cross except the BTC or the strongest swimmer.
- The BTC then removes all knots from the system.
- The BTC ties a fixed loop, inserts a carabiner, and attaches it to his rappel seat or harness. He then manages the rope as the slack is pulled to the far side.
- The BTC then moves across the obstacle while being belayed from the far side.

d. **Suspension Traverse**. The suspension traverse is used to move personnel and equipment over rivers, ravines, chasms, and up or down a vertical obstacle. By combining the transport tightening system used during the Rope bridge, an A-Frame used for the Vertical haul Line (Figure 10-25), and belay techniques device, Units can construct a suspension traverse (Figures 10-26 and 10-27). Installation of a suspension traverse can be time-consuming and equipment-intensive. All personnel must be well trained and rehearsed in the procedures.

(1) ***Construction***. The suspension traverse must be constructed using a static ropes. The max span that can be bridged is generally 75 percent length of the shortest rope. Planning considerations are the same as rope bridge and vertical haul line combined.

(2) ***Equipment***.

- Three static installation ropes.
- Seven sling Ropes.
- Nine carabiners.
- One heavy duty double pulley.
- One locking carabiner.
- One canvas pad.

Figure 10-25. ANCHORING THE TRAVERSE ROPE TO THE A-FRAME

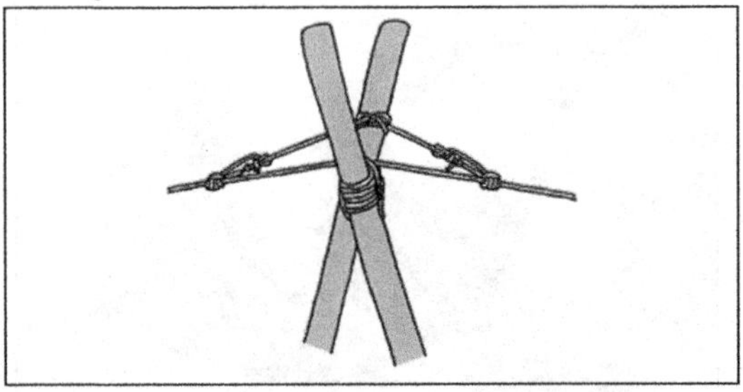

Figure 10-26. CARRYING ROPE FOR USE ON A SUSPENSION TRAVERSE

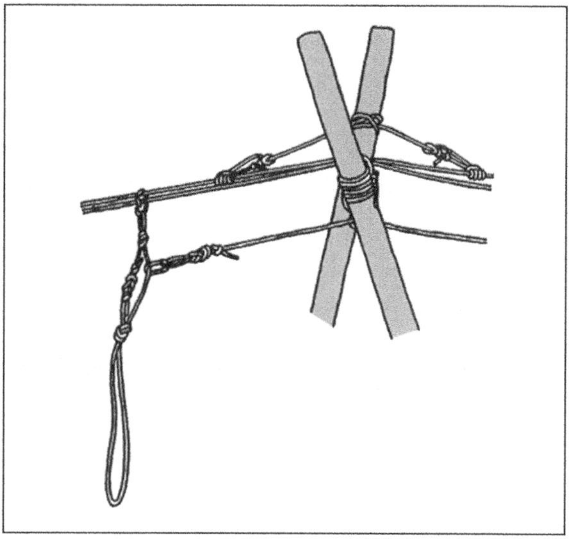

Figure 10-27. SUSPENSION TRAVERSE

10-9. **RAPPELLING**. Rappelling is a quick method of descent but can be extremely dangerous. These dangers include anchor failure, equipment failure, and individual error. Anchors in a mountainous environment should be selected carefully. Great care must be taken to load the anchor slowly and to ensure that no excessive stress is placed on the anchor. The best way to ensure this is to prohibit bounding rappels, and to use only walk-down rappels.

a. **Hasty and Body Rappels**. Hasty (Figure 10-28) and body rappels (Figure 10-29) are easier and faster than other methods, but should only be used on moderate pitches--never on vertical or overhanging terrain. Wear gloves to prevent rope burns.

Figure 10-28. Hasty Rappel

Figure 10-29. Body Rappel

b. **Seat-Hip Rappel**. The seat-hip rappel uses a mechanical rappel device that is inserted in a sling rope seat and is fastened to the rappeller to provide the necessary friction to remain in control. This method provides a faster and more controlled descent than any other method (Figure 10-30). There are several mechanical rappel devices (Figure 10-31 and 10-32) that can be used for the seat-hip rappel.

Figure 10-30. FIGURE EIGHT DESCENDER

Figure 10-30. FIGURE EIGHT DESCENDER (CONTINUED)

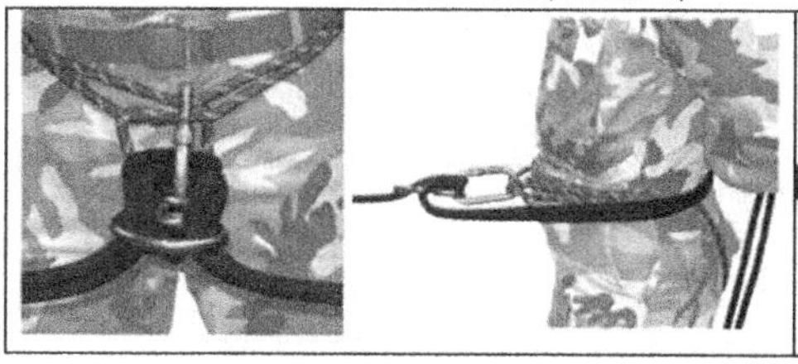

Figure 10-31. CARABINER WRAP

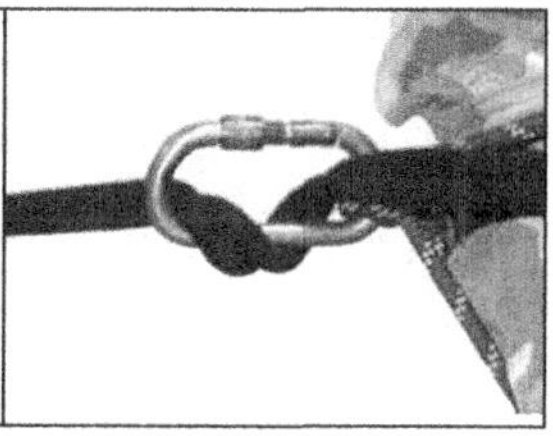

Figure 10-32. SEAT HIP RAPPEL WITH CARABINER WRAP DESCENDER

c. **Site Selection**. The selection of the rappel point depends on factors such as mission, cover, route, anchor points, and edge composition (loose or jagged rocks). There must be good anchors (primary and secondary). The anchor point should be above the rappeller's departure point. Suitable loading and off-loading platforms should be available.

- Each rappel point has primary and secondary anchors.
- Rappel point has equal tension between all anchor points.
- Double rope is used when possible.
- Ropes must reach the off loading platform.
- Site has suitable on and off loading platforms.
- Personnel working near the edge are tied in.
- Select a smooth route free of loose rock and debris.

Chapter 11
EVASION AND SURVIVAL

This chapter will cover basic material dealing with evading the enemy and survival techniques. Field expedient methods of acquiring food/ water, land navigation, constructing shelters, and making fires will be presented.

11-1. **EVASION**. When you become isolated or separated in a hostile area, either as an individual or as a group, your evasion and survival skills will determine whether or not you return to friendly lines.

a. When unable to continue the mission or unable to rejoin your unit, leave the immediate area, and move to your last rally point.

b. Observe activity in the area and form a plan.

c. Traveling alone offers the least possibility of detection, but traveling in groups of two to three is more desirable.

d. Plan a primary and alternate route. Consider distance, cover, food, and water. The easiest and shortest route may not be the best.

e. Food and water are daily requirements. You can do without food for several days; water, however, is essential.

f. Move at night. Use the daylight to observe, plan, and rest in a hide position.

g. Linkup only during daylight hours. Place friendly lines under observation.

h. Attempt to identify the unit you approach, note their movements and routine.

i. After carefully considering your approach route, make voice contact with the unit as soon as possible.

11-2. **SURVIVAL**.

With training, equipment, and the ***will to survive***, you can overcome *any* obstacle you may face. ***You will survive.*** Understand the emotional states associated with survival. "Knowing thyself" is extremely important in a survival situation. It bears directly on how well you cope with serious stresses, anxiety, pain, injury, illness; cold, heat, thirst, hunger, fatigue, sleep deprivation, boredom, loneliness and isolation.

a. **Memory Aid**. You can overcome and reduce the shock of being isolated behind enemy lines if you keep the key word S-U-R-V-I-V-A-L foremost in your mind. Its letters can help guide you in your actions.

(1) S - Size up the situation, the surroundings, your physical condition, and your equipment.

(2) U - Undue haste makes waste; don't be too eager to move. Plan your moves.

(3) R - Remember where you are relative to friendly and enemy units and controlled areas; water sources (most important in the desert); and good cover and concealment. This information will help you make intelligent decisions.

(4) V - Vanquish fear and panic.

(5) I – Improvise/Imagine. You can improve your situation. Learn to adapt what is available for different uses. Use your imagination.

(6) V - Value living. Remember your goal - to get out alive. Remain stubborn. Refuse to give in to problems and obstacles. This will give you the mental and physical strength to endure.

(7) A - Act like the natives; watch their daily routines and determine when, where, and how they get their food and water.

(8) L - Live by your wits. Learn basic skills.

b. **Navigation**. In a survival situation, you might find yourself without a compass. The ability to determine direction can enable you to navigate back to your unit or to a friendly sanctuary. In sunlight, there are two simple ways to determine direction: the shadow-tip and the watch.

(1) ***Shadow Tip***. Use the sun to find approximate true North. Use this in light bright enough to cast shadows. Find a fairly straight stick about 3 feet long, and follow the diagram below (Figure 11-l).

Figure 11-1. SHADOW-TIP METHOD

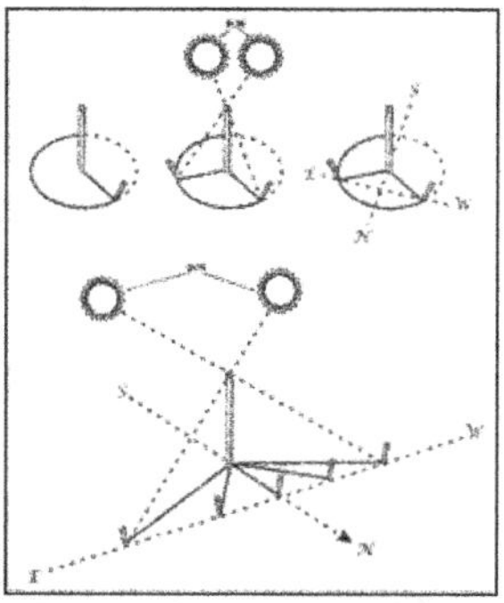

(2) ***Watch Method***. You can also determine direction using a watch (Figure 11-2). The steps you take will depend on whether you are in the northern Temperate Zone or in the southern Temperate Zone. The northern temperate zone is located between 23.4 north and 26.6 north. The southern Temperate Zone is located between 23.4 south and 66.6 south.

Figure 11-2. WATCH METHOD

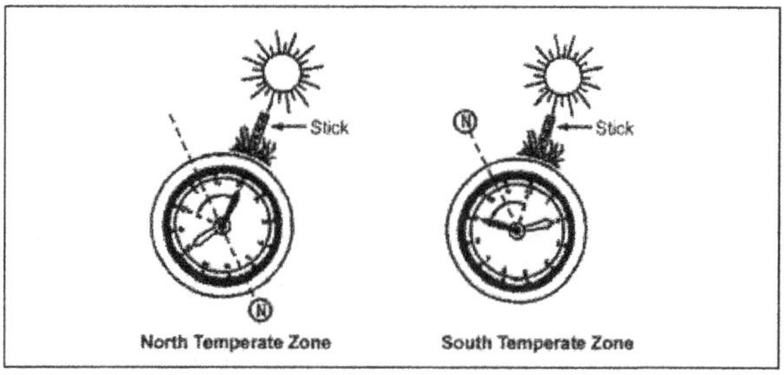

c. **Northern Temperate Zone**. Procedures in the northern temperate zone using a conventional watch are as follows:

(1) Place a small stick in the ground so that it casts a definite shadow.

(2) Place your watch on the ground so that the hour hand points toward and along the shadow of the stick.

(3) Find the point on the watch midway between the hour hand and 12 o'clock and draw an imaginary line from that point through and beyond the center of the watch. This imaginary line is a north-south line. You can then tell the other directions.

Note: If your watch is set on daylight savings time, then use the midway point between the hour hand and 1 o'clock to draw your imaginary line.

d. **Southern Temperate Zone**. Procedures in the southern temperate zone using a conventional watch are as follows:

(1) Place a small stick in the ground so that it casts a definite shadow.

(2) Place your watch on the ground so that 2 o'clock points to and along the shadow.

(3) Find the midway point between the hour and 12 o'clock and draw an imaginary line from the point through and beyond the center of the watch. This is a north-south line.

e. A hasty shortcut using a conventional watch is simply to point the hour hand at the sun in the northern temperate zone (or point the 12 at the sun in the southern temperate zone) and then follow the last step of the watch method above to find your directions. This shortcut, of course, is not as accurate as the regular method but quicker. Your situation will dictate which method to use.

11-3. **WATER**. Water is one of your most urgently needed resources in a survival situation. You can't live long without it, especially in hot areas where you lose so much through sweating. Even in cold areas, you need a minimum of 2 quarts of water a day to maintain efficiency. More than three-fourths of your body is composed of fluids. Your body loses fluid as a result of heat, cold, stress, and exertion. The fluid your body loses must be replaced for you to function effectively. So, one of your first objectives is to obtain an adequate supply of water.

a. **Purification**. Purify all water before drinking. Either--

(1) Boil it for at least one minute (plus 1 more minute for each additional 1,000 feet above sea level) or for a maximum of 10 minutes anywhere.

(2) Use water purification tablets.

(3) Add eight drops of 2-1/2 percent iodine solution to a quart (canteen full) of water. Let it stand for 10 minutes before drinking.

(4) Collect rain water directly in clean containers or on plants. This is generally safe to drink without purifying. Never drink urine or sea water; the salt content is too high. Use old, bluish sea ice, but newer, grayer ice may be salty. Glacier ice is safe to melt and drink.

b. **Desert Environment**. In a desert environment, water has a huge effect on Rangers. If a unit fails to plan properly for water, and resupply is unavailable, then they can run out of water. In the desert, look for four signs of water: animal trails, vegetation, birds, and civilization. Adequate water is critical in a hot environment if a unit is to survive and maintain the physical condition necessary to accomplish the mission. Unit leaders must enforce water discipline and plan for water resupply. The leader can use the following planning considerations for water resupply:

(1) Units average water consumption.

(2) Drop sites.

(3) Aviation support.

(4) DZ and LZ parties.

(5) Caches.

(6) Targets of opportunity (enemy).

c. **Survival Water Still**. Dig a belowground still (Figure 11-3). Select a site where you believe the soil will contain moisture (such as a dry stream bed or a spot where rain water has collected). The soil should be easy to dig, and be in sunlight most of the day:

(1) Dig a bowl-shaped hole about 3 feet across and 2 feet deep.

(2) Dig a sump in center of the hole. The depth and the perimeter of the sump will depend on the size of the container that you have to set inside of it. The bottom of the sump should allow the container to stand upright.

(3) Anchor the tubing to the bottom of the container by forming a loose overhand knot in the tubing.

(4) Place the container upright in the sump.

(5) Extend the unanchored end of the tubing up, over, and beyond the lip of the hole.

(6) Place plastic sheeting over the hole covering the edge with soil to hold it in place.

(7) Place a rock in the center of the plastic.

(8) Allow the plastic to lower into the hole until it is about 15 inches below ground level. The plastic now forms an inverted cone with the rock at its apex. Make sure that the apex of the cone is directly over your container. Also, make sure the plastic cone does not touch the sides of the hole, because the earth will absorb the condensed water.

(9) Put more soil on the edges of the plastic to hold it securely in place and to prevent loss of moisture.

(10) Plug the tube when not being used so that moisture will not evaporate.

d. You can drink water without disturbing the still by using the tube as a straw. You may want to use plants in the hole as a moisture source. If so, when you dig the hole you should dig out additional soil from the sides of the hole to form a slope on which to place the plants. Then proceed as above.

Figure 11-3. SURVIVAL WATER STILL

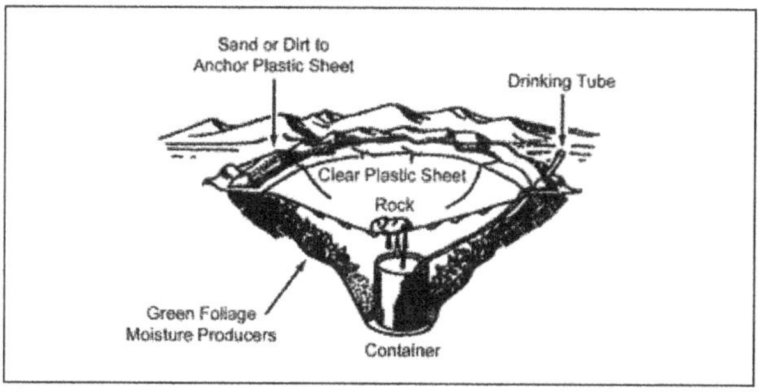

11-4. **PLANT FOOD**. There are many plants throughout the world. Tasting or swallowing even a small portion of some can cause severe discomfort, extreme internal disorders, or death. Therefore, if you have the slightest doubt as to the edibility of a plant, apply the universal edibility test described below before eating any part of it.

a. **Universal Edibility Test**. Before testing a plant for edibility, make sure there are a sufficient number of plants to make testing worth your time and effort. You need more than 24 hours to apply the edibility test outlined below:

(1) Test only one part of a potential food plant at a time.

(2) Break the plant into its basic components, leaves, stems, roots, buds, and flowers.

(3) Smell the food for strong or acid odors. Keep in mind that smell alone does not indicate a plant is edible.

(4) Do not eat for 8 hours before starting the test.

(5) During the 8 hours you are abstaining from eating, test for contact poisoning by placing a piece of the plant you are testing on the inside of your elbow or wrist. Usually 15 minutes is enough time to allow for reaction.

(6) During the test period, take nothing by mouth except purified water and the plant part being tested.

(7) Select a small portion and prepare it the way you plan to eat it.

(8) Before putting the prepared plant part in your mouth, touch a small portion (a pinch) to the outer surface of the lip to test for burning or itching.

(9) If after 3 minutes there is no reaction on your lip, place the plant part on your tongue, holding there for 15 minutes.

(10) If there is no reaction, thoroughly chew a pinch and hold it in your mouth for 15 minutes. *Do not swallow.*

(11) If no burning, itching, numbing, stinging, or other irritation occurs during the 15 minutes, swallow the food.

(12) Wait 8 hours. If any ill effects occur during this period, induce vomiting and drink a lot of water.

(13) If no ill effects occur, eat 1/2 cup of the same plant part prepared the same way. Wait another 8 hours. If no ill effects occur, the plant part as prepared is safe for eating.

b. **Poisonous Plants**. ***Do not*** eat unknown plants that have the below characteristics:

(1) Have a milky sap or a sap that turns black when exposed to air.

(2) Are mushroom like.

(3) Resemble onion or garlic.

(4) Resemble parsley, parsnip, or dill.

(5) Have carrot-like leaves, roots, or tubers.

11-5. **ANIMAL FOOD**. Animal food contains the most food value per pound. Anything that creeps, crawls, swims, or flies is a possible source of food, however you must first catch, kill and butcher it before this is possible. There are numerous methods for catching fish and animals in a survival situation. You can catch fish by using a net across a small stream, (Figure 11-4) or by making fish traps and baskets. Improvise fish hooks and spears as indicated in Figure 11-5, and use them for conventional fishing, spearing and digging.

Figure 11-4. SETTING A GILL NET IN THE STREAM

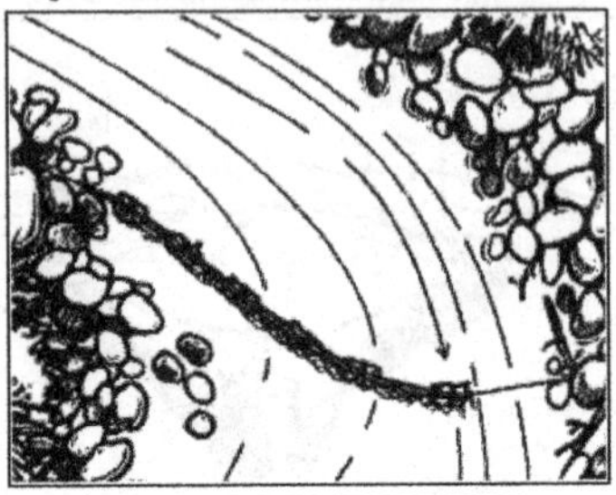

Figure 11-5. SPEARS AND FISH HOOKS

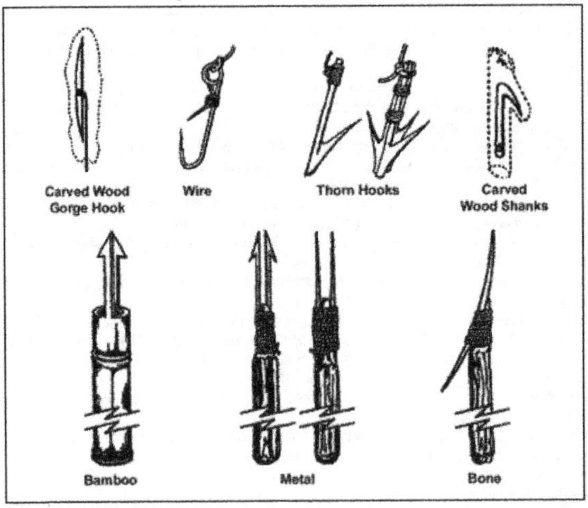

11-6. **TRAPS AND SNARES.** For an unarmed survivor or evader, or when the sound of a rifle shot could be a problem, trapping or snaring wild game is a good alternative. Several well-placed traps have the potential to catch much more game than a Ranger with a rifle is likely to shoot.

a. To be effective with any type of trap or snare, you must—

(1) Know the species of animal you intend to catch.

(2) Know how to construct a proper trap.

(3) Avoid alarming the prey with signs of your presence.

b. There are no catchall traps you can set for all animals. You must determine what species are in a given area and set your traps specifically with those animals in mind. Look for the following:

(1) Runs and trails.
(2) Tracks.
(3) Droppings.
(4) Chewed or rubbed vegetation.
(5) Nesting or roosting sites.
(6) Feeding and watering areas.

c. Position your traps and snares where there is proof that animals pass through. You must determine if it is a "run" or a "trail." A trail will show signs of use by several species and will be rather distinct. A run is usually smaller and less distinct and will only contain signs of one species. You may construct a perfect snare, but it will not catch anything if haphazardly placed in the woods. Animals have bedding areas, waterholes, and feeding areas with trails leading from one to another. You must place snares and traps around these areas to be effective.

d. An evader in a hostile environment must conceal traps and snares. It is equally important, however, to avoid making a disturbance that will alarm the animal and cause it to avoid the trap. Therefore, if you must dig, remove all fresh dirt from the area. Most animals will instinctively avoid a pitfall-type trap. Prepare the various parts of a trap or snare away from the site, carry them in, and set them up. Such actions make it easier to avoid disturbing the local vegetation, thereby alerting the prey. Do not use freshly cut, live vegetation to construct a trap or snare. Freshly cut vegetation will "bleed" sap that has an odor the prey will be able to smell. It is an alarm signal to the animal.

e. You must remove or mask the human scent on and around the trap you set. Although birds do not have a developed sense of smell, nearly all mammals depend on smell even more than on sight. Even the slightest human scent on a trap will alarm the prey and cause it to avoid the area. Removing the scent from a trap is difficult but masking it is relatively easy. Use the fluid from the gall and urine bladders of previous kills. Do not use human urine. Mud, particularly from an area with plenty of rotting vegetation, is also good. Use it to coat your hands when handling the trap and to coat the trap when setting it. In nearly all parts of the world, animals know the smell of burned vegetation and smoke. It is only when a fire is actually burning that they become alarmed. Therefore, smoking the trap parts is an effective means to mask your scent. If one of the above techniques is not practical, and if time permits, allow a trap to weather for a few days and then set it. Do not handle a trap while it is weathering. When you position the trap, camouflage it as naturally as possible to prevent detection by the enemy and to avoid alarming the prey.

f. Traps or snares placed on a trail or run should use canalization. To build a channel, construct a funnel-shaped barrier extending from the sides of the trail toward the trap, with the narrowest part nearest the trap. Canalization should be inconspicuous to avoid alerting the prey. As the animal gets to the trap, it cannot turn left or right and continues into the trap. Few wild animals will back up, preferring to face the direction of travel. Canalization does not have to be an impassable barrier. You only have to make it inconvenient for the animal to go over or through the barrier. For best effect, the canalization should reduce the trail's width to just slightly wider than the targeted animal's body. Maintain this constriction at least as far back from the trap as the animal's body length, then begin the widening toward the mouth of the funnel.

(1) Use a treadle snare against small game on a trail (Figure 11-6). Dig a shallow hole in the trail. Then drive a forked stick (fork down) into the ground on each side of the hole on the same side of the trail. Select two fairly straight sticks that span the two forks. Position these two sticks so that their ends engage the forks. Place several sticks over the hole in the trail by positioning one end over the lower horizontal stick and the other on the ground on the other side of the hole. Cover the hole with enough sticks so that the prey must step on at least one of them to set off the snare. Tie one end of a piece of cordage to a twitch-up or to a weight suspended over a tree limb. Bend the twitch-up or raise the suspended weight to determine where you will tie a 5 centimeter or so long trigger. Form a noose with the other end of the cordage.

(2) Route and spread the noose over the top of the sticks over the hole. Place the trigger stick against the horizontal sticks and route the cordage behind the sticks so that the tension of the power source will hold it in place. Adjust the bottom horizontal stick so that it will barely hold against the trigger. As the animal places its foot on a stick across the hole, the bottom horizontal stick moves down, releasing the trigger and allowing the noose to catch the animal by the foot. Because of the disturbance on the trail, an animal will be wary. You must therefore use canalization.

Figure 11-6. TREADLE SNARE

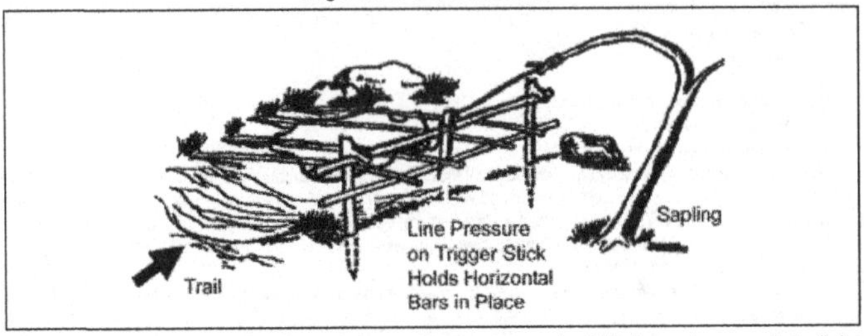

g. Trapping game can be accomplished through the use of snares, traps, or deadfalls. A snare is a noose that will slip and strangle or hold any animal caught in it. You can use inner core strands of parachute suspension lines, wire, bark of small hardwood saplings as well as hide strips from previously caught animals to make snares.

(1) The drag noose snare, Figure 11-7, is usually the most desirable in that it allows you to move away from the site, plus it is one of the easiest to make and fastest to set. It is especially suitable for catching rabbits. To make the drag noose snare, make a loop in the string using a bowline or wireman's knot. (When using wire, secure the loop by intertwining the end of the wire with the wire at the top of the loop). Pull the other end of the string (or wire) through the loop to form a noose that is large enough for the animal's head but too small for its body; tie the string (or attach the wire) to a sturdy branch. The branch should be large enough to span the trail and rest on the bush or support (two short forked sticks) you have selected. A snared animal will dislodge the drag stick, pulling it until it becomes entangled in the brush. Any attempt to escape tightens the noose, strangling or holding the animal.

Figure 11-7. DRAG NOOSE SNARE

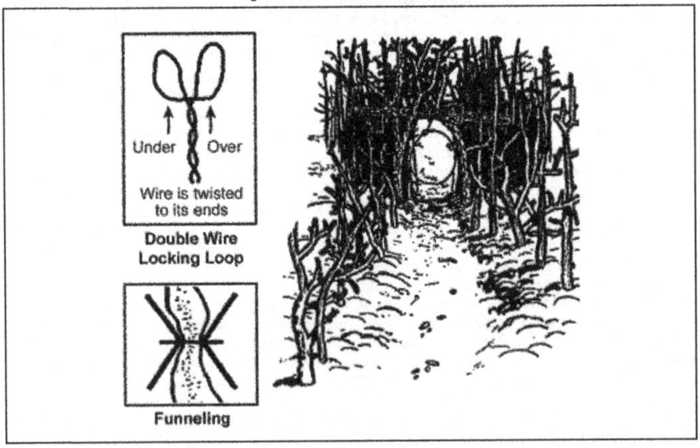

(2) Another type snare is the locking type snare loop (Figure 11-8) that will lock when pulled tight, ensuring the snared animal cannot escape. Use lightweight wire to make this snare, such as, trip wire or the wire from a vehicle or aircraft electrical system. To construct this snare, cut a piece of wire twice the length of the desired snare wire. Double the wire and attach the running ends to a securely placed object, such as the branch of a tree. Place a stick about 1/2 inch in diameter through the loop end of the wire; holding the wire taut, turn the stick in a winding motion so that the wire is twisted together. You should have four to five twists per inch. Detach the wire from the branch and then remove the loop from the stick; make a Figure Eight in the l/2-inch loop by twisting the loop over itself then fold the Figure Eight so the small loops are almost overlapping; run the loose wire ends through these loops. This forms a stiff noose that is strong. Tie the loose end to the stick (for a drag noose square) or branch you are using to complete the snare. This is an excellent snare for catching large animals.

Figure 11-8. FORMING A LOCKING-TYPE SNARE LOOP

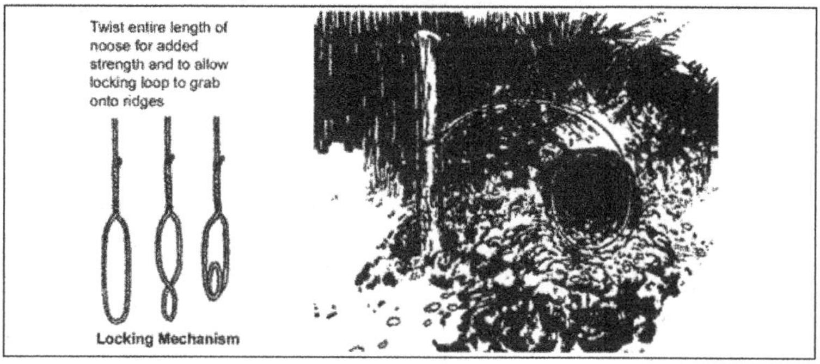

(3) Another means of obtaining game is the use of the deadfall trap (Figure 11-9 and Figure 11-10).

Figure 11-9. TRIGGER WITH DEADFALL

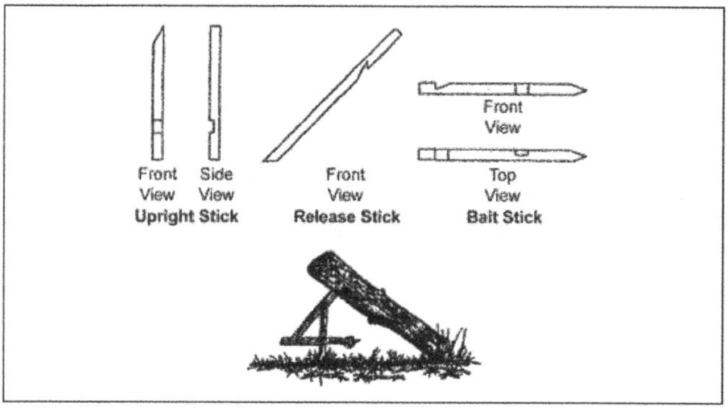

Figure 11-10. TRIP-STRING DEADFALL TRAP

h. Once you have obtained your fish or game, you must clean/butcher and cook/store it. Improper cleaning storing can result in inedible fish and game.

(1) ***Fish***. You must know how to tell if fish are free of bacterial decomposition that makes the fish dangerous to eat. Although cooking may destroy the toxin from bacterial decomposition, do not eat fish that appear spoiled.

(a) Signs of spoilage are:

- A peculiar odor.
- A suspicious color. (Gills should be red or pink. Scales should be a pronounced-not faded shade of gray).
- A dent remaining after pressing the thumb against the flesh.
- A slimy rather than moist or wet body.
- A sharp or peppery taste.

(b) Eating spoiled or poisoned fish may cause diarrhea, nausea, cramps, vomiting, itching; paralysis, or a metallic taste in the mouth. These symptoms appear suddenly 1 to 6 hours after eating. If you are near the sea, drink sea water immediately upon on set of such symptoms and force yourself to vomit.

(c) Fish spoil quickly after death, especially on a hot day, so prepare fish for eating as soon as possible after you catch them.

(d) Cut out the gills and large blood vessels that lie next to the backbone. (You can leave the head if you plan to cook the fish on a spit).

(e) Gut fish that are more than 4 inches long cut along the abdomen and scrape out the intestines.

(f) Scale or skin the fish.

(g) You can impale a whole fish on a stick and cook it over an "open fire." However, boiling the fish with the skin on is the best way to get the most food value. The fats and oil are under the skin, and by boiling the fish, you can save the juices for broth. Any of the methods used for cooking plant food can be used for cooking fish. Fish is done when the meat flakes off.

(h) To dry fish in the sun, hang them from branches or spread them on hot rocks. When the meat has dried, splash it with sea water, if available, to salt the outside. Keep seafood only if it is well dried or salted.

(2) ***Snakes***. All poisonous and nonpoisonous fresh water and land snakes are edible.
To prepare snakes for eating use the following steps (Figure 11-11):

(a) Grip the snake firmly behind the head and cut off the head with a knife.

(b) Slit the belly and remove the innards. (You can use the innards for baiting traps and snares).

(c) Skin the snake. (You can use the skin for improvising, belts, straps, or similar items).

CAUTION

Take extreme care in securing snakes--the bite of some poisonous snakes can be fatal. Even after a snake's head is cut off, its reflex action can cause it to bite, injecting poison. The best time to capture snakes is in the early morning or late evening when temperatures are low and they move slow. Kill or use a long stick to pin down its head and capture it. To pick up a snake, place the index finger on the top rear of its head with your thumb and middle finger on either side of the head behind the jaws. Keep your index finger on top of snake's head to prevent it from turning inside its skin and biting you.

Figure 11-11. CLEANING A SNAKE

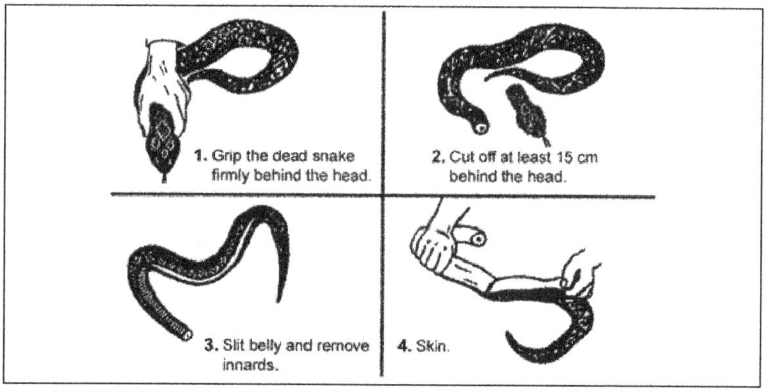

(3) ***Fowl.*** Your first step after killing a fowl for eating or preserving is to pluck its feathers. If plucking is impractical, you can skin the fowl. Keep in mind, however, that a fowl cooked with the skin on retains more food value. Waterfowl are easier to pluck while dry, but other fowl are easier to pluck after scalding. After you pluck the fowl--

(a) Cut off its neck close to the body.

(b) Cut an incision in the abdominal cavity and clean out the insides. Save the neck, liver, and heart for stew. Thoroughly clean and dry the entrails to use for cordage.

(c) Wash out the abdominal cavity with fresh clean water. You can boil fowl or cook it on a spit over a fire. You should boil scavenger birds such as vultures and buzzards for at least 20 minutes to kill any parasites. Use the feathers from fowl for insulating your shoes clothing, or bedding. You can also use feathers for fish lures.

(4) **Medium-Sized Mammals**. The game you trap or snare will generally be alive when you find it and therefore dangerous. Be careful when you approach a trapped animal. Use a spear or club to kill it so you can keep a safe distance from it. After you kill an animal, immediately bleed it by cutting its throat. If you must drag the carcass any distance, do so before you cut off the hide so that the carcass is protected from dirt and debris that might contaminate it. Clean the animal near a stream if possible so that you can wash and cool the carcass and edible parts. Fleas and parasites will leave a cooled body so if the situation allows, wait until the animal cools before cleaning and dressing the carcass. To skin and dress the animal (Figure 11-12 and 11-13).

Figure 11-12. SKINNING AND BUTCHERING LARGE GAME

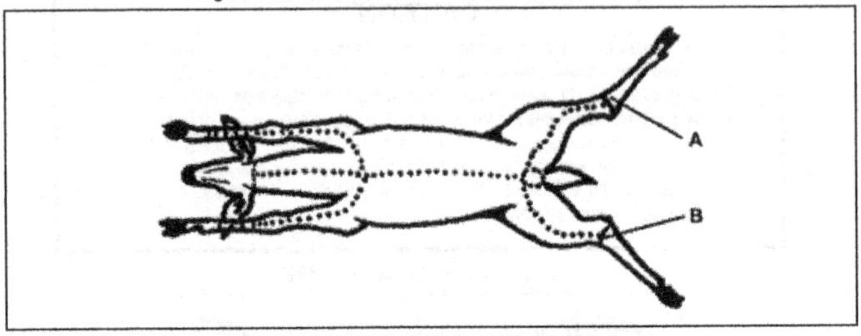

Figure 11-13. SKINNING SMALL GAME

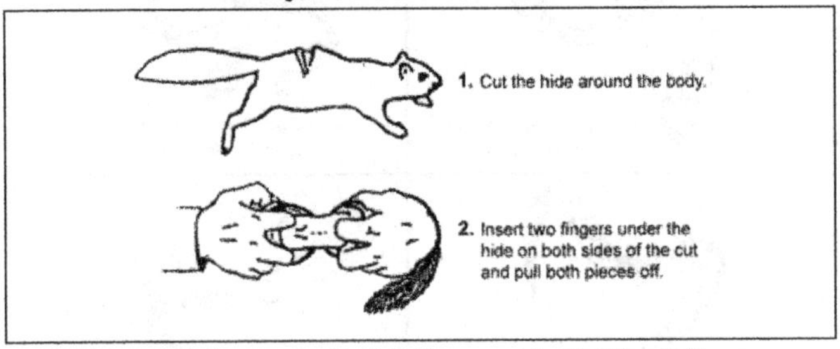

(a) Place carcass, belly up, on a slope if available. You can use rocks or brush to support it.

(b) Remove genitals or udder.

(c) Remove musk glands to avoid tainting meat.

(d) Split hide from tail to throat. Make the cut shallow so that you do not pierce the stomach.

(e) Insert your knife under the skin, taking care not to cut into the body cavity. Peel the hide back several inches on each side to keep hair out of the meat.

(f) Open the chest cavity by splitting the sternum. You can do this by cutting to one side of the sternum where the ribs join.

(g) Reach inside and cut the windpipe and gullet as close to the base of the skull as possible.

(h) With the forward end of the intestinal tract free, work your way to the rear lifting out internal organs and intestines. Cut only where necessary to free them.

(i) Carefully cut the bladder away from the carcass so that you do not puncture the bladder (urine can contaminate meat). Pinch the urethra tightly and cut it beyond the point you are pinching.

(j) Remove the bladder.

(k) From the outside of the carcass, cut a circle around the anus.

(l) Pull the anus into the body cavity and out of the carcass.

(m) Lift or roll the carcass to drain all blood.

Note: Try to save as much blood as you can as it is a vital source of food and salt. Boil the blood.

(n) Remove the hide, and make cuts along the inside of the legs to just above the hoof or paw. Peel the skin back, using your knife in a slicing motion to cut the membrane between the skin and meat. Continue this until the entire skin is removed.

(o) Most of the entrails are usable. The heart, liver, and kidneys are edible. Cut open the heart and remove the blood from its chambers. Slice the kidneys and if enough water is available, soak or rinse them. In all animals except those of the deer family, the gall bladder (a small, dark-colored, clear-textured sac) is attached to the liver.

(p) Sometimes, the sac looks like a blister on the liver. To remove the sac, hold the top portion of it and cut the liver around and behind the sac. If the gall bladder breaks and gall gets on the meat, wash it off immediately so the meat will not become tainted. Dispose of the gall.

(q) Clean blood splattered on the meat will glaze over and help preserve the meat for a short time. However, if an animal is not bled properly, the blood will settle in the lowest part of its body and will spoil in a short time. Cut out any meat that becomes contaminated.

(r) When temperatures are below 40 degrees, you can leave meat hanging for several days without danger of spoilage. If maggots get on the meat, remove the maggots and cut out the discolored meat. The remaining meat is edible. Maggots, which are the larvae of insects, are also edible.

(s) Blood, which contains salts and nutrients, is a good base for soups.

(t) Thoroughly clean the intestines and use them for storing or smoking food or lashings for general use. Make sure they are completely dry to preclude rotting.

(u) The head of most animals contains a lot of meat, which is relatively easy to get. Skin the head, saving the skin for leather. Clean the mouth thoroughly and cut out the tongue. Remove the outer skin from the tongue after cooking. Cut or scrape the meat from the head. If you prefer, you can roast the head over an open fire before cutting off the meat. Eyes are edible. Cook them but discard the retina (this is a plastic like disc). The brain is also edible; in fact, some people consider it a delicacy. The brain is also used to tan leather, the theory being that the brain of an animal is adequate to tan its hide.

(v) Use the tendons and ligaments of the body of large animals for lashings.

(w) The marrow in bones is a rich-food source. Crack the bones and scrap out the marrow, and use bones to make weapons.

(x) If the situation and time allow, you should preserve the extra meat for later use. If the air is cold enough, you can freeze the meat. In warmer climates however, you will need to use a drying or smoking process to preserve it. One night of heavy smoking will make meat edible for about 1 week. Two nights will make it remain edible for 2 to 4 weeks. To prepare meat for drying or smoking, cut it with the grain in quarter inch strips. To air dry the meat, hang it in the wind and hot sun out the reach of animals; cover it so that blow flies cannot land on it.

(y) To smoke meat, you will need an enclosed area – for instance, a teepee (Figure 11-14) or a pit. You will also need wood from deciduous trees, preferably green. Do not use conifer trees such as pines, firs, spruces, or cedars, as the smoke from these trees give the meat a disagreeable taste.

(z) When using the para-teepee or other enclosed area with a vent at the top, set the fire in the center and let it burn down to coals, then stoke it with green wood. Place the strips of meat on a grate or hang them from the top of the enclosure so that they are about 2 feet above the smoking coals. To use the pit method of smoking meat dig, a hole about 3 feet (1 meter) deep and 1 1/2 feet (1/2 meter) in diameter. Make a fire at the bottom of the hole. After it starts burning well, add chipped green wood or small branches of green wood to make it smoke. Place a wooden grate about 1 1/2 feet (1/2 meter) above the fire and lay the strips of meat on the grate. Cover the pit with poles, boughs, leaves, or other material.

11-7. **SHELTERS**. A shelter can protect you from the sun, insects, wind, rain, snow, hot or cold temperatures, and enemy observation. In some areas your need for shelter may take precedence over your need for food, possibly even your need for water. After determining your shelter site, you should keep in mind the type of shelter (protection) you need. You need to know how to make different types of shelters. Only two are described in this handbook. Additional information is available in FM 21-76.

a. **Planning Considerations**.

- How much time and effort are needed to build the shelter?
- Will the shelter adequately protect you from the elements (rain, snow, wind, sun, and so on)?
- Do you have tools to build it? If not, can you improvise tools from materials in the area?
- Do you have the type and amount of manmade materials needed to build it? If not, are there sufficient natural materials in the area?

(f) If you plan to use the lean-to for more than one night, or if you expect rain, make a center support to the lean-to. You can do this by stretching a rope between two upright poles or trees that are in line with the center of the poncho.

(g) Tie another rope to the poncho hood; pull it upward so that it lifts the center of the poncho, and tie it firmly to the rope stretched between the two uprights.

(h) Another method is to cut a stick to place upright under the center of the lean-to. This method, however, will restrict your space and movements in the shelter.

(i) To give additional protection from wind and rain, place boughs, brush, your rucksack, or other equipment at the sides of the lean-to.

(j) To reduce heat loss to the ground, place some type of insulating material, such as leaves or pine needles, inside your lean-to.

Note: When at rest, as much as 80 percent of your body heat can be lost to the ground.

(l) To increase your security from enemy observation, lower the silhouette of the lean-to by making two modifications.

- Secure the support lines to the trees knee-high rather than waist-high.
- Use two knee-high sticks in the two center grommets (sides of lean-to), and angle the poncho to the ground, securing it with sharpened sticks as above.

(2) ***Field Expedient Lean-To***. If you are in a wooded area and have sufficient natural materials, you can make an expedient lean-to (Figure 11-16) without the aid of tools or with only a knife. You need more time to make it than the shelter previously mentioned, but it will protect you from most environmental elements. You will need two trees, (or two upright poles), about 6 feet apart; one pole about 7 feet long and 1 inch in diameter. Five to eight poles about 10 feet long and 1 inch in diameter for beams, cord or vines for securing, the horizontal support to the trees and other poles, saplings, or vines to crisscross the beams. To make this lean-to:

Figure 11-16. FIELD-EXPEDIENT LEAN-TO

(a) Tie the 7-foot pole to the two trees at point about waist to chest high. This is your horizontal support. If there is a fork in the tree, you can rest the pole in it instead of tying the pole in place. If a standing tree is not available, construct a bipod using an Y-shaped sticks or two tripods.

(b) Place one end of the beams (10-foot poles) one side of the horizontal support. As with all lean-to type shelters, make sure the backside of the lean-to is placed into the wind.

(c) Crisscross sapling or vines on the beams.

(d) Cover the framework with brush, leaves, pine needles, or grass, starting at the bottom and working your way up like shingling.

(f) If you plan to use the lean-to for more than one night, or if you expect rain, make a center support to the lean-to. You can do this by stretching a rope between two upright poles or trees that are in line with the center of the poncho.

(g) Tie another rope to the poncho hood; pull it upward so that it lifts the center of the poncho, and tie it firmly to the rope stretched between the two uprights.

(h) Another method is to cut a stick to place upright under the center of the lean-to. This method, however, will restrict your space and movements in the shelter.

(i) To give additional protection from wind and rain, place boughs, brush, your rucksack, or other equipment at the sides of the lean-to.

(j) To reduce heat loss to the ground, place some type of insulating material, such as leaves or pine needles, inside your lean-to.

Note: When at rest, as much as 80 percent of your body heat can be lost to the ground.

(l) To increase your security from enemy observation, lower the silhouette of the lean-to by making two modifications.

- Secure the support lines to the trees knee-high rather than waist-high.
- Use two knee-high sticks in the two center grommets (sides of lean-to), and angle the poncho to the ground, securing it with sharpened sticks as above.

(2) ***Field Expedient Lean-To***. If you are in a wooded area and have sufficient natural materials, you can make an expedient lean-to (Figure 11-16) without the aid of tools or with only a knife. You need more time to make it than the shelter previously mentioned, but it will protect you from most environmental elements. You will need two trees, (or two upright poles), about 6 feet apart; one pole about 7 feet long and 1 inch in diameter. Five to eight poles about 10 feet long and 1 inch in diameter for beams, cord or vines for securing, the horizontal support to the trees and other poles, saplings, or vines to crisscross the beams. To make this lean-to:

Figure 11-16. FIELD-EXPEDIENT LEAN-TO

(a) Tie the 7-foot pole to the two trees at point about waist to chest high. This is your horizontal support. If there is a fork in the tree, you can rest the pole in it instead of tying the pole in place. If a standing tree is not available, construct a bipod using an Y-shaped sticks or two tripods.

(b) Place one end of the beams (10-foot poles) one side of the horizontal support. As with all lean-to type shelters, make sure the backside of the lean-to is placed into the wind.

(c) Crisscross sapling or vines on the beams.

(d) Cover the framework with brush, leaves, pine needles, or grass, starting at the bottom and working your way up like shingling.

(e) Place straw, leaves, pine needles, or grass inside the shelter for bedding.

(f) In cold weather, you can add to the comfort of your lean-to by building a fire-reflector wall (Figure 11-16). Drive four stakes about 4 feet long into the ground to support the wall. Stack green logs on top of one another between the support stales. Bind the top of the support stakes so the green logs will stay in place. Fill in the spaces between the logs with twigs or small branches. With just a little more effort, you can have a drying rack. Cut a few 3/4-inch diameter poles. The length depends on distance between the lean-to support and the top of the fire-reflector wall. Lay one end of the poles on the lean-to horizontal support and the other ends on top of the reflector wall. Place and tie into place smaller sticks across these poles. You now have a place to dry clothes, meat, or fish.

11-8. **FIRES**. A fire can full fill several needs. It can keep you warm, it can keep you dry: you can use it to cook food, to purify water, and to signal. It can also cause you problems when you are in enemy territory: it creates smoke, which can be smelled and seen from a long distance: It causes light which can be seen day or night and it leaves signs of your presence. Remember you should always weigh your need for a fire against your need to avoid enemy protection. When operating in remote areas you should always take a supply of matches in a waterproof case and keep them on your person.

a. **Selection**. When selecting a site to build a fire, you should consider the following:

(1) The area (terrain and climate) in which you are operating.
(2) The material and tools available.
(3) How much time you have.
(4) Why you need a fire.
(5) The nearness of the enemy.

b. **Preparation**. If you are in a wooded or brush-covered area, clear brush away, and scrape the surface soil from the spot you selected. The cleared circle should be at least 3 feet (1 meter) in diameter so that there is little chance of the fire spreading. To prepare the site for a fire, ensure that it is dry and that it--look for a dry spot that--

(1) Offers protection from the wind.
(2) Is suitably placed in relation to your shelter (if any).
(3) Concentrates the heat in the direction you desire.
(4) Has a supply of wood or other fire burning material.

c. **Dakota Fire Hole**. In some situations, an underground fireplace will best meet your needs. It conceals the fire to some extent and serves well for cooking food. To make an underground fireplace or Dakota fire hole (Figure 11-17):

Figure 11-17. DAKOTA FIRE HOLE

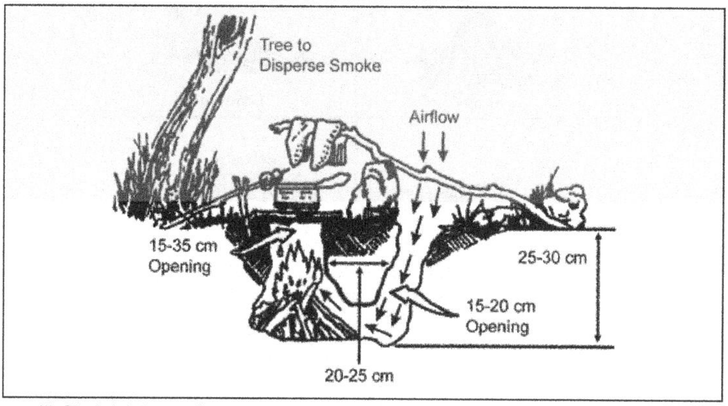

(1) Dig a hole in the ground.

(2) On the upwind side of this hole, poke one large connecting hole for ventilation.

d. **Aboveground fire**. If you are in a snow covered or wet area, you can use green logs to make a dry base for your fire (Figure 11-18). Trees with wrist-size trunks are easily broken in extreme cold. Cut or break several green logs and lay them side by side on top of the snow. Add one or two more layers, laying the top layer logs in a direction opposite those of the layer below it.

Figure 11-18. BASE FOR FIRE IN SNOW COVERED AREA

e. **Methods**. There are several methods for laying a fire for quick fire making. Three easy methods are Tepee, lean-to, and cross-ditch. Tepee (Figure 11-19). Arrange tinder and a few sticks of kindling in the shape of a cone. Fire the center. As the cone burns away, the outside logs will fall inward, feeding the heart of the fire. This type of fire burns well even with wet wood.

(1) ***Lean-To*** (Figure 11-19). Push a green stick into the ground at a 30 degree angle. Point the end of the stick in the direction of the wind. Place some tinder (at least a handful) deep inside this lean-to stick. Light the tinder. As the kindling catches fire from the tinder, add more kindling.

Figure 11-19. METHODS FOR LAYING A FIRE

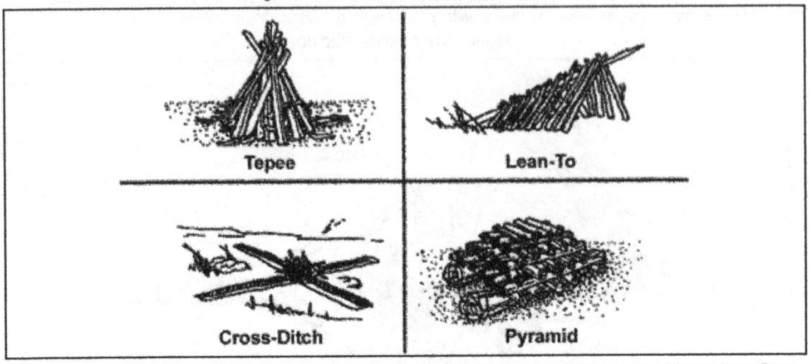

(2) ***Cross-Ditch*** (Figure 11-19). Scratch a cross about 1 foot in size in the ground. Dig the cross 3 inches deep. Put a large wad of tinder in the middle of the cross. Build a kindling pyramid above the tinder. The shallow ditch allows air to sweep under the fire to provide a draft.

NOTES

Chapter 12

FIRST AID

Patrolling, more than some other types of missions, puts Rangers in harms way. CASEVAC planning is vital. Also, because trained medical personnel might be unavailable at the initial point of injury, everyone must know how to diagnose and treat injuries, wounds, and common illnesses. The unit should also have a plan for handling KIAs.

12-1. **LIFESAVING STEPS**. Whatever the injury, (1) open the airway and restore breathing; (2) stop the bleeding and protect the wound; (3) check, treat and monitor for shock; and (4) MEDEVAC the casualty.

12-2. **CARE UNDER FIRE**. When still under fire, (1) maintain situational awareness; (2) return fire; (3) protect the casualty; (4) Identify and control severe bleeding with bandage or tourniquet; (5) and move the casualty to cover.

12-3. **PRIMARY SURVEY**. Use the ABC's to help you remember how to identify and deal with life-threatening injuries, airway complications, breathing problems, or uncontrolled hemorrhaging (bleeding).

A	Airway. Open airway by patient position or with airway adjuncts.
B	Breathing. Seal open chest wounds with occlusive dressing.
C	Circulation. Identify uncontrolled bleeding and control with pressure or tourniquet. Start IV if needed.
D	Disability. Determine Level of consciousness.
E	Exposure. Fully expose patient. (Environment dependent)

12-4. **AIRWAY MANAGEMENT**. The *airway* is most often obstructed at the base of the tongue.

a. If this happens, open the airway using the chin lift (nontrauma, Figure 12-1) or jaw thrust (trauma, Figure 12-2).

Figure 12-1. CHIN LIFT

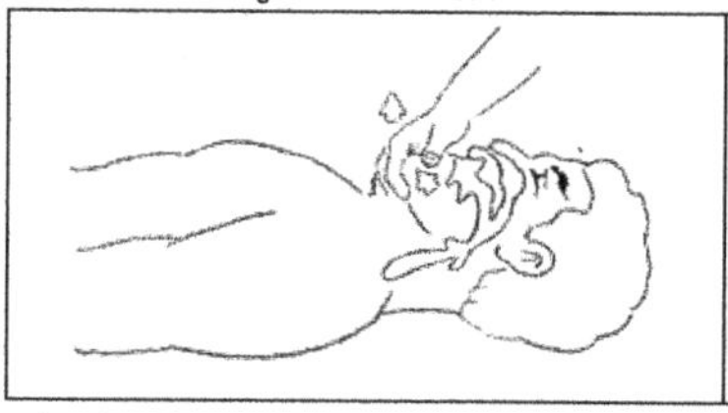

Figure 12-2. JAW THRUST

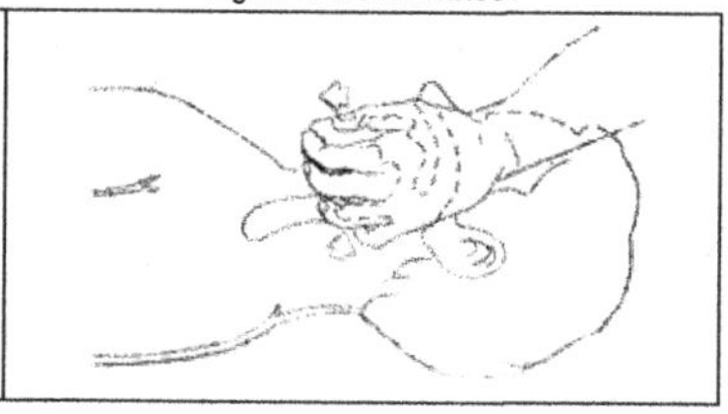

b. **Remove** debris (teeth, blood clots, bone) from the oral cavity; use suction if you have it; and place airway adjuncts to allow the victim to breathe through their nose (Figure 12-3) or mouth (Figure 12-4).

Figure 12-3. NASAL AIRWAY

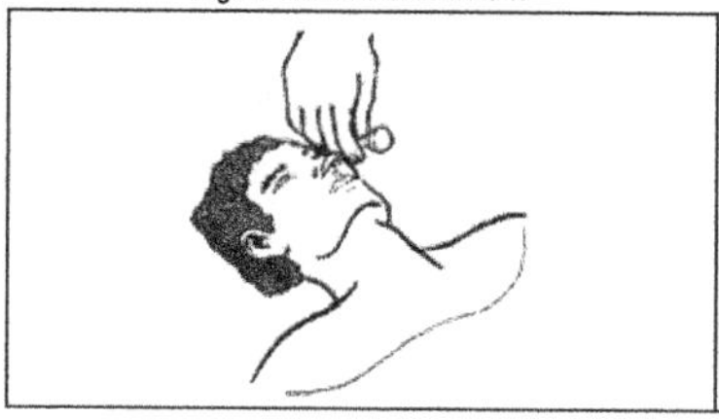

Figure 12-4. MOUTH AIRWAY

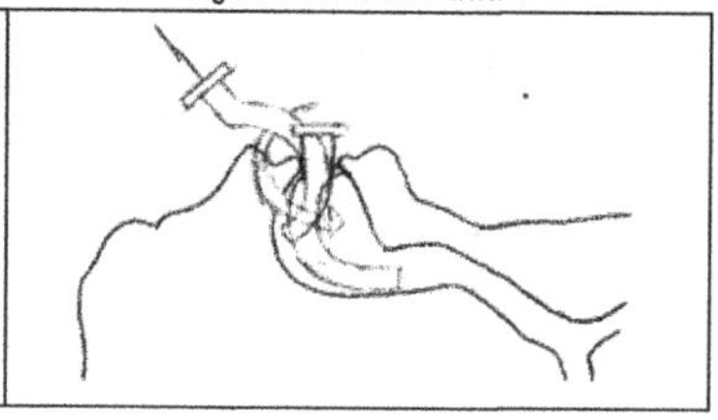

12-5. **BREATHING**. If the patient is having difficulty breathing, (1) expose the chest and identify open chest injuries; (2) place occlusive dressing (plastic and tape, or a COTS chest dressing) over open entry and exit chest wounds; and (3) place the patient on his injured side, or position him where he can breathe most comfortably.

12-6. **BLEEDING**. Quickly identify and control bleeding. (1) Apply a standard or pressure dressing. (2) If this does not control the bleeding, immediately apply a tourniquet. (3) Reassess dressings frequently to ensure bleeding has been controlled.

12-7. **SHOCK**. Shock causes an inadequate flow of oxygen to the tissues. Shock can cause bleeding (hemorrhagic) or not (nonhemorrhagic). Signs and symptoms of shock can include increased pulse, increased respiration; and decreased level of consciousness. (1) Open the victim's airway; (2) restore his breathing; (3) control the bleeding; (4) initiate an IV or a saline lock; and (5) monitor his condition.

12-8. **EXTREMITY INJURIES**. Identify and control bleeding. If you suspect a fracture, splint it as it lies. Do not reposition injured extremity.

12-9. **ABDOMINAL INJURIES**. Identify and control bleeding. Treat for shock. (1) If internal organs are exposed, cover them with dry, sterile dressing. Do not place organs back in wound area. Do not handle internal organs. (2) Place patient in position of comfort. Flex knees to relax abdomen. (3) Do not give anything by mouth to the patient.

12-10. **BURNS**. Remove patient from burn source. Then (1) remove all clothing and jewelry from the area; (2) cover with dry, sterile dressings, and ensure fingers and toes have dressings between them before covering entire area; and (3) Evacuate immediately any casualties with burns of the face, neck, hands, genitalia, or over 20 percent of his body surface.

12-11. **WEATHER (HEAT AND COLD) INJURIES**. Tables 12-1, 12-2, and 12-3 show first aid for (respectively) heat injuries, cold injuries, and environmental injuries.

Table 12-1. HEAT INJURIES

INJURY	*SIGNS/SYMPTOMS*	*FIRST AID*
Heat Cramps	Casualty experiences muscle cramps in arms, legs and/or stomach, may also have wet skin and extreme thirst.	1. Move the casualty to a shaded area and loosen clothing. 2. Allow casualty to drink 1 quart of cool water slowly per hour. 3. Monitor casualty and provide water as needed. 4. Seek medical attention if cramps persist.
Heat Exhaustion	Casualty experiences loss of appetite, headache, excessive sweating, weakness or faintness, dizziness, nausea, muscle cramps. The skin is moist, pale, and clammy.	1. Move the casualty to a cool, shaded area and loosen clothing. 2. Pour water on casualty and fan to increase cooling effect of evaporation. 3. Provide at least one quart of water to replace lost fluids. 4. Elevate legs. 5. Seek medical aid if symptoms continue.
Heatstroke (Sunstroke)	Casualty stops sweating (hot, dry skin), may experience headache, dizziness, nausea, vomiting, rapid pulse and respiration, seizures, mental confusion. Casualty may suddenly collapse and lose consciousness. THIS IS A MEDICAL EMERGENCY!	1. Move casualty to a cool, shaded area, loosen clothing, and remove outer clothing if the situation permits. 2. Immerse in cool water. If cool bath is not available, massage arms and legs with cool water. Fan casualty to increase the cooling effect of evaporation. 3. If conscious, slowly consume one quart of water. 4. SEEK MEDICAL AID AND EVACUATE AS SOON AS POSSIBLE. Perform any lifesaving measures.

Table 12-2. COLD INJURIES

INJURY	***SIGNS/SYMPTOMS***	***FIRST AID***
Chilblain	Red, swollen, hot, tender, itching skin. Continued exposure may lead to infected (ulcerated bleeding) skin lesions.	1. Area usually responds to locally applied warming (body heat). 2. Do Not rub or massage area. 3. Seek medical treatment.
Immersion (trench) foot	Affected parts are cold, numb, and painless. As parts warm, they may be hot, with burning and shooting pains. Advanced stage: skin pale with bluish cast: pulse decreases, blistering, swelling, heat hemorrhages, and gangrene may follow.	1. Gradual warming by exposure to warm air. 2. DO NOT massage or moisten skin. 3. Protect affected parts from trauma. 4. Dry feet thoroughly: avoid walking. 5. Seek medical treatment.
Frostbite	***SUPERFICIAL*** Redness, blisters in 24-36 hours and sloughing of the skin. ***DEEP*** Preceded by superficial frostbite; skin is painless, pale- yellowish, waxy, "wooden" or solid to touch, blisters form in 12-36 hours	***SUPERFICIAL*** 1. Keep casualty warm; gently warm affected parts. 2. Decrease constricting clothing, increase exercise and insulation. ***DEEP*** 1. Protect the part from additional injury. 2. Seek medical treatment as fast as possible.
Snow Blindness	Red, scratchy, or watery eyes; headache; increased pain in eyes with exposure to light.	1. Cover the eyes with a dark cloth. 2. Seek medical treatment.
Dehydration	Similar to heat exhaustion.	1. Keep warm, loosen clothes. 2. Replace lost fluids, rest, and additional medical treatment.
Hypothermia	Casualty is cold, shivers uncontrollably until shivering stops. Rectal (core) temp less 95° F can affect consciousness. Uncoordinated movements, shock, and coma may occur as body temperature drops.	***MILD HYPOTHERMIA*** 1. Warm body evenly and without delay. (Heat source must be provided.) 2. Keep dry, protect from elements. 3. Warm liquids may be given to conscious casualty only. 4. Be prepared to start CPR. 5. Seek medical treatment immediately. ***SEVERE HYPOTHERMIA*** 1. Quickly stabilize body temperature. 2. Attempt to prevent further heat loss. 3. Handle the casualty gently. 4. Evacuate to nearest medical treatment facility as soon as possible.

Table 12-3. ENVIRONMENTAL INJURIES

TYPE	*FIRST AID*
Snake bite	1. Get the casualty away from the snake. 2. Remove all rings and bracelets from the affected extremity. 3. Reassure the casualty and keep him quiet. 4. Apply constricting band(s) 1 to 2 finger widths close to the bite. You should be able to slip 1 finger between the band and skin. ***ARM OR LEG BITE*** - Place one band above and one band below the bite site. ***HAND OR FOOT BITE*** - Place one band above the wrist or ankle. 5. Immobilize the affected limb below the level of the heart. 6. Kill the snake, if possible, (without damaging its head or endangering yourself) and send it with the casualty. 7. Seek medical treatment immediately.
Brown recluse or black widow, spider bite	1. Keep the casualty calm. 2. Wash the area. 3. Apply ice or a freeze pack, if available. 4. Seek medical treatment.
Tarantula bite, scorpion sting, ant bite	1. Wash the area. 2. Apply ice or a freeze pack, if available. 3. Apply baking soda, calamine lotion, or meat tenderizer to the bite site to relieve pain and itching. 4. If site of bite(s) or sting(s) is on the face, neck (possible airway blockage), or genital area, or if reaction is severe, or if the sting is by the dangerous Southwestern scorpion, keep the casualty as quiet as possible, administer the epinephrine auto-injector (EAI) if needed, and then seek immediate medical aid.
Wasp or bee sting	1. If the stinger is present, remove by scraping with a knife or finger nail. DO NOT squeeze venom sack on stinger, more venom may be injected. 2. Wash the area. 3. Apply ice or freeze pack, if available. 4. If allergic signs or symptoms appear, be prepared to administer the EAI, and then seek medical assistance.
Human and other animal bites	1. Cleanse the wound thoroughly with soap or detergent solution. 2. Flush bite well with water. 3. Cover bite with a sterile dressing. 4. Immobilize injured extremity. 5. Transport casualty to a medical treatment facility. 6. For animal bites, without endangering yourself or damaging the animal's head, kill the animal and send its head with the casualty. 7. For human bites, try to extract some of the attacker's saliva from the wound and send that in a sealed, identified container with the casualty.

Table 12-3. ENVIRONMENTAL INJURIES (CONTINUED)

Poison Ivy, Oak, Sumac	1. Gently clean affected area. Clean areas 2-3 times daily. Wash clothing. 2. Apply topical anti-itch lotion or ointment as needed, and cover. 3. Avoid scratching or itching the area. 4. Observe for signs of infection (increasing redness, tenderness, warm to touch). 5. Seek medical attention if rash persists or signs of infection develop.

12-12. **POISONOUS PLANT IDENTIFICATION**

Poison plants include, among others, poison ivy, oak, and sumac, as well as a few more such as stinging nettles which are not discussed here.

a. **Poison Ivy**. Poison ivy is grows in two forms, a vine or shrub. The compound leaves of poison ivy consist of three pointed leaflets; the middle leaflet has a much longer stalk than the two side ones. The leaflet edges can be smooth or toothed but are rarely lobed. The leaves vary greatly in size, from 8 to 55 mm (0.31" to 2.16") in length. They are reddish when they emerge in the spring, turn green during the summer, and become various shades of yellow, orange, or red in the autumn. Small greenish flowers grow in bunches attached to the main stem close to where each leaf joins it. Later in the season, clusters of poisonous, berrylike drupes form. They are whitish, with a waxy look.

b. **Poison Oak**. Poison oak is a widespread deciduous shrub throughout mountains and valleys of North America, generally below 5,000 feet elevation. It commonly grows as a climbing vine with aerial (adventitious) roots that adhere to the trunks of oaks and sycamores. Poison oak also forms dense thickets in chaparral and coastal sage scrub. Poison oak leave typically have three leaflets (sometimes five), the terminal one on a slender rachis (also called a stalk or petiolule). Eastern poison ivy often has a longer rachis and the leaflet margins tend to be less lobed and serrated (less "oak-like"). Like many members of the sumac family (Anacardiaceae) new foliage and autumn leaves often turn brilliant shades of pink and red.

c. **Poison Sumac**. Poison Sumac is a woody perennial shrub or small tree growing from 5 - 25 feet tall that grows in peat bogs and swamps. To identify Poison Sumac, look for the fruit that grows between the leaf and the branch. Leaves are arranged in an alternate pattern on the vine. There are about 7-13 leaflets forming a feather-like appearance. The foliage has brilliant orange or scarlet coloring in the fall.

Figure 12-5. POISONOUS PLANTS

12-13. **FOOT CARE.** Use moleskin to prevent blisters prior to movement or foot march. Drain large blisters. Clean area, puncture with needle, drain blister. Place moleskin over area. Observe for signs of infection. Keep feet as clean and dry as possible. Use foot powder and change socks. Let feet air dry as mission permits.

12-14. **CASUALTY RESCUE AND TRANSPORT SYSTEM.** Remove the hoistable CRTS litter from its pack, and place the litter on the ground. Unfasten the retainer strap, step on the foot end of the litter, and unroll the litter completely to the opposite end. Then--

a. Bend the CRTS litter in half and backroll it. Repeat with the opposite end of the litter. The litter will now lay flat. Lift the sides of the litter, and fasten the four cross-straps to the buckles directly opposite the straps.

b. Place the CRTS litter next to the casualty. Insure the head end of the stretcher is adjacent to the head of casualty. Place the cross-straps under the litter. Logroll the casualty and slide the litter as far under the casualty as possible. Gently roll the casualty down on the litter. Slide the casualty to the center of the litter. Be sure to keep his spinal column as straight as possible.

c. Pull the straps out from under the CRTS litter and fasten it to the buckles. Position the foot end of the litter at the head of the casualty. Have one rescuer straddle the litter and support the casualty's head, neck, and shoulders.

d. Grasp the foot straps of the litter, and slide the litter under the casualty. Center the casualty on the litter, and feed foot straps thru the unused grommets at the foot end of the litter. Fasten the straps to the buckles.

12-15. **HYDRATION AND ACCLIMATIZATION**

a. Table 12-4 shows strategies for minimizing dehydration and increasing acclimatization and good hydration practices.

Table 12-4. HYDRATION MANAGEMENT AND ACCLIMATIZATION

STRATEGY	***SUGGESTIONS FOR IMPLEMENTATION***
Start early	1. Start at least 1 month prior to School 2. Be flexible and patient: performance benefits take longer than the physiological benefits
Mimic the training environment climate	1. In warm climates, acclimatize in the heat of day. 2. In temperate climates workout in a warm room wearing sweats.
Ensure adequate heat stress	1. Induce sweating. 2. Work up to 100 minutes of continuous physical exercise in the heat. Be patient. The first few days, you may not be able to go 100 minutes without resting. 3. Once you can comfortably exercise for 100 minutes in the heat, then continue for at least 7-14 days with added exercise intensity (loads or training runs).
Teach yourself to drink and eat	1. Your thirst mechanism will improve as you become heat acclimatized, but you will still under-drink if relying on thirst sensation. 2. Heat acclimatization will increase your water requirements. 3. Dehydration will negate most benefits of physical fitness and heat acclimatization. 4. You will sweat out more electrolytes when not acclimatized, so add salt to your food, or drink electrolyte solutions during the first week of heat acclimatization. 5. A convenient way to learn how much water your body needs to replace is to weigh yourself before and after the 100 minutes of exercise in the heat. For each pound lost, you should drink about one-half quart of fluid. 6. Do not skip meals, as this is when your body replaces most of its water and salt losses.

12-16. **WORK, REST, AND WATER CONSUMPTION.** Table 12-5 shows a work, rest, and water consumption table. The guidance applies to the average size, heat acclimated Soldier wearing ACU, not hot weather gear, except as specified otherwise. The work and rest times and fluid replacement volumes shown will help the Soldier sustain his performance and hydration for at least 4 hours of work in the specified heat category. Fluid needs can vary based on individual differences (±1 quart per hour).

a. "NL" means that there is no limit to work time per hour. "Rest" means minimal physical activity such as sitting or standing, preferably in the shade.

b. Consume no more than 1.5 quarts of fluid per hour, and no more than 12 quarts per day.

c. If you are wearing body armor in a humid climate, then add 5° F to the WBGT. if wearing MOPP 4 clothing, add 10° F to the WBGT.

d. Work categories include easy, moderate, and hard.

1) ***Easy Work***. This includes, for example, maintaining weapons; walking on hard surfaces at 2.5 mph with a load weighing no more than 30 pounds; participating in marksmanship training; and participating in drills or ceremonies.

2) ***Moderate Work***. This includes, for example, walking in loose sand at 2.5 mph (maximum) or under with no load; walking on a hard surface at 3.5 mph (maximum) with a load weighing no more than 40 pounds; performing calisthenics; patrolling; or conducting individual movement techniques such as the low or high crawl.

3) ***Hard Work***. This includes, for example, walking on a hard surface at 3.5 mph with a load weighing 40 or more pounds; walking in loose sand at 2.5 mph while carrying a load; and conducting field assaults.

Table 12-5. WORK, REST, AND WATER CONSUMPTION TABLE

HEAT CATEGORY	WBGT INDEX, F°	***EASY WORK***		***MODERATE WORK***		***HARD WORK***	
		WORK/ REST	**WATER INTAKE (QT/H)**	**WORK/ REST**	**WATER INTAKE (QT/H)**	**WORK/ REST**	**WATER INTAKE (QT/H)**
1	78° to 81.9°	NL	0.50	NL	0.75	40/20	0.75
2 (Green)	82° to 84.9°	NL	0.50	50/10	0.75	30/30	1.00
3 (Yellow)	95° to 87.9°	NL	0.75	40/20	0.75	30/30	1.00
4 (Red)	88° to 89.9°	NL	0.75	30/30	0.75	20/40	1.00
5 (Black)	90° or more	50/10 min	1.00	20/40	1.00	10/50	1.00

NOTES

Chapter 13
DEMOLITIONS

This chapter introduces Rangers to the characteristics of explosives, to initiation systems, MDI components, detonation systems, safety considerations, expedient explosives, breaching charges, and timber cutting charges (FM 5-250). The two categories of explosives are low and high (Table 13-1).

- *Low explosives* have a detonating velocity up to 1,300 feet per second, which produces a pushing or shoving effect.
- *High explosives* have a detonating velocity of 3,280 to 27,888 feet per second, which produces a shattering effect.

Table 13-1. CHARACTERISTICS OF US DEMOLITIONS EXPLOSIVES

NAME	APPLICATIONS	DETONATION VELOCITY MIN/SEC	DETONATION VELOCITY FT/SEC	RE FACTOR*	FUME TOXICITY	WATER RESISTANCE
AMMONIUM NITRATE	CRATERING CHG	2,700	8.800	0.42	DANGEROUS	POOR
PETN	DET CORD BLASTING CAPS DEMOLITION CHGS	8,300	27,200	1.66	SLIGHT	EXCELLENT
RDX	BLASTING CAPS COMPOSITION EXPLOSIVE	8,350	27,400	1.60	DANGEROUS	EXCELLENT
TRINITROTOLUENE (TNT)	DEMOLITION CHG COMPOSITION EXPLOSIVE	6,900	22,600	1.00	DANGEROUS	EXCELLENT
TETRYL	BOOSTER CHG COMPOSITION EXPLOSIVE	7,100	23,300	1.25	DANGEROUS	EXCELLENT
NITROGLYCERIN	COMMERCIAL DYNAMITE	7,700	25,200	1.50	DANGEROUS	GOOD
BLACK POWDER	TIME FUSE	400	1,300	0.55	DANGEROUS	POOR
AMATOL 80/20	BURSTING CHG	4,900	16,000	1.17	DANGEROUS	POOR
COMPOSITION A3	BOOSTER CHG BURSTING CHG	8,100	26,500	---	DANGEROUS	GOOD
COMPOSITION B	BURSTING CHG	7,800	25,600	1.35	DANGEROUS	EXCELLENT
COMPOSITION C4 (M112)	CUTTING CHG BREACHING CHG	8.040	26,400	1.34	SLIGHT	EXCELLENT
COMPOSITION H6	CRATERING CHG	7,190	23,600	1.33	DANGEROUS	EXCELLENT
TETRYTOL 75/25	DEMOLITION CHG	7,000	23,000	1.20	DANGEROUS	EXCELLENT
PENTOLITE 50/50	BOOSTER CHG BURSTING CHG	7,450	24,400	---	DANGEROUS	EXCELLENT

NAME	APPLICATIONS	DETONATION VELOCITY		RE FACTOR*	FUME TOXICITY	WATER RESISTANCE
		MIN/SEC	FT/SEC			
M1 DYNAMITE	DEMOLITION CHG	6,100	20,000	0.92	DANGEROUS	FAIR
DET CORD	PRIMING DEMOLITION CHG	6,100 TO 7,300	20,000 TO 24,000	---	SLIGHT	EXCELLENT
SHEET EXPLOSIVE M118 AND M186	CUTTING CHG	7,300	24,000	1.14	DANGEROUS	EXCELLENT
BANGALORE TORPEDO M1A2	DEMOLITION CHG	7,800	25,600	1.17	DANGEROUS	EXCELLENT
SHAPED CHARGES M2A3, M2A4, AND M3A1	CUTTING CHG	7,800	25,600	1.17	DANGEROUS	EXCELLENT
* TNT = 1.00 RELATIVE EFFECTIVENESS						

13-1. **INITIATING (PRIMING) SYSTEMS**. The best way to prime demolition systems is with modernized demolition initiators (MDI). These consist of blasting caps attached to various lengths of time fuse or shock tube. MDI can be used with a fuse igniter and detonating cord to create many firing systems. In the absence of MDI, field expedient methods may be used.

a. **Shock Tube**.

(1) Thin, plastic tube of extruded polymer with a layer of special explosive material on the interior surface.

(2) Explosive material propagates a detonation wave which moves along the shock tube to a factory crimped and sealed blasting cap.

(3) Detonation is normally contained within the plastic tubing. However, burns may occur if the shock tube is held.

(4) Advantages of shock tube.

(a) The shock tube offers the instantaneous action of electric initiation without the risk of accidental initiation caused by radio transmitters, static electricity, and so on.

(b) Extremely reliable.

(c) May be extended using left over pieces from previous operations.

b. **Blasting Caps**. Five MDI blasting caps are available which replace the M6 electric and M7 Non-electric blasting cap. Three of these are high-strength caps and two are low-strength. High-strength blasting caps can be used to prime all standard military explosives (including detonating cord) or to initiate the shock tube of other MDI blasting caps.

(1) ***M11***.

- Factory crimped to 30 feet of shock tube.
- A movable "J" hook is attached for quick and easy attachment to detonating cord.
- A red flag is attached 1 meter from the blasting cap and a yellow flag 2 meters from the blasting cap.

(2) ***M14***.

- Factory crimped to 7.5 feet of time fuse.
- May be initiated using a fuse igniter or match.
- Burn-time for total length is about five minutes.
- Yellow bands indicate calibrated one-minute time intervals.

(3) ***M15***.

- Two blasting caps factory crimped to 70 feet of shock tube.
- Each blasting cap has delay elements to allow for staged detonations.
- Low-strength blasting caps. Used as a relay device to transmit a shock tube detonation impulse from an initiator to a high strength-blasting cap.

(4) ***M12***. Factory crimped to 500 feet of shock tube on a cardboard spool.

(5) ***M13***. Factory crimped to 1,000 feet of shock tube.

c. **Matches**. If fuse igniter is unavailable, light the time blasting fuse with a match. Split the fuse at the end (Figure 13-1), and place the head of an unlit match in the powder train. Light the inserted match head with a flaming match, or rub the abrasive on the match box against it. It may be necessary to use two match heads during windy conditions. Burn time will increase with altitude and colder temperatures.

d. **M81 Fuse Igniter**. This is used to ignite time blasting fuse or to initiate the shock tube of MDI blasting caps.

Figure 13-1. LIGHTING TIME FUSE WITH A MATCH

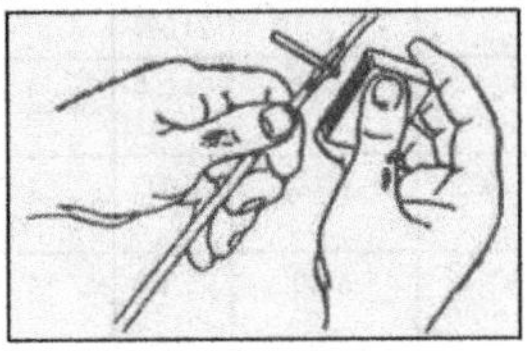

Notes: Low strength blasting caps cannot reliably set off explosives. They should only be used to set off additional shock tubes. The M60 fuse igniter may still be used to ignite time blasting fuse. However, it will not reliably initiate the shock tube.

13-2. **DETONATION (FIRING) SYSTEMS**. The two types of firing systems are MDI alone, or MDI plus detonating cord.

a. **MDI Alone**. An MDI firing system is one in which the initiation set, transmission and branch lines are constructed using MDI components and the explosive charges are primed with MDI blasting caps. Construct the charge in the following manner.

(1) Emplace and secure explosive charge (C4, TNT, cratering charge, and so on.) on target.

(2) Place a sandbag or other easily identifiable marker over the M11, M14, or M15 blasting cap to be used.

(3) Connect to an M12 or M13 transmission line if desired.

(4) Connect blasting cap with shock tube to an M14 cap with time fuse. Cut the time blasting fuse to desired delay time.

(5) Prime the explosive charge by inserting the blasting cap into the charge.

(6) Visually inspect firing system for possible misfire indicators such as cracks, bulges, or corrosion.

(7) Return to the firing point and secure a fuse igniter to the cut end of the time fuse.

(8) Remove the safety cotter pin from the igniter's body.

(9) Actuate the charge by grasping the igniter body with one hand while sharply pulling the pull ring.

b. **MDI and Detonating Cord**. Construct the charge using the above steps for MDI stand-alone system. Incorporate detonating cord branch lines into the system using the "J" hooks of the M11 shock tube. Taping the ends of the Detonation Cord reduces the effect of moisture on the system.

13-3. **SAFETY**. MDI is not recommended for below ground use, except in quarry operations with water-gel or slurry explosives. Use detonating cord when it is necessary to bury primed charges.

a. Do not handle misfires downrange until the required 30 minute waiting period for both primary and secondary initiation systems has elapsed and other safety precautions have been accomplished.

b. Never yank or pull hard on the shock tube. This may actuate the blasting cap.

c. Do not dispose of used shock tubes by burning because of potentially toxic fumes given off from the burning plastic.

d. Do not use M1 dynamite with the M15 blasting cap. The M15 delay-blasting cap should be used only with water-gel or slurry explosives.

e. Always use protective equipment when handling demolitions. Minimum protection consists of leather gloves, ballistic eye protection, and helmet.

13-4. **EXPEDIENT EXPLOSIVES--IMPROVISED SHAPED CHARGE**. An improvised shaped charge (Figure 13-2) concentrates the energy of the explosion released on a small area, making a tubular or linear fracture in the target.

a. The versatility and simplicity of these charges make them effective against targets, especially those made of concrete or those with armor plating.

(1) Bowls, funnels, cone shaped glasses, (champagne glasses with stem removed) used as cones. Champagne or cognac bottles are excellent.

(2) Charge characteristics.

(a) *Cavity Liners.* These are made of copper, tin, or zinc. If none is available, cut a cavity out of the plastic explosive.

(b) *Cavity Angle.* This will work with 30 to 60 degree angles. The cavity angle in most high-explosive antitank (HEAT) ammunition is 42 to 45 degrees.

(c) *Explosive Height (In Container).* 2 x the height of the cone measured from the base of the cone to the top of the explosive.

(d) *Standoff.* Normal standoff is one and one half the cone's diameter.

(e) *Detonation Point.* The exact top center of the charge is the detonation point. Cover the blasting cap with a small amount of C4 if any part of the blasting cap is exposed.

b. Remove the narrow neck of a bottle or the stem of a glass by wrapping it with a piece of soft, absorbent twine or by soaking the string in gasoline and lighting it. Place two bands of adhesive tape, one on each side of the twine, to hold the twine firmly in place. The bottle or stem must be turned continuously with the neck up, to heat the glass uniformly.

c. A narrow band of plastic explosive placed around the neck and burned gives the same result. After the twine or plastic has burned, submerge the neck of the bottle in water and tap it against some object to break it off. Tape the sharp edge of the bottle to prevent cutting hands while tamping the explosive in place.

d. Do not immerse the bottle in water before the plastic has been completely burned, or it could detonate.

Figure 13-2. IMPROVISED SHAPED CHARGE

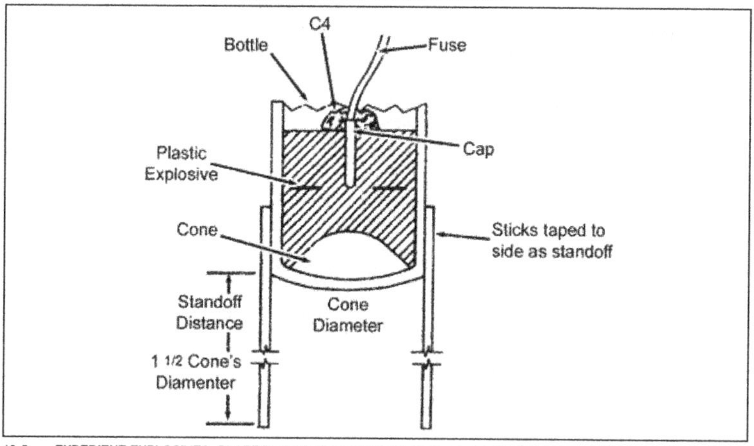

13-5. **EXPEDIENT EXPLOSIVES--PLATTER CHARGE**. This device turns a metal plate into a powerful blunt-nosed projectile (Figure 13-3). The plate should be steel, preferably round but square will work, and should weigh from 2 to 6 pounds.

a. The weight of the explosive should equal the weight of the platter.

b. Uniformly pack the explosive behind the platter. You will only need a container if the explosives fail to remain firmly against the platter. You can use tape to anchor the explosives, if needed.

c. Prime the charge at the exact, rear center of the charge. If any part of the blasting cap is exposed, cover the blasting cap with a small quantity of C4.

d. Aim charge at the direct center of the target, and ensure that the charge is on the opposite side of the platter from the target. Effective range is 35 yards for a small target. With practice, you might hit a 55-gallon drum at 25 yards 90 percent of the time. A gutted fuse igniter can serve as an expedient aiming device.

Figure 13-3. PLATTER CHARGE

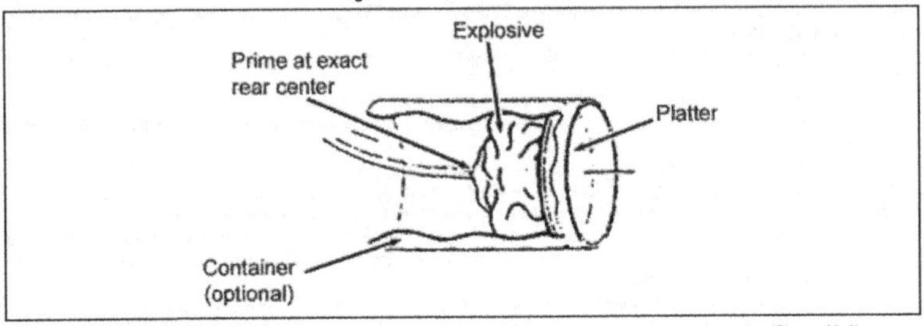

13-6. **EXPEDIENT EXPLOSIVES--GRAPESHOT CHARGE**. To use this antipersonnel fragmentation mine (Figure 13-4)--

a. **Hole**. Create a hole in the center, bottom of the container, for the blasting cap.

b. **Explosives**. Place explosives evenly on the bottom of the container. Remove all voids and air pockets by pressing the C4 into place using a non-sparking instrument.

c. **Buffer**. Place buffer material directly over top of the explosives.

d. **Projectiles**. Place projectiles over top of the buffer materials, then cover to prevent spilling from movement.

e. **Aim**. Aim at target from about 100 feet. Use a small amount of C4 on any exposed portion of the blasting cap.

Figure 13-4. GRAPESHOT CHARGE

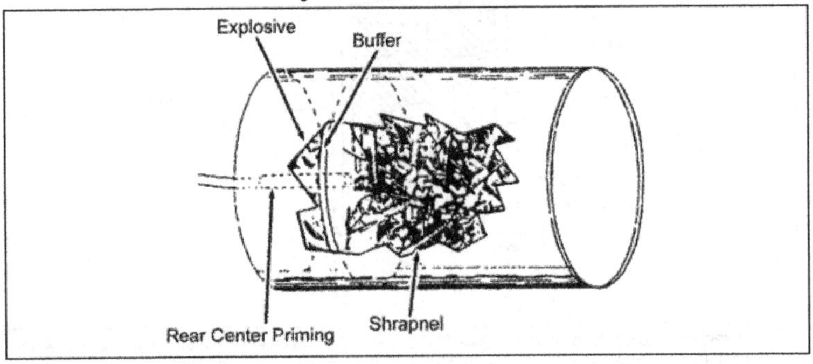

13-7. **DEMOLITION KNOTS.** Several knots are used in demolitions. Figures 13-5 and 13-6 show a few simple knots that can join demolitions to detonation cord.

Figure 13-5. VARIOUS JOINING KNOTS USED IN DEMOLITIONS

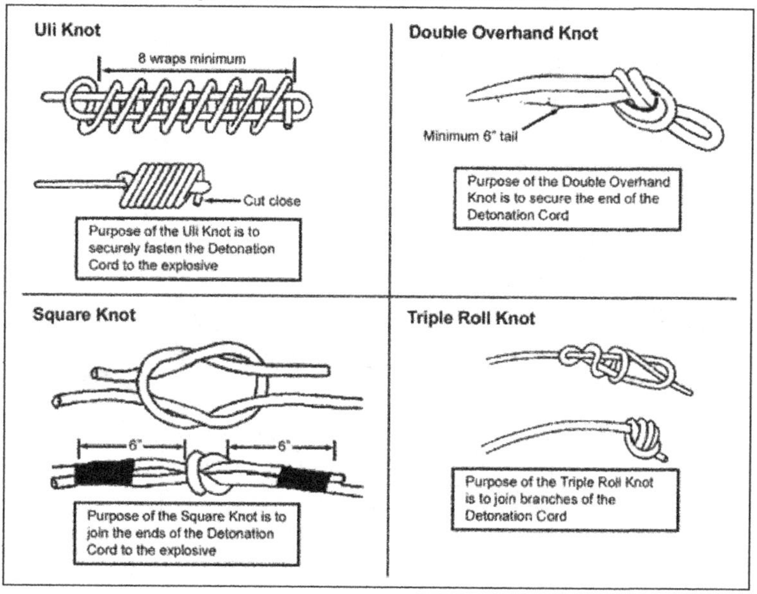

Figure 13-6. BRITISH JUNCTION

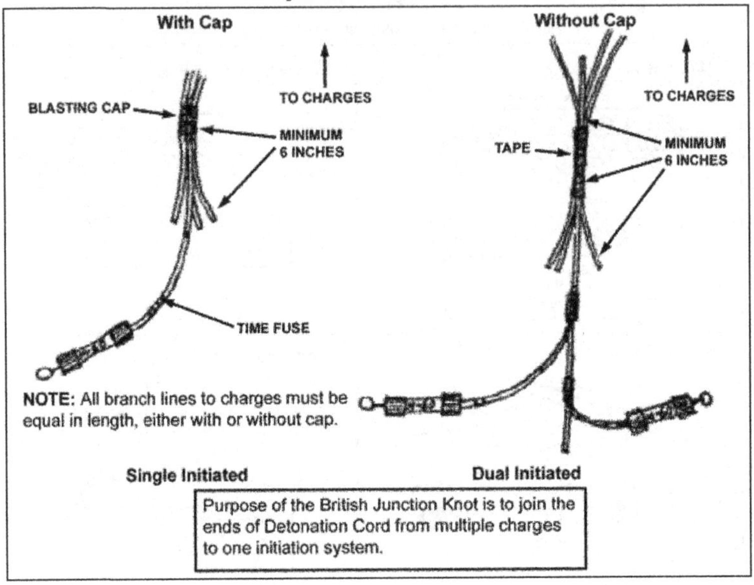

13-8. **MINIMUM SAFE DISTANCES**. Rangers must remain especially aware of their situation when using demolitions. (Table 13-2 shows minimum safe distances for employing demolitions.) For charges over 500 pounds, use the formulas shown in Figure 13-7.

Table 13-2. MINIMUM SAFE DISTANCE FOR PERSONNEL IN OPEN (BARE CHARGE)

EXPLOSIVE WEIGHT (LB)	SAFE DISTANCE		EXPLOSIVE WEIGHT (LB)	SAFE DISTANCE	
	FEET	METERS		FEET	METERS
27 OR LESS	985	300	175	1,838	560
30	1,021	311	200	1,920	585
35	1,073	327	225	1,999	609
40	1,123	342	250	2,067	630
45	1,168	356	275	2,136	651
50	1,211	369	300	2,199	670
60	1,287	392	325	2,258	688
70	1,355	413	350	2,313	705
80	1,415	431	375	2,369	722
90	1,474	449	400	2,418	737
100	1,526	465	425	2,461	750
125	1,641	500	500	2,625	800
150	1,752	534	---	---	---

Figure 13-7. MINIMUM SAFE DISTANCES FOR CHARGES OVER 500 POUNDS

$$\textit{safe distance (meters)} = 100\sqrt[3]{\textit{pounds of explosive}}$$

$$\textit{safe distance (feet)} = 300\sqrt[3]{\textit{pounds of explosive}}$$

Notes: For charges on targets, the minimum radius of danger is 1,000 meters. Minimum safe distance when in a missile-proof shelter from the point of detonation is 100 meters.

13-9. **BREACHING CHARGES**. For Table 13-3, the left column represents the thickness of reinforced concrete wall. The remaining 7 columns show the number of packages of C4 required to breach the wall using the charge placements shown in the drawings above the columns.

a. Use Table 13-3, 13-4 and 13-5 for breaching charges.

b. Use the formula in Figure 13-8 to calculate the charges.

c. Multiply number of packages of C4 from Table 13-3 by conversion factor from Table 13-4 for materials other than reinforced concrete.

Table 13-3. BREACHING CHARGES FOR REINFORCED CONCRETE

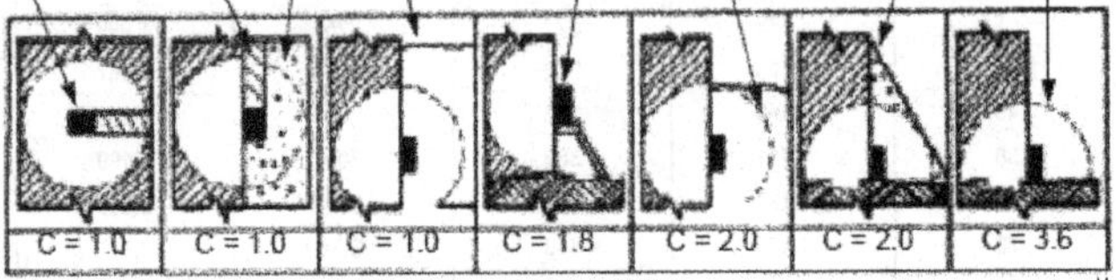

Rein-forced-Concrete Thick-ness (ft)	Placement Methods						
	Placed in center of mass (C = 1.0)	Tamped or stemmed, Fill (C = 1.0)	Deep water (C = 1.0)	Elevated untamped (C = 1.8)	Shallow water (C = 2.0)	Earth tamping (C = 2.0)	Ground placed, untamped (C = 3.6)
	Packages of M112 (C4)						
2.0	1	5	5	9	10	10	17
2.5	2	9	9	17	18	18	33
3.0	2	13	13	24	26	26	47
3.5	4	21	21	37	41	41	74
4.0	5	31	31	56	62	62	111
4.5	7	44	44	79	88	88	157
5.0	9	48	48	85	95	95	170
5.5	12	63	63	113	126	126	226
6.0	13	82	82	147	163	163	293
6.5	17	104	104	186	207	207	372
7.0	21	111	111	200	222	222	399
7.5	26	137	137	245	273	273	490
8.0	31	166	166	298	331	331	595

NOTE: The results of all calculations for this table have been rounded UP to the next whole package.

Table 13-4. CONVERSION FACTORS FOR MATERIALS OTHER THAN REINFORCED CONCRETE

MATERIAL	*CONVERSION FACTOR*
• EARTH	0.1
• ORDINARY MASONRY • HARDPAN • SHALE • ORDINARY CONCRETE • ROCK • GOOD TIMBER • EARTH CONSTRUCTION	0.5
• DENSE CONCRETE • FIRST-CLASS MASONRY	0.7

Table 13-5. MATERIAL FACTOR (K) FOR BREACHING CHARGES

Material	R	K
Earth	All values	0.07
Poor masonry, shale, hardpan, good timber, and earth construction	Less than 1.5 m (5 ft)	0.32
	1.5 m (5 ft) or more	0.29
Good masonry, concrete block, and rock	0.3 m (1 ft) or less	0.88
	Over 0.3 m (1 ft) to less than 0.9 m (3 ft)	0.48
	0.9 m (3 ft) to less than 1.5 m (5 ft)	0.40
	1.5 m (5 ft) to less than 2.1 m (7 ft)	0.32
	2.1 m (7 ft) or more	0.27
Dense concrete and first-class masonry	0.3 m (1 ft) or less	1.14
	Over 0.3 m (1 ft) to less than 0.9 m (3 ft)	0.62
	0.9 m (3 ft) to less than 1.5 m (5 ft)	0.52
	1.5 m (5 ft) to less than 2.1 m (7 ft)	0.41
	2.1 m (7 ft) or more	0.35
Reinforced concrete (factor does not consider cutting steel)	0.3 m (1 ft) or less	1.76
	Over 0.3 m (1 ft) to less than 0.9 m (3 ft)	0.96
	0.9 m (3 ft) to less than 1.5 m (5 ft)	0.80
	1.5 m (5 ft) to less than 2.1 m (7 ft)	0.63
	2.1 m (7 ft) or more	0.54

Figure 13-8. FORMULA FOR COMPUTING SIZES OF CHARGES TO BREACH CONCRETE, MASONRY, AND ROCK

$$P = R^3KC$$

where—

P = *TNT required, in pounds*

R = *breaching radius, in feet*

K = *material factor, which reflects the strength, hardness, and mass of the material to be demolished (Table 3-5)*

C = *tamping factor, which depends on the location and tamping of the charge (Figure 3-15)*

13-10. **TIMBER CUTTING CHARGES**. Table 13-6 shows timber-cutting charge sizes. Figures 13-9 through 13-15 show the types of charges and the formulas to use with each.

Table 13-6. TIMBER-CUTTING CHARGE SIZE

	Packages of C4 Required (1.25-lb Packages) by Timber Diameter (in)											
Charge Type	**6**	**8**	**10**	**12**	**15**	**18**	**21**	**24**	**27**	**30**	**33**	**36**
Internal	1	1	1	1	1	1	2	2	2	3	3	4
External	1	1	2	3	4	5	7	9	11	14	17	20
Abatis	---	---	---	---	---	---	---	7	9	11	14	16
NOTE: **Packages required are rounded UP to the next whole package.**												

Figure 13-9. ABATIS

Figure 13-10. FORMULA FOR FALLEN TREE OBSTACLES OR TEST SHOT

$$P = \frac{D^2}{50} = P = 0.02D^2$$

where—

P = *TNT required per tree, in pounds*

D = *diameter or least dimension of dimensioned timber, in inches*

Figure 13-11. TIMBER-CUTTING RING CHARGE

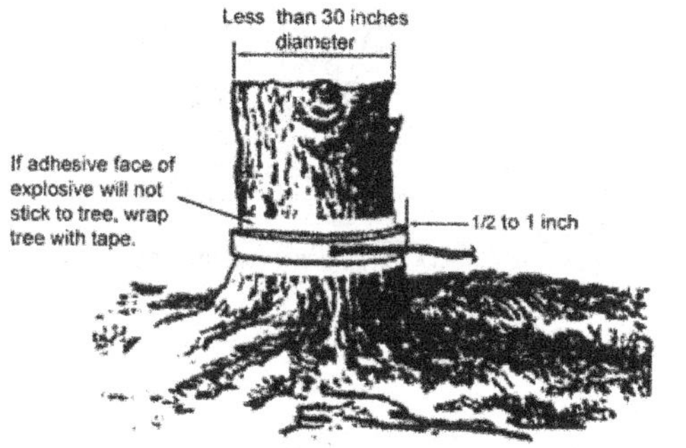

Figure 13-12. TIMBER-CUTTING CHARGE (EXTERNAL)

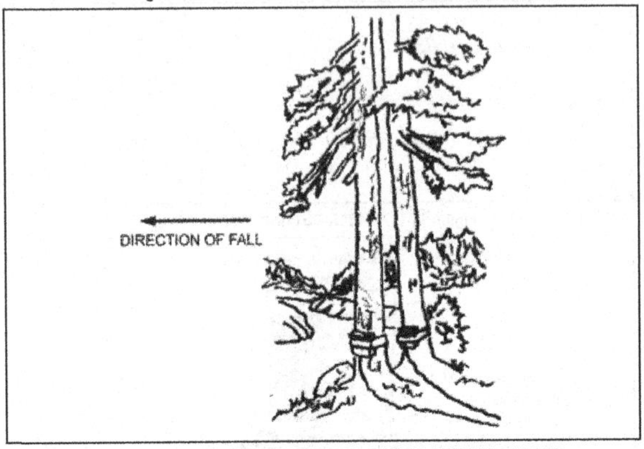

Figure 13-13. FORMULA FOR EXTERNAL TIMBER-CUTTING CHARGE

$$P = \frac{D^2}{40} \text{ or } P = 0.025D^2$$

where—

P = TNT required per target, in pounds

D = diameter or least dimension of dimensioned timber, in inches

Figure 13-14. TIMBER-CUTTING CHARGE (INTERNAL)

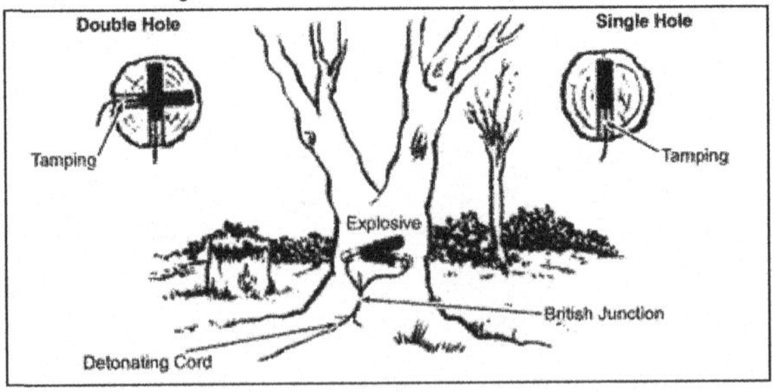

Figure 13-15. FORMULA FOR INTERNAL TIMBER-CUTTING CHARGE

$$P = \frac{D^2}{250} \text{ or } P = 0.004D^2$$

where—

P = TNT required per tree, in pounds

D = diameter or least dimension of dressed timber, in inches

NOTES

Chapter 14
RANGER URBAN OPERATIONS

Urban operations are planned and conducted in an area of operations (AO) that includes one or more urban areas. An urban area is a topographical complex where man-made construction or high population density is the dominant feature (FM 30-06.11). Urban terrain is likely to be one of the most significant future areas of operations for American forces throughout the world. Expanding urban development affects military operations as the terrain is altered. The increasing focus on stability and support operations, urban terrorism, and civil disorder emphasizes that combat in urbanized areas is unavoidable. Urban areas are the power centers, the centers of gravity, and thus the future battlefield. For further study, see FM 30-06.11, FM 90-10, FM 90-1, FM 7-8, 75th Ranger Regiment Advanced UO SOP, and Ranger Training Circular 350-1-2.

14-1. **TYPES**. Infantry units must be trained to conduct urban combat under high-intensity conditions. High-intensity urban combat requires the employment of combat power of the joint combined arms team. An Infantry unit's mission is normally to recon, isolate, penetrate, systematically clear, defend the urban area, and engage and defeat the enemy with decisive combat power. Although the changing world situation may have made urban combat under high-intensity conditions less likely for US forces, it represents the high end of the combat spectrum, and units must be trained for it. High-intensity urban operations can be casualty-intensive for both sides. With the integrated firepower of the joint, combined arms team, leaders must make every attempt to limit unnecessary destruction of critical infrastructure and casualties among noncombatants.

a. **Precision Conditions**. Infantry units train to defeat an enemy that is mixed with non-combatants in precision urban combat. Leaders plan to limit civilian casualties and collateral damage through the establishment of strict rules of engagement (ROE) and the employment of precision weapons and munitions. The ROE provides the focus for the use and restraint of combat power. The ROE may be significantly more restrictive than under high-intensity conditions.

b. **Surgical Conditions**. Operations conducted under surgical conditions include special-purpose raids, small precision strikes, or small-scale personnel seizure or recovery operations in an urban environment (for example, hostage rescue). Joint special operation forces usually conduct these operations. They may closely resemble US police operations performed by Special Weapons and Tactics (SWAT) teams. They may even involve cooperation between US forces and host nation police. Though regular units may not usually be involved in the actual surgical operation, they may support it by isolating the area, by providing security or crowd control, or providing search and rescue teams.

c. **Transitions**. Leaders must always be prepared to transition rapidly from one type of urban combat to another, and back. Real-world combat shows us that urban operations can deteriorate rapidly and without warning. A force involved in stability and support operations can suddenly find itself in a high-intensity combat situation.

14-2. **PRINCIPLES**

a. **Surprise**. Strike the enemy at a time or place or in a manner for which he is unprepared. Key to success: gives the assaulting element the advantage.

b. **Security**. Never permit the enemy to acquire unexpected advantage.

(1) Maintain during all phases of the operation.

(2) Four-dimensional battlefield (height, depth, width, subterranean).

(3) Always maintain 360 degree security (include elevated and subterranean areas).

(4) Mission is never complete as long as you remain in the urban environment. The status of actors in the urban environment does not afford the sense of security offered by "open" terrain. The key to survivability is a constant state of situational awareness.

c. **Simplicity**. Prepare clear, uncomplicated plans, and provide subordinates with concise orders to ensure thorough understanding.

(1) Always keep plans simple.

(2) Ensure everyone understands the mission and the commander's intent.

(3) Plan and prepare for the worst.

d. **Speed**. Rate of military action.

(1) Act as security.

(2) Move in a careful hurry.

(3) Smooth is fast and fast is smooth.

(4) Never move faster than you can accurately engage targets.
(5) Exercise tactical patience.

e. **Violence of Action**. Eliminate the enemy with sudden, explosive force.
(1) Combined with speed gives surprise.
(2) Prevents enemy reaction.
(3) Both physical and mental.

14-3. **METT-TC**. To effectively plan combat operations in urban environments, leaders must use Troop Leading Procedures and conduct a thorough analysis utilizing METT-TC factors. The following lists specific planning guidance that must be incorporated when planning for urban operations. For more specifics on mission planning, refer to Chapter 2, Ranger Handbook.

a. **Mission**. Know correct Task Organization to accomplish the mission (Offense, Defense, or Stability and Support Operations).

b. **Enemy**.

(1) ***Disposition***. Analyze the arrayal of enemy forces in and around your objective, known and suspected. Example: Known or suspected locations of minefields, obstacles, and strong points.

(2) ***Composition and Strength***. Analyze the enemy's task organization, troops available, suspected strength, and amount of support from local civilian populace based on intelligence estimates. Is the enemy a conventional or unconventional force?

(3) ***Morale***. Analyze the enemy's current operational status based on friendly intelligence estimates. For example, is the enemy well supplied; have they had recent success against friendly forces, taken many casualties; and what is the current weather?

(4) ***Capabilities***. Determine what the enemy can employ against your forces. Example: Enemy's weapons, artillery assets, engineer assets, air defense assets, NBC threats, thermal/NVG capabilities, close air support, armor threat, and so on.

(5) ***Probable Course(s) of Action***. Based on friendly intelligence estimates, determine how the enemy will fight within his area of operation (in and around your area of operation).

c. **Terrain**. Leaders conduct a detailed terrain analysis of each urban setting, considering the types of built-up areas and composition of existing structures. They use OCOKA when analyzing terrain, in and around the area of operation.

(1) ***Observation and Fields of Fire***. Always be prepared to conduct urban operations under limited visibility conditions. This includes the effects of reduced illumination, as well as natural and manmade obscuration. Leaders should ensure that Rangers are equipped with adequate resources, which allow them to successfully operate in the urban environment under these types of conditions. (2) ***Cover and Concealment***. Leaders must perform a thorough analysis of peripheral as well as intra-urban areas. Leaders should identify routes to objectives, which afford assault forces with the best possible cover and concealment. Additionally, leaders should take advantage of limited visibility conditions, which would allow forces to move undetected to their final assault / breaching positions. When in the final assault position, forces should move as rapidly as tactically possible to access structures, which afford additional cover and concealment. Leaders must learn to properly employ obscurants and exercise "tactical patience" to fully take advantage of these effects. Finally, all members of the urban force must practice noise and light discipline. Soldiers must avoid unnecessary voice communications, learn the proper use of white light, and limit contact with surfaces that may alert the enemy of their presence.

(3) ***Obstacles***. There are many manmade and natural obstacles on the periphery, as well as within the urban environment. Leaders should conduct a detailed reconnaissance of routes and objectives (this must include subterranean complexes), taking into consideration route adjustments and special equipment needs.

(4) ***Key Terrain***. Analyze which buildings, intersections, bridges, LZ/PZ, airports, and elevated areas that provide a tactical advantage to you or the enemy. Additionally, the leader must identify critical infrastructure within his area of operations, which would provide the enemy with a tactical advantage on the battlefield. These may include, but are not limited to, communication centers, medical facilities, governmental facilities, and facilities that are of psychological significance.

(5) ***Avenues of Approach***. Consider roads, intersections, inland waterways, and subterranean constructions (subways, sewers, and basements). Leaders should classify areas as go, slow go, or no-go based on the navigability of the approach.

Note: Military maps may not provide enough detail for urban terrain analysis or reflect the underground sewer systems, subways, water systems, or mass transit routes.

(6) ***Troops***. Analyze your forces utilizing their disposition, composition, strength, morale, capabilities, and so on. Leaders must also consider the type and size of the objective to plan effective use of troops available.

(7) ***Time***. Operations in an urban environment have a slower pace and tempo. Leaders must consider the amount of time required to secure, clear, or seize the urban objective and stress and fatigue Rangers will encounter. Additional time must also be allowed for area analysis efforts; these may include, but are not limited to--

- Maps and urban plans Recon and analysis.
- Hydrological data analysis.
- Line-of-sight surveys.
- Long Range Surveillance and Scout reconnaissance.
- Artillery.
- Armor.
- Aviation.
- Armor.
- Engineers.

(8) ***Civilians***. Authorities such as the National Command will establish the Rules of Engagement. Commanders at all levels may provide further guidance regarding civilians occupying the area of operations (AO). Leaders must daily reiterate the ROE to subordinates, and immediately inform them of any changes to the ROE. Rangers must have the discipline to identify the enemy from noncombatants and ensure civilians understand and follow all directed commands.

Note: Civilians may not speak English, may be hiding (especially small children), or dazed from a breach. Civilians must not be given the means to resist. Rehearse how clearing/search teams will react to these variables. Never compromise the safety of your Rangers.

14-4. **CLOSE QUARTERS COMBAT.** Due to the very nature of a CQC encounter, engagements will be very close (within 10 meters) and very fast (targets exposed for only a few seconds). Most close quarter's engagements are won by who hits first and puts the enemy down. It is more important to knock a Ranger down as soon as possible than it is to kill him. In order to win a close quarters engagement, Rangers must make quick, accurate shots by mere reflex. This is accomplished by reflexive fire training. Remember, no matter how proficient you are, always fire until the enemy goes down. All reflexive fire training is conducted with the eyes open.

Note: Research has determined that, on average, only three out of ten people actually fire their weapons when confronted by an enemy during room clearing operations. Close quarters combat success for the Ranger begins with the Ranger being psychologically prepared for the close quarter's battle. The foundation for this preparedness begins with the Ranger's proficiency in basic rifle marksmanship. Survival in the urban environment does not depend on advanced skills and technologies. Rangers must be proficient in the basics.

14-5. **REHEARSALS**. Similar to the conduct of other military operations, leaders need to designate time for rehearsals. Urban operations require a variety of individual, collective, and special tasks, which are not associated with operations on less complex terrain. These task require additional rehearsal time for clearing, breaching, obstacle reduction, casualty evacuation, and support teams. Additionally, rehearsal time must be identified for rehearsals with combined arms elements. These may include, but are not limited to--

a. **Stance**. Feet are shoulder width apart, toes pointed straight to the front (direction of movement). The firing side foot is slightly staggered to the rear of the non-firing foot. Knees are slightly bent and the upper body leans slightly forward. Shoulders are not rolled or slouched. Weapon is held with the butt stock in the pocket of the shoulder. The firing side elbow is kept in against the body. The stance should be modified to ensure the Ranger maintains a comfortable boxer stance.

(1) ***Low-Carry Technique***. The butt stock of the weapon is placed in the pocket of the shoulder. The barrel is pointed down so the front sight post and day optic is just out of the field of vision. The head is always up identifying targets. This technique is safest and is recommended for use by the clearing team once inside the room.

(2) ***High-Carry Technique***. The butt stock of the weapon is held in the armpit. The barrel is pointed slightly up with the front sight post in the peripheral vision of the individual. Push out on the pistol grip and thrust the weapon forward and pull straight back into the pocket of the shoulder to assume the proper firing position. This technique is best suited for the line-up outside the door. Exercise caution with this technique always maintaining situational awareness, particularly in a multi-floored building.

Note: Muzzle awareness is critical to the successful execution of close quarter's operations. Rangers must never point their weapons or cross the bodies of their fellow Rangers at any time. Additionally Rangers should always avoid exposing the muzzle of their weapons around corners; this is referred to as "flagging."

b. **Malfunction**. If a Ranger has a malfunction with his weapon during any CQC training, he will take a knee to conduct immediate action. Once the malfunction is cleared, there is no need to immediately stand up to engage targets. Rangers can save precious seconds by continuing to engage from one knee. Whenever other members of the team see a Ranger down, they must automatically clear his sector of fire. Before rising to his feet, the Ranger warns his team members of his movement and only rises after they acknowledge him. If a malfunction occurs once committed to a doorway, the Ranger must enter the room far enough to allow those following him to enter and move away from the door. This drill must be continually practiced until it is second nature.

c. **Approach to a Building or Breach Point**. One of the trademarks of ranger operations is the use of limited visibility conditions. Whenever possible, breaching and entry operations should be executed during hours and conditions of limited visibility. Rangers should always take advantage of all available cover and concealment when approaching breach and entry points. When natural or manmade cover and concealment is not available, Rangers should employ obscurants to conceal their approach. There are times when Rangers will want to employ obscurants to enhance existing cover and concealment. Members of the breach / entry team should be numbered for identification, communication, and control purposes.

(1) The # 1 Ranger should always be the most experienced and mature member of the team, usually the team leader. The # 1 Ranger is responsible for frontal and entry and breach point security. The team leader is responsible for initiating all voice and physical commands. The team leader must exercise situational awareness at all time with respect to the task, friendly force, and enemy activity.

(2) The # 2 Ranger is directly behind the # 1 Ranger in the order of movement, and he moves through the breach point in the opposite direction from the #1 Ranger.

(3) The # 3 Ranger will simply go opposite the # 2 Ranger inside the room at least 1 meter from the door.

(4) The # 4 Ranger moves the opposite the # 3 Ranger and is responsible for rear security (and is normally the last Ranger into the room). An additional duty of the # 4 Ranger is breaching.

Note: Consider how much firepower each Ranger delivers. Where do you put the SAW gunner in the order? You must weigh the benefits of firepower against those of quick, accurate shots. If the # 4 Ranger has breaching responsibilities, it should not be the SAW gunner, because this would reduce your firepower.

d. **Actions Outside the Point of Entry**. Entry point position and individual weapon positions are important. The clearing team members should stand as close to the entry point as possible, ready to enter. Weapons are oriented in such a manner that the team provides itself with 360 degree security at all times. Team members must signal to one another that they are ready at the point of entry. This is best accomplished by sending up a "squeeze." If a tap method is used, an inadvertent bump may be misunderstood as a tap.

e. **Enter Building/Clear Room**. See Battle Drill 5, Chapter 6.

f. **Locking Down the Room**.

(1) Control the situation within the room.

(2) Use clear, concise arm and had signals. Voice commands should be kept to a minimum to reduce the amount of confusion and to prevent the enemy--who might be in the next room--from discerning what is going on. This enhances the opportunity for surprise and allows the assault force the opportunity to detect any approaching force.

(3) Physically and psychologically dominate.

(4) Establish security and report status.

(5) Cursory search of the room to include the ceiling (3 Dimensional Fight).

(6) Identify the dead using reflexive response techniques (Eye thump method).

(7) Search the room for PIR, precious cargo as per the mission and time available.

(8) Evacuate personnel.

(9) Mark room clear using chemical lights, engineer tape, chalk, paint, VS-17 panels, and so on.

14-6. **TTPS FOR MARKING BUILDINGS AND ROOMS**. Units have long identified a need to mark specific buildings and rooms during UO. Sometimes rooms need to be marked as having been cleared, or buildings need to be marked as containing friendly forces. The US Army Infantry School is currently testing a remote marking device that can be used to mark doors from as far away as across a wide street. In the past, units have tried several different field-expedient marking devices; some with more success than others. Chalk has been the most common. It is light and easily obtained but not as visible as other markings. Some of the other techniques have been to use spray paint, and paintball guns.

a. **NATO Standard Marking SOP**. The North Atlantic Treaty Organization (NATO) has developed a standard marking SOP for use during urban combat. It uses a combination of colors, shapes, and symbols. These markings can be fabricated from any material available. (Figure 14-1 shows examples.)

Figure 14-1. EXAMPLE NATO STANDARD MARKINGS

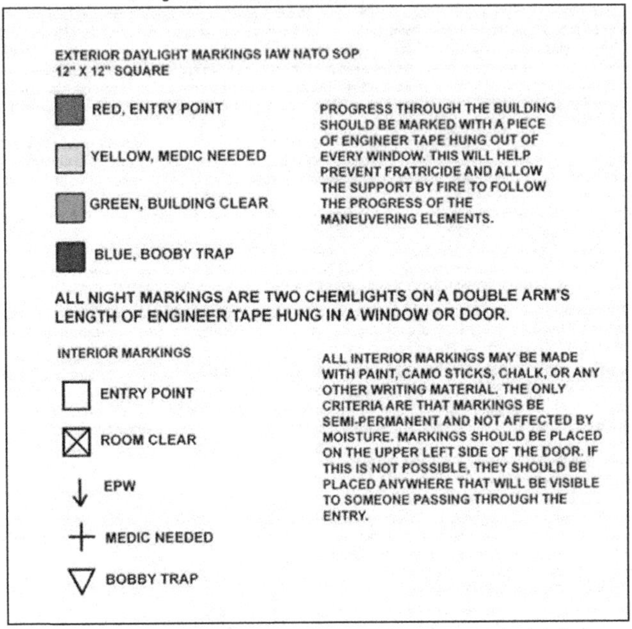

b. **Spray Paint**. Canned spray paint is easily obtained and comes in a wide assortment of colors including florescent shades that are highly visible in daylight. It cannot be removed once used. Cans of spray paint are bulky and hard to carry with other combat equipment. Paint is not visible during darkness nor does it show up well through thermal sights.

c. **Paintball Guns**. Commercial paintball guns have been purchased by some units and issued to small unit leaders. Some models can be carried in standard military holsters. They can mark a building or door from about 30 meters. The ammunition and propellant gas is not easily obtainable. The ammunition is fragile and often jams the gun if it gets wet. The available colors are not very bright, and just like spray paint, cannot be seen at night or through thermal sights.

d. **Wolf Tail**. A simple, effective, easy-to-make, lightweight device called a "wolf tail" can be fabricated to mark buildings, doorways, and windows (Figure 14-1). A unit has changed its tactical SOP to require that each Infantryman carry one of these devices in his ACU cargo pocket. Wolf tails, when used IAW a simple signaling plan understood by all members of the unit, can aid in command and control, reduce the chances of fratricide, and speed up casualty collection during urban combat.

(1) The wolf tail marking device is simple to make and versatile. It can be used together with the NATO marking scheme. Rolled up, it makes a small, easily accessible package that can be carried in the cargo pocket of the ACUs. It can be recovered easily and used again if the situation changes. All its components can be easily obtained through unit supply. It combines a variety of visual signals (colored strapping and one or more chemical lights of varying colors) with a distinctive heat signature that is easily identified through a thermal weapon sight. An infrared chemlight can be used either as a substitute for the colored chemlight(s) or in addition to them.

(2) Constructing the wolf-tail marking device requires the following material:

- A 2-foot length of nylon strap (the type used for cargo tie-downs) (engineer tape can be substituted).
- About 5 feet of 550 cord.
- A small weight such as a bolt or similar object.
- Duct tape.
- Chemical lights (colored or IR).
- Two 9-volt batteries.

(3) Assemble the items by tying or taping the cord to the small weight. Attach the other end of the cord to the nylon strapping, securing it with duct tape. Attach the 9-volt batteries in pairs to the lower end of the strapping with several wraps of duct tape, making sure that the negative terminals are opposite the positive, but not actually touching. Use more duct tape to attach the chemical lights, about 2 inches above the batteries, to the strapping.

(4) When you want to mark your position, push the batteries together firmly until the male and female plugs lock. This shorts out the battery, causing it to heat up rapidly. The hot battery is easily identified through the thermal sights of tanks or BFVs. The batteries stay visible for about 45 minutes. Activating the chemical lights provides an easily identified light source visible to the naked eye. You can use infrared chemical lights if you want them to be seen through night vision devices but not with the naked eye.

Figure 14-2. EXAMPLE OF A WOLF-TAIL MARKING DEVICE

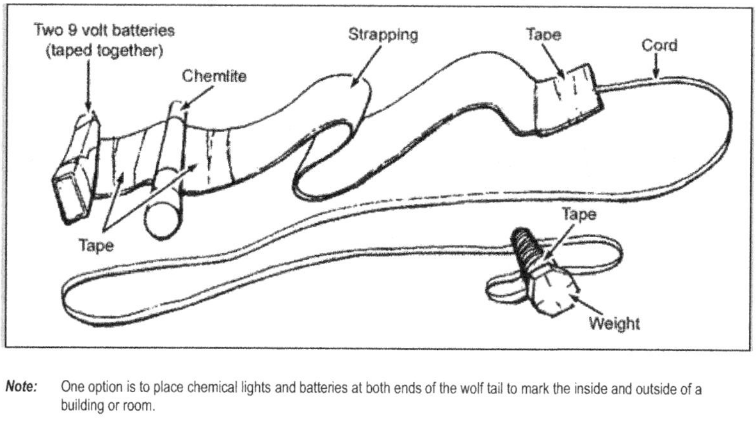

Note: One option is to place chemical lights and batteries at both ends of the wolf tail to mark the inside and outside of a building or room.

(5) Use the cord and the small weight to hold the wolf tail in position by tying or draping it out a window or hanging it on a door, wherever it is best seen by other friendly troops. Squads or platoons can vary the numbers and colors of chemical lights, or use multiple battery sets to identify precisely what unit is in which building.

(6) Medics and combat lifesavers can carry a standardized variation that can be used to clearly identify a building as containing wounded personnel needing evacuation. This could be a white strap with multiple red chemical lights, or any other easily identified combination.

14-7. **URBAN ASSAULT BREACHES**. Understanding how to employ and incorporate breaching as part of a leader's planning process is an important part of urban operations. It is imperative that elements of your patrol be skilled in the art of breaching. Whether infiltrating or exfiltrating from an objective, leaders must plan for either option. One constant disadvantage with the employment of explosives is that placement requires Rangers to expose themselves to possible enemy fires. Breach teams need to be supported by fires or obscurants and breaching operations should be performed during hours of limited visibility whenever possible. Breaching classes follow:

a. **Mechanical**. Mechanical breaching should be an important part of a leader's breaching training program because it is usually an option. Mechanical breaching is best described as gaining access by the use of tools or saws. Although most tools and saws used are recognizable and self explanatory to the individual Ranger, one must practice on various techniques to increase speed an effectiveness. This reduces fatigue and expedites the actual assault.

(1) **Tools**.

(a) Hooligan tools..........................Doors/windows of all types.
(b) Sledge hammer........................Heavy duty doors, locks, and window frames.
(c) Picket pounder.........................Doors of all types, light walls.
(d) Bolt Cutters..............................Chain link fence, locks, and wire obstacles.
(e) Pick Ax.....................................Lightweight doors and locks.
(f) Saws...Fences, light doors, locks.

b. **Ballistic**. Ballistic breaching is defined as a forced entry or exit by the use of weapons. Whether using shotguns, M16A2/M4, M249 SAW, specific considerations must be addressed:

(a) Type of round and ricochet factor.
(b) Composition of the breaching point.
(c) Composition of the floor beyond the door.
(d) Personnel behind the door (friendly/enemy).
(e) Always shoot at a 45-degree angle.

c. **Explosive**. Explosive breaching is the most viable because it is the most effective. When employing explosives during breaching operations, leaders must consider three major factors.

(1) ***Overpressure***. The amount of PSI released from the concussion of the blast.

(2) ***Missile Hazard***. Fragmentation or projectiles sent at tremendous speed from the explosion area. This occurs from either the charge or target being breached.

(3) ***Minimum Safe Distance Requirements***. Use of explosives in the urban environment must consider the presence of noncombatants and friendly forces. Additionally, the are many hazardous materials located in the urban environment, these may include chemicals as well as construction materials. There is always a risk of secondary explosions and fires, when employing explosive breaching techniques.

(4) ***Charges***. Various charges can be used for explosive breaching. Leaders must conduct extensive training on the use of the charges to get proper target feedback. Figure 14-3 shows examples of charges used for explosive breaching.

Figure 14-3. EXAMPLE CHARGES FOR EXPLOSIVE BREACHING

Charge	Target
Water Impulse................	Steel/wood doors
Flexible linear.................	Wood doors
Ranger wall breach........	Masonry/brick walls
Chain link ladder............	Chain link fence
E-Type silhouette...........	Wood doors
Brashier breacher..........	Concertina wire

Chapter 15
VEHICLE CONVOY OPERATIONS

This chapter outlines a technique for conducting vehicle convoy operations. Convoy operations present a challenge to the Ranger leader. Trucks and other combat vehicles produce a large signature on the battlefield and increase your unit's value as a target. Vehicle movement is restricted to roads and terrain that they can traverse; therefore, a sound plan must be implemented to minimize the possibility of compromise.

15-1. **PLANNING CONSIDERATIONS**. This paragraph explains how to plan a vehicle convoy IAW the eight steps of the TLP.

a. **Receive the Mission**. (PL will extract the following information from Company OPORD).
   (1) Truck Support (number and type of Truck, ACL).
   (2) Weather. Road conditions.
   (3) Truck pickup and drop-off location / marking.
   (4) Truck movement Timeline (Pick-up time, roll time, H-hour).
   (5) Truck Routes (primary and alternate, Check points).
   (6) Drop-off locations (primary and alternate).
   (7) IED/Contact compromise and contingencies.

b. **Issue the Warning Order**. During the WARNO the PL gives a basic overview of the Truck Movement and gives specific Task to Maneuver Units.
   (1) Manifest Personnel (PSG).
   (2) Pick up point security and set-up (PSG).
   (3) Primary and alternate routes (Security SL).

c. **Make a tentative Plan**. PL considers METT-TC.
   (1) ***Mission***. Develop your Ground Tactical Plan and a then develop a truck plan that best supports it.
   (2) ***Enemy***. Consider enemy patrols and activity in AO.
   (3) ***Terrain***. Evaluate routes and dropoff and pickup points using OCOKA.
   (4) ***Troops***.
   (a) Truck ACL.
   (b) Chalks and chalk leaders identified.
   (c) Chalks planed with tactical integrity & self sufficiency.
   (d) Key leaders and weapons are cross-loaded.
   (e) Pickup point marking team identified.
   (f) Pickup point security teams identified.
   (5) ***Time***. PL considers the following when evaluating time available.
   (a) Movement to Pickup point.
   (b) Recon of Pickup point.
   (c) Emplacement of security.
   (d) Marking the Pickup point.
   (e) Pickup point Posture time.

d. **Initiate Necessary Movement**. PL decides whether to move the platoon to the pickup location or to conduct a map reconnaissance and complete the plan.

e. **Conduct Reconnaissance**. (Is not necessary if pickup point is secured and established by Higher HQ's).
   (1) Leader's Recon consists of PL, RTO, Chalk Leaders and Pickup point Security element.
   (2) PL determines suitability of Pickup point.
   (3) PL determines security plan (either overwatch or strong point).

f. **Complete the Plan**. Complete Truck Movement Annex on Page ??? the Ranger Handbook.

g. **Issue the OPORD**.

h. **Supervise**.
   (1) Back brief from chalk leaders.
   (2) Rock drill of truck movement and contingency plans.

15-2. **FIVE PHASES OF TRUCK MOVEMENT**. Each phase must support the ground tactical plan, which specifies actions in the objective area to accomplish the commander's intent for the assigned mission, be it a raid, ambush, recon or other follow-on missions.

a. **Staging Plan**.

(1) Secure the pickup point (strongpoint or overwatch).

(2) Mark pickup point (day/night).

b. **Loading Plan**. Each Ranger is assigned a truck.

c. **Movement Plan**.

(1) Troops awake and alert during movement.

(2) Platoon leader and chalk leaders tracking route progress.

(3) Compromise and contingency plan.

(a) React to IED.

(b) React to ambush.

(c) Truck breakdown.

d. **Dropoff Plan**.

(1) Establish security of dropoff point (overwatch or strongpoint).

(2) Dismount vehicles.

(3) Recon and secure assembly area.

(4) Adjust perimeter as chalks arrive.

(5) PSG clears the trucks when last chalk departs.

e. **Ground Tactical Plan**. Execute after platoon is reconsolidated or minimum force required is assembled.

15-3. **CONVOY TECHNIQUES**. The following convoy techniques have been included for your convenience:

- Convoy warning order
- Convoy brief
- React to Ambush Near/Far
- Convoy forced to stop (Methods 1 and 2)
- Break contact
- Recovery/CASEVAC Operations

Convoy Warning Order

1. SITUATION: A brief statement of the enemy and friendly situation. Who, What, Where
2. MISSION: Who, What, When, Where, Why
3. TASK ORGANIZATION: Based on tasking from higher WARNO.
4. INITIAL TIME SCHEDULE:

When	Who	What

5. SPECIAL INSTRUCTIONS: PCC/PCI Guidance, Rehearsals, additional tasks to be accomplished.
6. SERVICE SUPPORT:

Class I: (Rations & Water)

Class III: (POL)

Class V: (Ammunition /Pyrotechnics)

Weapon System	Rounds	Type

Pyrotechnic Device	Number	Location

Class VIII: (Medical)

7. UNIFORM AND EQUIPMENT COMMON TO ALL.

CONVOY BRIEF (Modified from FM 55-30)

Movement Order No. ____

References: A. ______________________________ *(Maps, tables and relevant documents)*

B. ______________________________

TASK ORGANIZATION: (Internal organization for convoy – Manifest)

1. **SITUATION:**

a. **Enemy Forces:**

(1) Weather. General forecast.

(2) Light Data (EENT, % Illumination, MR, MS, BMNT)

(3) Discuss Enemy.

Identification of enemy (If known).

Composition / capabilities / strength / equipment

Location

(Hot Spots highlighted on map)

b. **Friendly Forces:**

(1) Operational support provided by higher headquarters.

Helicopter / Gunships

Quick Reaction Forces (QRFs)

MP Escorts / Rat Patrols

Fire Support elements

c. **Attachments:** (From outside the organization)

2. **MISSION: (WHO, WHAT, WHEN, WHERE, & WHY)**

3. **Employment Tactics, Techniques and Procedures**

a. Readiness Level
- i. Ride / Scanning (Observation)
- ii. At the Ready

b. Scanning / Sector of Fire

i. Driver
1. Sector of Scan is 9 – 1 clock position
2. Observation with Mirrors
3. Sector of Fire is 9 – 11 o'clock

ii. TC
4. Sector of Scan is 11 – 3 clock position
5. Sector of Fire is 1 – 3 clock position

iii. Other Systems are based off vehicle type and load

b. Target ID
- i. Communicate / Signal

c. Body Positioning
- i. Engage as you Train (Right or Left handed Firing)
- ii. Firing-side Shoulder Down
- iii. Maintain your Body Position
- iv. Weapon to Head

d. Rules of Engagement Concerns – Employ the appropriate systems based off the threat.

e. Point of Aim
- i. Moving Platform – Stationary Target requires aim to the rear (trail) and low dependant upon speed.

CONVOY BRIEF (Modified from FM 55-30) (Continued)

3. Employment Tactics, Techniques and Procedures

e. Point of Aim
   - ii. Moving Platform – Moving Target requires aim directly on and low.
   - iii. Stationary Platform – Moving Target requires aim to the front (lead) and low dependent upon speed.

f. Rate of Fire
   - i. Burst Mode
   - ii. Steady Suppression (ROE)

g. Magazine Awareness
   - i. Serviceability / maintenance
   - ii. Tracer Mix
   - iii. Magazine Storage / Placement
   - iv. Mounted, Reload when time is available

Dismounted, Seek cover prior to the need to Reload

REACT TO AMBUSH (NEAR)

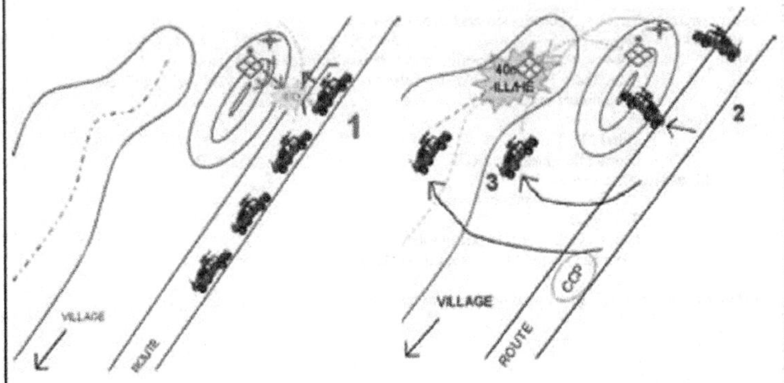

(Enemy Initiates Weapons Fire / RPG / IED)

1. Lead element of convoy reacts to contact and immediately returns fire. Lead fire team fires 40mm HE (at night also white star if NVGs drown out by vehicle headlights – leader calls it. FYI, most travel at night will be with headlights on).
2. Lead elements assualt through ambush – fire team dismounts while gun trucks use M249, MK-19, and .50 cal in close support.
3. Trail fire team and trail gun truck move to block most likely enemy escape route. Fire team uses 40mm HE/Illumination if needed. (Use of 40mm becomes less likely as units get closer to urban areas due to collateral damage risk.) Establish CCP upon action completion or enemy neutralized.

REACT TO AMBUSH (FAR)

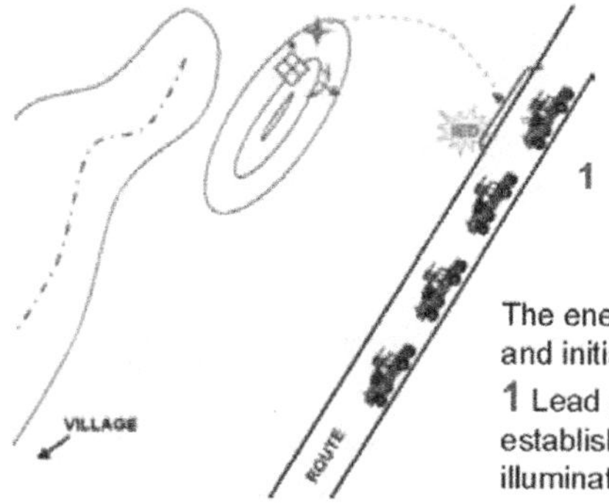

The enemy is outside 100 meters and initiates with IED/RPG

1 Lead gun truck and lead fire team establish SBF—fire team fires 40mm illumination

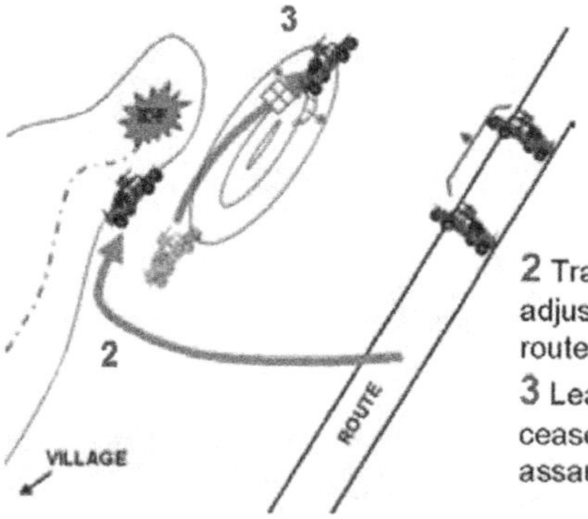

2 Trail element flanks—SL/PL adjusts IDF into enemy escape route(if in range)

3 Lead element shifts fire—arty cease loading—trail element assaults across—enemy destroyed

CONVOY FORCED TO STOP

(Weapons Fire / RPG / IED / Indirect)

1. Vehicles forced to stop. Activate vehicle turn signal to indicate direction of contact.
2. Vehicle(s) / Personnel not in direct contact, report on internal communication, identifying truck number, type of contact, clock direction and eight digit grid if available.

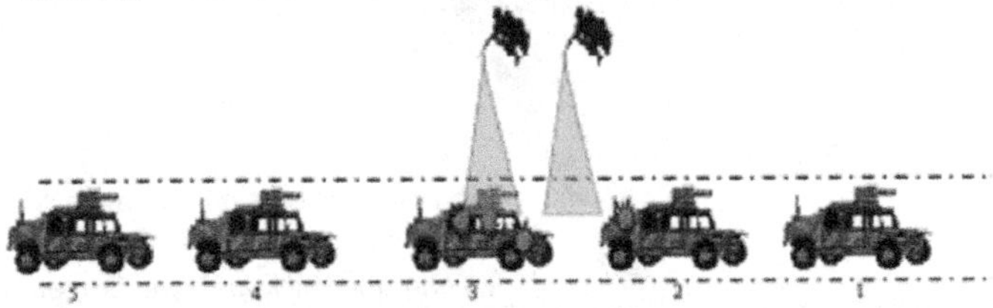

3. Personnel on vehicle(s) forced to stop dismount on the non-contact side, assume covered position(s) and provide initial base of fire.
4. The <u>entire</u> convoy halts; personnel will dismount vehicle(s) on the non-contact side and provide additional base of fire on the enemy if in range. Vehicle(s) not in contact will reposition to attack the attackers. M16's should now be on <u>Semi-Automatic</u> in order to conserve ammunition.
5. PL/CC/SC will move to better assess the situation and position the Gun Truck(s) in order to best suppress the enemy while maintaining <u>standoff</u>. Gun trucks will close with and engage the enemy.
6. Once the PL/CC/SC determines the convoy has either gained fire superiority or defeated the enemy contact, Recovery / CASEVAC operations will begin.
7. If the PL/CC/SC determines the convoy cannot gain fire superiority, leader will then conduct Break Contact procedures

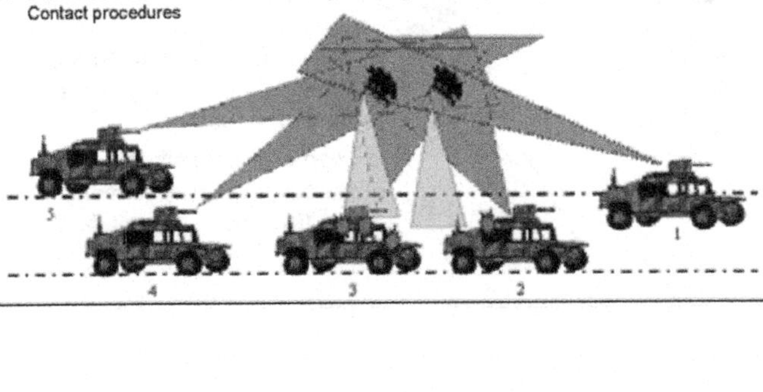

CONVOY FORCED TO STOP - METHOD 2

(Weapons Fire /RPG / IED / Indirect)

1. Vehicle(s) forced to stop. Activate vehicle turn signal to indicate direction of contact.

*** ALL PERSONNEL STAY IN VEHICLES ***

2. All vehicles immediately drive forward out of the kill zone.

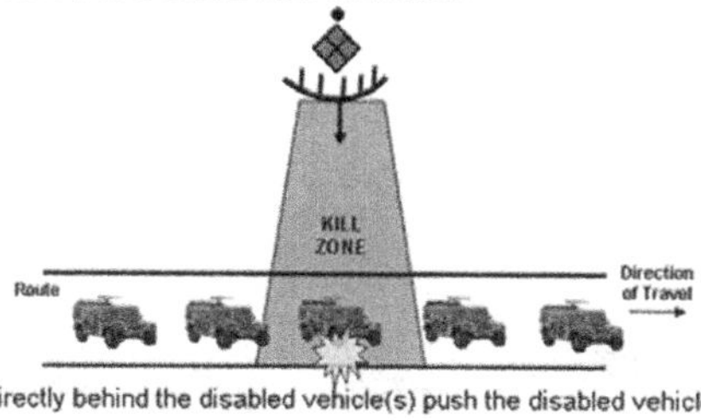

3. The vehicle(s) directly behind the disabled vehicle(s) push the disabled vehicle(s) out of the kill zone.
4. The vehicle(s) that are not disabled will establish a base of fire towards the suspected/known contact.
5. If fire superiority can be gained the PL will use the minimum force necessary to destroy the enemy.
6. If the PL determines the convoy cannot gain fire superiority, the leader will conduct break contact.

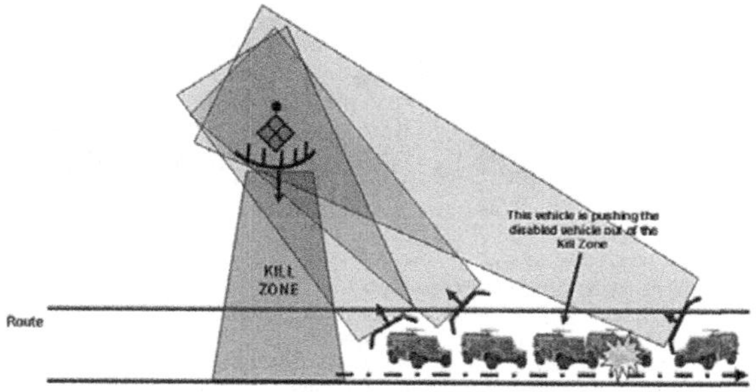

BREAK CONTACT

Always try to close with and destroy the enemy first. This way he cannot come back later on to attack you or another convoy again. However – if you must...

1. The PL/CC/SC determines the convoy cannot gain fire superiority and the decision has been made to Break Contact.
2. The PL/CC/SC designates that either Rally Point "Rear" or "Forward" will be used. If necessary, both Rally Points may be used. Communication systems and appropriate pyrotechnic signals will be used to communicate Break Contact and Rally Point.
3. Personnel will deploy obscuration measures if available. Utilizing cover and concealment, Aid & Litter team(s) will evacuate all casualty(s) under support of Gun Truck and other protective fire(s).
4. Personnel will maintain position and suppression in contact zone and assist Aid & Litter team(s) as necessary.
5. Disabled vehicle(s) will be hauled back or destroyed as directed by leaders. (Thermite or explosives)

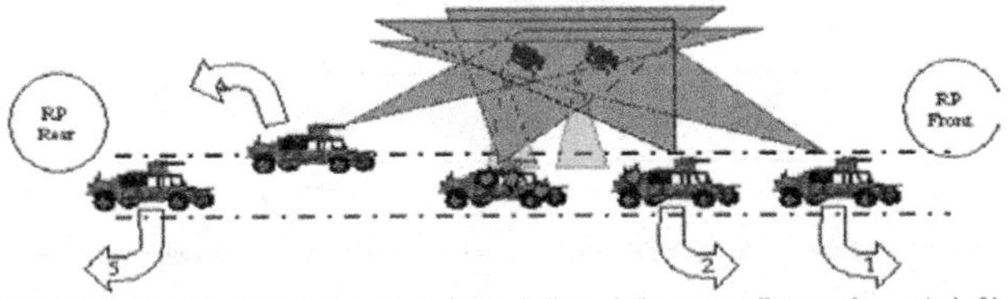

6. Vehicles will displace either backwards or forwards through the convoy lines under control of leaders. The most forward vehicle in the contact zone moves first, then the next most forward vehicle moves second. Vehicles will continue to displace. As vehicles displace, Gun Truck repositions as necessary until contact is broken.

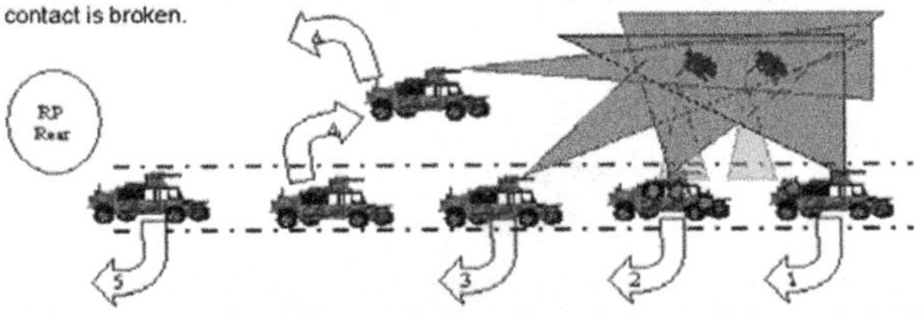

7. If Break Contact occurs with vehicles on both sides of the contact zone, displacement of vehicles will occur using an alternating displacement technique.
8. Upon occupation of the Rally Point, leaders will immediately position vehicles, setting 360 degree security, and conduct Consolidation and Reorganization.
9. If the convoy vehicles get separated when not in contact with the enemy, personnel and vehicles stay together and move to the closest Rally Point or Check Point.

RECOVERY / CASEVAC OPERATIONS

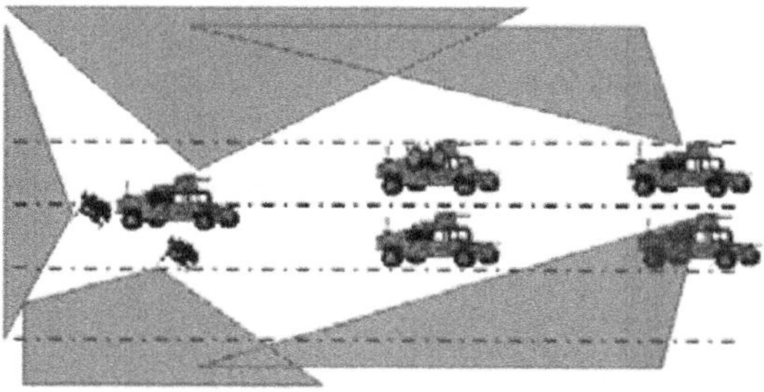

1. Once the leader assesses the enemy threat to be destroyed, neutralized, defeated and the area secured – Recovery / CASEVAC operations will begin. This keeps soldiers focused first on defeating and destroying the threat.

2. CASEVAC:

   a. Aid & Litter team will position on the safe side of the vehicle and extract casualty(s) and personnel.

   b. Treatment of casualty(s) will occur once they are safely removed from the contact area.

3. Vehicle Recovery Procedures:

   a. Recovery team will position on the safe side of the disabled vehicle.

   b. TC will dismount and assess the disabled vehicle.

   c. If determined the vehicle can be safely recovered, TC guide the recovery vehicle into position and conduct a hasty hook-up. TC will operate the disabled vehicle is necessary.

   d. Upon exiting the contact area, complete and correct hook-up procedures will occur.

   e. If assessment results in outside support necessary for recovery, leader will contact higher for guidance.

4. Once recovery operations are complete, the team will displace and conduct link-up with the convoy at the Rally Point.

5. Disabled vehicle(s) will be abandoned or destroyed as directed by leaders. (Thermite or explosives)

GLOSSARY

1SG	first sergeant
5 W's	Who, What, Where, When, and Why?
5-S Rule	Search, Silence, Segregate, Safeguard, and Speed to rear
AA	avenue(s) of approach
AAR	after-action review
AATF	air assault task force
ABCDE	method of identification and response to life-threatening conditions: Airway, Breathing, Circulation, Disability, Exposure
ACE	ammunition, casualties, and equipment
ACL	allowable combat load
ACP	aerial checkpoint
ACU	Army combat uniform
ADA	air defense artillery
AG	assistant gunner
ATC	air traffic controller; a mechanical belay device that locks down on itself when tension is applied in opposite directions
ALT	alternate
AMC	air mission commander
ammo	ammunition
ANCD	automated net-control device
AO	area of operations
AOO	actions on the objective
AR	automatic rifleman
ABF	attack by fire
ATL	Alpha team leader
ATM	Alpha Team
bangalore torpedo	A manually emplaced, 1.5-meter long explosive-filled tube used to breach wire and detonate simple, pressure-activated antipersonnel mines. Ten tubes clear a 1- by 15-meter lane.
belay	any action taken to stop a climber's fall or to control the rate a load descends
binos	binoculars
BMNT	begin morning nautical twilight
BN	battalion
body belay	belay that uses the belayer's body to apply friction by routing the rope around the his body
bowline on a coil	knot used to secure a climber to the end of a climbing rope
BP	battle position
British junction knot	knot used to join the ends of detonation cords from multiple charges to one initiation system
BTC	bridge team commander
BTL	Bravo Team leader
BTM	Bravo Team

C2	command and control
CAS	close air support
CASEVAC	casualty evacuation
CCA	close combat attack
CCIR	commander's critical information requirements
CCP	casualty collection point
CDR	commander
CDS	Camp Darby Special (map)
CLS	combat lifesaver
CO	company
COA	course of action
commo	communications
COMSEC	communications security
cordelette	short section of static rope or static cord. Also called "sling rope"
COTS	commercial off the shelf
CP	command post
CPR	cardiopulmonary resuscitation
CQC	close quarters combat
CS	combat support
CSS	*obsolete: now referred to as* sustainment
CTT	common task test
DAR	designated area of recovery
DOL	direction of landing
double figure eight knot	knot used to form a fixed loop in the end of the rope; loops are large enough to insert a carabiner
double overhand knot	knot used to secure the end of detonation cord
DP	duty position; decision point (depending on context)
DST	distance
DTG	date-time group
dynamic ropes	one of two classifications of kernmantle rope; used for climbing; *see also* static rope
DZ	drop zone
EA	engagement area
EEFI	essential elements of friendly information
EENT	end evening nautical twilight
end-of-the-rope clove hitch	intermediate anchor knot that requires constant tension
end-of-the-rope Prusik	knot used to attach a movable rope to a fixed rope; *see also* middle-of-the-rope Prusik
ENY	enemy
EPW	enemy prisoner of war
FA	field artillery
FDC	fire direction center
FFIR	friendly force information requirements

figure 8 slip knot	knot used to form an adjustable bight in the middle of a rope
FLIR	forward looking infrared
FLOT	forward line of own troops
FO	forward observer
FPF	final protective fires
FRAGO	fragmentary order
FSC	fire support coordinator
FSO	fire support officer
GOTWA	G - Where leader is GOING O - Others he is taking with him T - Time he plans to go W - WHAT to do if the leader does not return in time A - The unit's and the leader's ACTIONS on chance contact while the leader is gone
GPS	global positioning system
GRN	grenadier
HDG	heading
HE	high explosive
H-Hour	hit hour (the time the unit plans to accomplish the mission)
HI	high temperature (weather)
HPT	high-payoff target
HQ	headquarters
IAW	in accordance with
ICM	improved conventional munitions
ID	identification
IP	initial point
IR	intelligence requirements
ERRP	en route to release point
JAAT	joint air attack team
JD	Julian date
KIA	killed in action
LAW	light antiarmor weapon
LBV	load-bearing vest
LD	line of departure
LDA	linear danger area
LO	low temperature (weather)
LOGSTAT	logistical status
LP	listening post
LZ	landing zone
material factor	the strength, hardness, and mass of the material to be demolished
mb	millibar (a metric unit used to measure air pressure)

MDI	modernized demolition initiator
ME	main effort
mechanical belay	a belay that uses mechanical devices to help the belayer control the rope, as in rappelling
MEDEVAC	medical evacuation
METL	mission-essential task list
METT-TC	mission, enemy, terrain (and weather), troops (and support) available, time available, and civil considerations
middle of the rope clove hitch	knot that secures the middle of a rope to an anchor
middle-of-the-rope Prusik	knot that attaches a movable rope to a fixed rope, anywhere along the length of the fixed rope; *see also* end-of-the-rope Prusik
MG	machine gun/ner
MOPP	mission-oriented protective posture
MR	moonrise
MRE	meal, ready to eat
MS	moonset
MSD	minimum safe distance
MSL	mean sea level
munter hitch	commonly used belay that requires little equipment
NATO	North Atlantic Treaty Organization
NAV	navigation
NFA	no-fire area
NLT	no later than
NVD	night-vision device
NVG	night vision goggles
NVS	night vision system
OAKOC	observation and fields of fire, avenues of approach, key terrain, observation, and cover and concealment
OBJ	objective
OD	olive drab
OOM	order of movement
OP	observation post
OPORD	operations order
OPSKED	operational schedule
ORP	objective rally point
OT	observer-target
PB	patrol base
PCC	precombat checks
PCI	precombat inspection
PDF	principal direction of fire

PI	probability of incapacitation
PIR	priority intelligence requirements
PL	platoon leader
PLD	probable line of deployment
PLOT-CR	purpose, location, observer, trigger, communication method, resources (a format for planning fire support)
PLT	platoon
POL	petroleum, oils, and lubricants
PRI	primary
protection	a piece of equipment, natural or artificial, that is used to construct an anchor
PSG	platoon sergeant
PSI	pounds per square inch
PZ	pickup zone
R	rifleman
R&S	reconnaissance and surveillance
RACO	rear area combat operations
RAP	rocket-assisted projectile
rappel seat	a rope harness used in rappelling and climbing
RED	risk-estimate distance
REQ	required
rerouted figure 8 knot	anchor knot that also attaches a climber to a climbing rope
RFA	restrictive fire area
RFL	restrictive fire line
RFLM	rifleman
RHB	Ranger Handbook
ROE	rules of engagement
round-turn with two half hitches	a constant tension anchor knot
RP	release point
RTO	radio operator
S-2	intelligence staff officer
S-3	operations staff (and training) officer
SALUTE	Size, Activity, Location, Unit/Uniform, Time, and Equipment
SAW	squad automatic weapon
SBF	support by fire (position)
SDT	self-development test
SE	supporting effort
SEAD	suppression of enemy air defenses
SITREP	Situation Report
SITTEMP	situational template

SL	squad leader
sling rope	short section of static rope or static cord. Also called "cordelette"
SLLS	Stop, Look, Listen, Smell
SOC	succession of command
SOI	signal operating instructions
SOP	standing operating procedures
SP	start point
square knot	knot used to join two ropes of equal diameter; used to join the ends of the detonation cord to the explosive
SR	sunrise
SS	sunset
STANO	surveillance, target acquisition, and night observation
static ropes	one of two classifications of kernmantle rope; used where rope stretch is undesired, and when the rope is subjected to heavy static weight. *See also* dynamic rope
SURVIVAL	S Size up the situation, your surroundings, your physical condition, and your equipment. U Undue haste makes waste; don't be too eager to move. Plan your moves. R Remember where you are in relation to important friendly and locations and critical resources V Vanquish fear and panic. I Improvise. You can improve your situation. Use what you have. Use your Imagination. V Value your life. Remember your goal: to get out alive. Remain stubborn. Refuse to give in to problems and obstacles that face you. This will give you the mental and physical strength to endure. A Act like the natives; watch their daily routines. When, where, and how do they get food? Where they get water? L Live by your wits. Learn basic skills.
suspension traverse	used to move personnel and equipment over rivers, ravines, chasms, and up or down a vertical obstacle
TAC	tactical air controller
tamping factor	depends on the location and tamping of the charge
technical climbing	using safe and proper equipment and techniques to climb on a rock formation in parties of two or more
tensionless anchor	used to anchor rope on high-load installations such as bridging
TL	team leader
TLP	troop-leading procedures
TL	team leader
TOC	tactical operations center
triple roll knot	knot used to join branches of detonation cord
TTP	tactics, techniques, and procedures
uli knot	knot used to securely fasten detonation cord to explosive
VIXL	video image crosslink
WARNO	warning order
WBGT	wet bulb globe temperature

WFFs	warfighting functions (fire support, movement and maneuver, protection, command and control, and sustainment)
WIA	wounded in action
XO	executive officer
WSL	weapons squad leader

NOTES

INDEX

Appendix A

RESOURCES

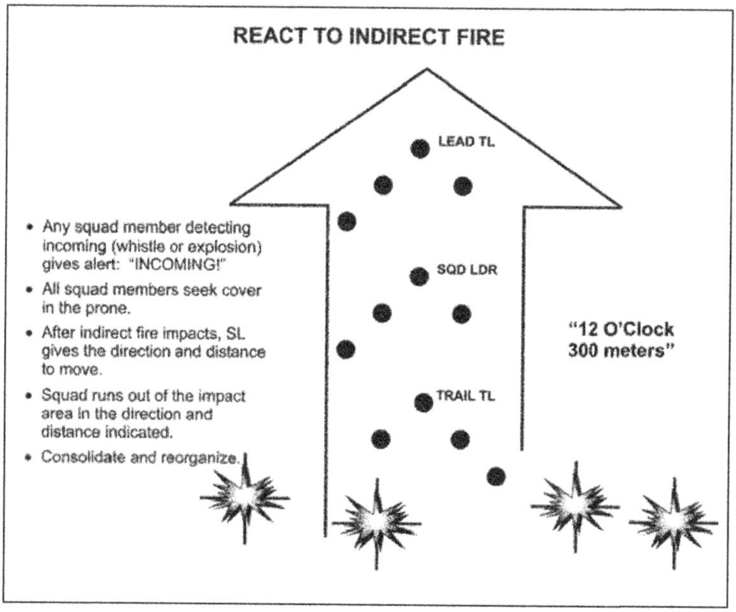

REACT TO CONTACT

- Seek nearest cover.
- Return fire (known or suspected enemy position).
- TM LDRs control fire by using fire commands.
- Report enemy situation (number/position).
- Maintain contact (visual/oral) with team members.
- SQD LDR moves to team in contact and makes an assessment of the situation.
- Factors of his assessment:
  - Can squad move out to engagement area.
  - Can squad gain and maintain suppressive fire.
  - Location of enemy.
  - Size of enemy.
  - Vulnerable flanks.
  - Covered/concealed flanking routes.
- SQD LDR determines COA (break contact, squad attack, etc.)
- Report situation to PL.

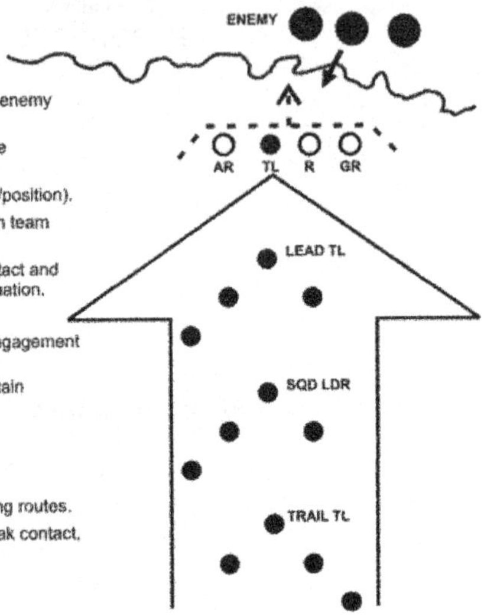

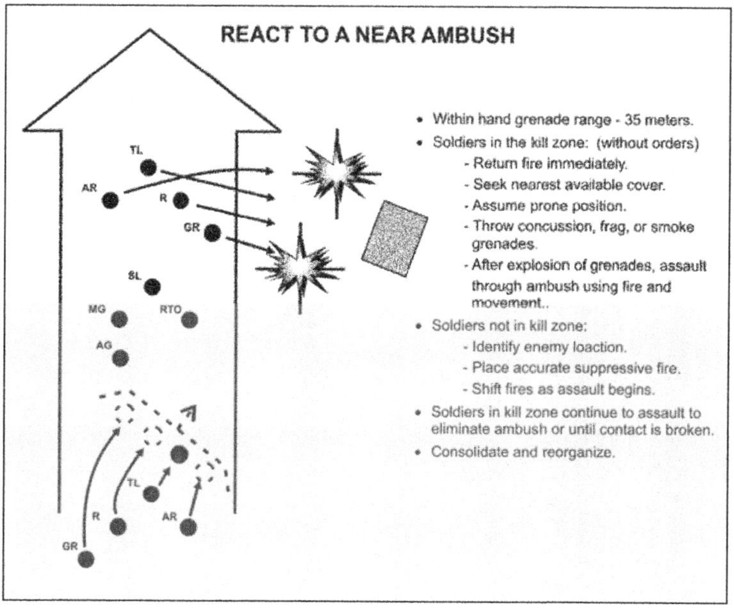
REACT TO A NEAR AMBUSH
TL
AR
R
GR
SL
MG
RTO
AG
TL
R
AR
GR
• Within hand grenade range - 35 meters.
• Soldiers in the kill zone: (without orders)
- Return fire immediately.
- Seek nearest available cover.
- Assume prone position.
- Throw concussion, frag, or smoke grenades.
- After explosion of grenades, assault through ambush using fire and movement..
• Soldiers not in kill zone:
- Identify enemy loaction.
- Place accurate suppressive fire.
- Shift fires as assault begins.
• Soldiers in kill zone continue to assault to eliminate ambush or until contact is broken.
• Consolidate and reorganize.

REACT TO A FAR AMBUSH

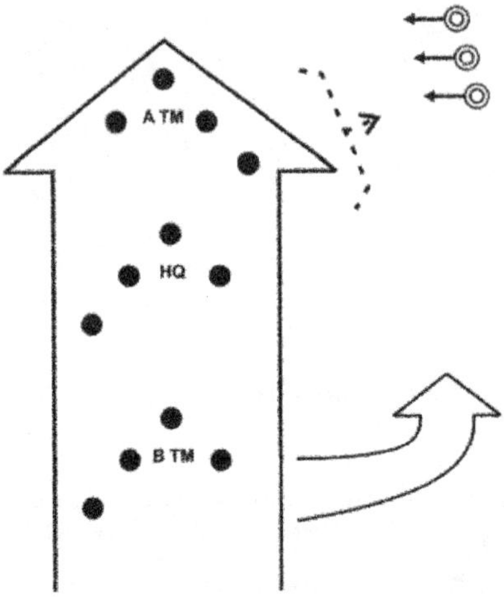

- More than 35 meters.
- TM in the kill zone: (without orders)
  - Return fire immediately.
  - Seek cover and concealment.
  - Suppress enemy (overwatch).
- SL assess situation.
  - Determines COA (flank).
- TM not in contact:
  - Moves along covered and concealed route.
  - Assaults enemy on weak flank.
  - Suppress enemy (overwatch).
- Overwatch TM continues to suppress shifts/cease fire as bounding team enters sector.
- Bounding team continues to assault through enemy.
- SL may request indirect fire.
- Consolidate and reorganize.

BREAK CONTACT

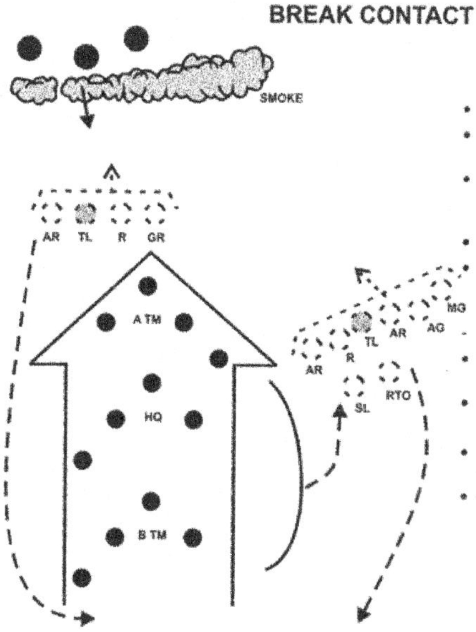

- Squad Leader orders: "Break Contact".
- Squad Leader designates SPT element and maneuver element.
- SL issues distance and direction or a terrain feature for the maneuver element.
- SBF suppresses enemy position.
- Maneuver uses smoke to mask movement.
  - Takes up overwatch position.
  - Begins to suppress enemy
- Squad Leader directs SBF to break contact.
- SBF uses smoke to screen movement.
  - Takes up overwatch position.
- Squad continues to bound away until contact is broken.
- Consolidate / reorganize.

FORMATIONS AND ORDERS OF MOVEMENT

I. Movement Formation: Fire Team Wedge; MG Team attached.

II. Three Movement Techniques utilized:

A. Traveling technique used behind FFL when contact is not likely.

B. Traveling Overwatch forward of the FFL when enemy contact is possible.

C. Bounding Overwatch used forward of the FFL when enemy contact is expected.

III. Distances are based on but not dictated by visibility, terrain, and vegetation.

IV. Actions at Night: Modified Wedge

V. Actions at the Halt: Short and Long Halt (GV/LV)

VI. Leader Location: Fixed/Unfixed

FIRE TEAM WEDGE

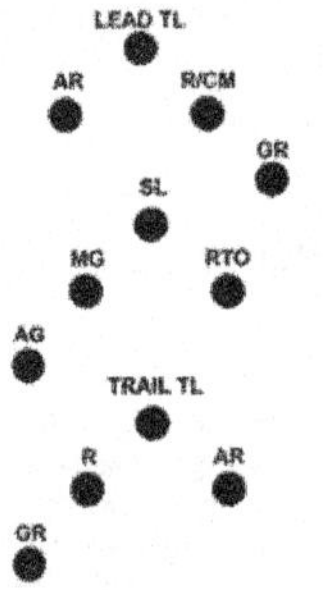

MODIFIED WEDGE

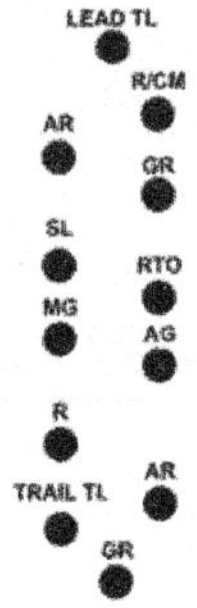

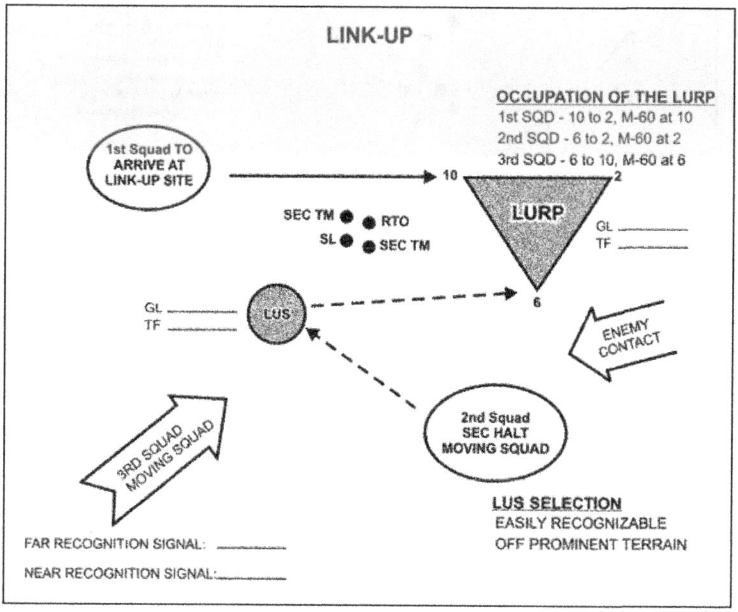
LINK-UP
OCCUPATION OF THE LURP
1st SQD - 10 to 2, M-60 at 10
2nd SQD - 6 to 2, M-60 at 2
3rd SQD - 6 to 10, M-60 at 6
1st Squad TO ARRIVE AT LINK-UP SITE
10
2
LURP
6
GL
TF
SEC TM
RTO
SL
SEC TM
LUS
GL
TF
ENEMY CONTACT
3RD SQUAD MOVING SQUAD
2nd Squad SEC HALT MOVING SQUAD
LUS SELECTION
EASILY RECOGNIZABLE
OFF PROMINENT TERRAIN
FAR RECOGNITION SIGNAL:
NEAR RECOGNITION SIGNAL:

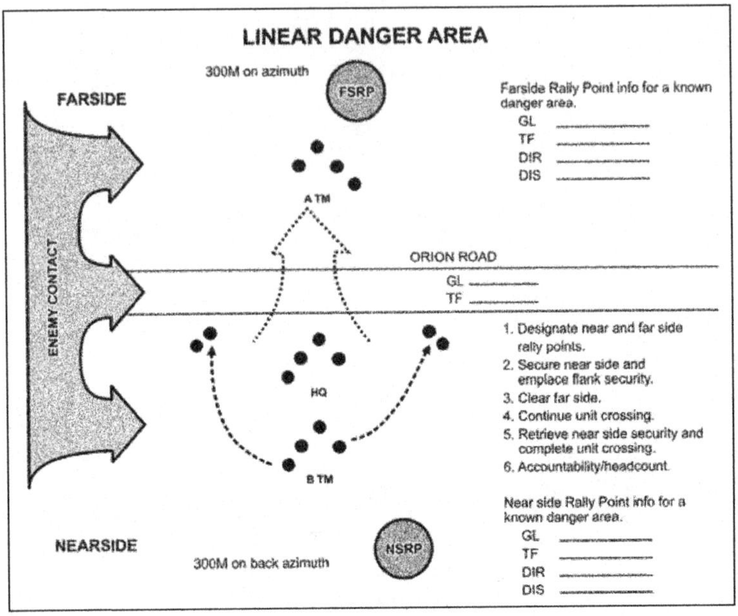
LINEAR DANGER AREA
300M on azimuth
FSRP
FARSIDE
Farside Rally Point info for a known danger area.
GL
TF
DIR
DIS
A TM
ENEMY CONTACT
ORION ROAD
GL
TF
1. Designate near and far side rally points.
2. Secure near side and emplace flank security.
3. Clear far side.
4. Continue unit crossing.
5. Retrieve near side security and complete unit crossing.
6. Accountability/headcount.
HQ
B TM
Near side Rally Point info for a known danger area.
GL
TF
DIR
DIS
NEARSIDE
300M on back azimuth
NSRP

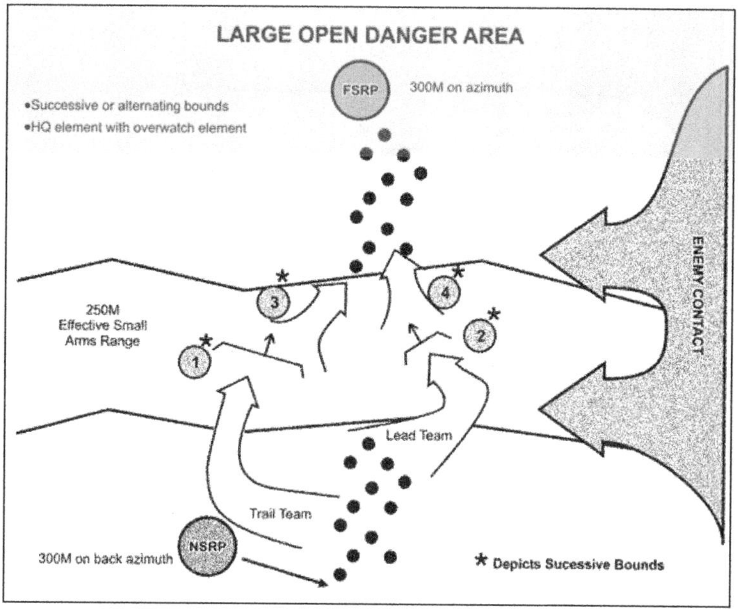
LARGE OPEN DANGER AREA
FSRP
300M on azimuth
•Successive or alternating bounds
•HQ element with overwatch element
ENEMY CONTACT
250M
Effective Small
Arms Range
1
2
3
4
Lead Team
Trail Team
NSRP
300M on back azimuth
* Depicts Sucessive Bounds

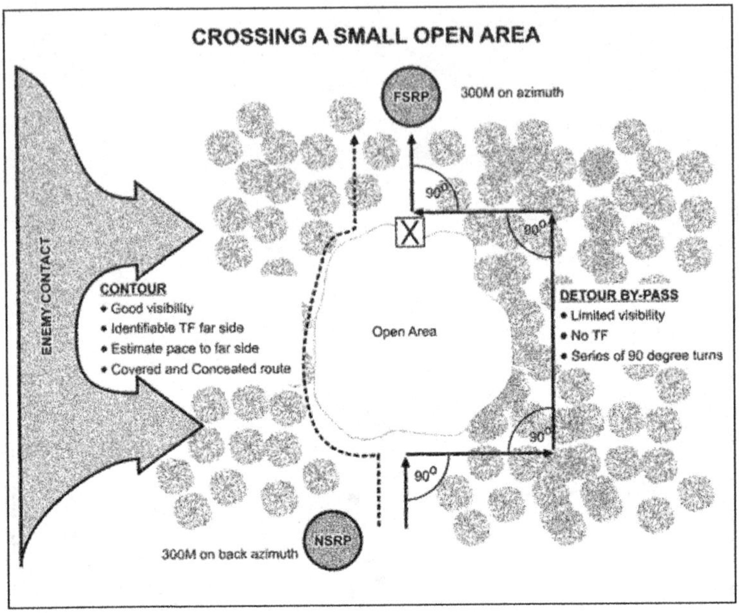

CROSSING A SMALL OPEN AREA
FSRP
300M on azimuth
ENEMY CONTACT
CONTOUR
• Good visibility
• Identifiable TF far side
• Estimate pace to far side
• Covered and Concealed route
Open Area
DETOUR BY-PASS
• Limited visibility
• No TF
• Series of 90 degree turns
90°
90°
90°
90°
NSRP
300M on back azimuth

SQUAD ATTACK

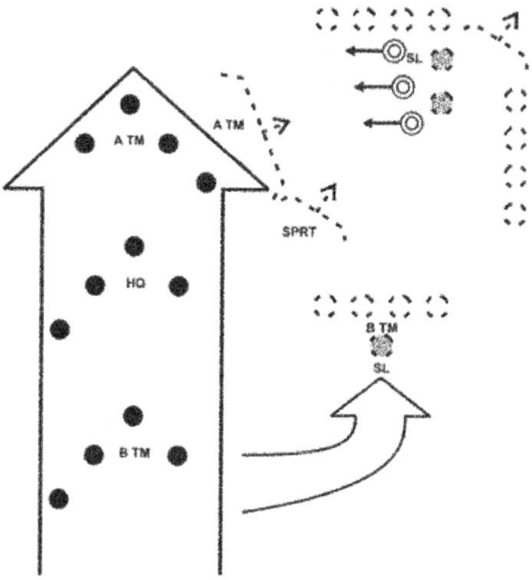

- React to contact.
- TM in the kill zone: (without orders)
  - Return fire.
  - Seek cover and concealment.
  - Suppress enemy calls out 3 D's becomes overwatch.
- SL assess situation.
  - Determines COA (flank/attack).
- TM not in contact: (with SL)
  - Moves along covered and concealed route.
  - Assaults enemy on weak flank.
- Overwatch TM continues to suppress shifts/cease fire as bounding team enters sector.
- Bounding team continues to assault through enemy.
- SL may request indirect fire.
- Consolidate and reorganize.

RAID BOARDS (LEFT)
GENERAL INFO

Raid References:
- RHB Chapter 5
- FM 3-21.8

Purposes:
1. Destroy
2. Liberate
3. Collect Intel

Planning Considerations:
- Mission
- Enemy
- Troops
- Terrain / OAKOC
- Time
- Civilians

STUDENT INPUT WRITTEN HERE

Definition:
A surprise attack on a fixed position or installation ending in a planned withdrawal.

Characteristics:
1. Surprise - Gain
2. Coordinated fires - Maintain
3. Violence of action - Retain
4. Planned withdrawal

INITIATIVE

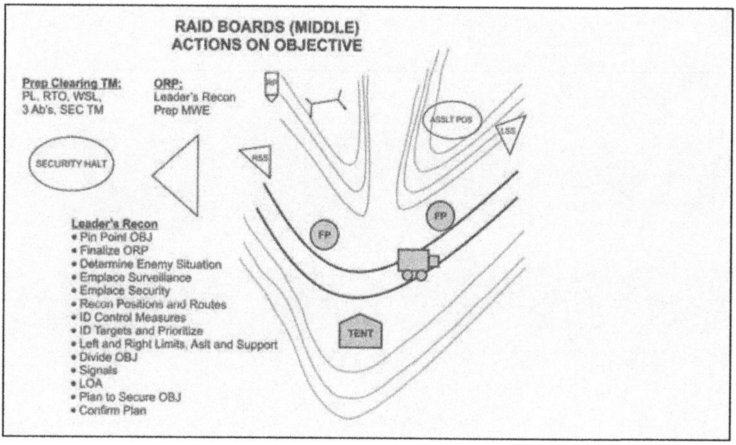
RAID BOARDS (MIDDLE)
ACTIONS ON OBJECTIVE
Prep Clearing TM:
PL, RTO, WSL,
3 Ab's, SEC TM
ORP:
Leader's Recon
Prep MWE
SECURITY HALT
RSS
ASSLT POS
LSS
FP
FP
TENT
Leader's Recon
• Pin Point OBJ
• Finalize ORP
• Determine Enemy Situation
• Emplace Surveillance
• Emplace Security
• Recon Positions and Routes
• ID Control Measures
• ID Targets and Prioritize
• Left and Right Limits, Aslt and Support
• Divide OBJ
• Signals
• LOA
• Plan to Secure OBJ
• Confirm Plan

RAID BOARDS (MIDDLE)
TASK ORGANIZATION

ELEMENT	WHO	WHY	WHAT
HQ	PL, APL RTO, FO, MEDIC	C2/FACILITATE SIGNALS, PEQ-2, STANO	1x119, ICOM, SKEDCO, PLUGGER
ASSLT 1	SQUAD (+)(-)	DESTROY	ICOM, STANO, M18A1, AT4, SPECIAL TEAM KITS
ASSLT 2	SQUAD (+)(-)	DESTROY	ICOM, STANO, M18A1, AT4, SPECIAL TEAM KITS
SUPPORT	WPN S SQD (+)(-)	SUPPRESS	3xM240B COMPLETE, ICOM, PEQ-2, PVS-4, AT, STANO, SIGNALS, MAP
SECURITY	SQUAD (+)(-)	DELAY	ICOMs, M18A1, AT, STANO, MAP SIGNALS

RAID BOARDS (RIGHT)
SOP'S

CONTINGENCIES

1. COMPROMISE
   a. ORP
   b. LEADERS RECON
   c. OCCUPATION
2. MASS CAL
3. COUNTERATTACK

SOP's

1. EPW SEARCH
2. AID LITTER
3. MED EVAC
4. CCP
5. WITHDRAW PLAN

AMBUSH SOPS (LEFT)

References:
- RHB Chapter 5
- FM 3-21.8

Ambush:
A surprise attack from a covered and concealed position on a moving or temporarily halted target.

Types:
- Point
- Area

Fundamentals:
1. Surprise - Gain
2. Fire superiority - Maintain
3. Violence of action - Retain

Purposes:
1. Disrupt/Destroy
2. Collect Intelligence
3. Block or Deny Access
4. Canalize

Categories:
Deliberate: has specific target at a predetermined time and location.

Hasty: Platoon makes visual contact with the enemy and has time to establish an ambush without being detected.

Planning Considerations:
- Mission - Task/Purpose
- Enemy - MPCOA, MDCOA
- Terrain - Map Recon/Ldrs Recon
- Time - 1/3 2/3 Rule
- Troops - Task Org
- Civilians -

AMBUSH BOARDS (MIDDLE)

HQ	PL, PL RTO, PSG, MEDIC, FO	COMMAND AND CONTROL	PRC-119s, ICOM, NVGs, BINO's, SKEDCO, PEQ-2, SIGNALS, MAP
SECURITY	MINIMUM OF 3-2 MAN TEAMS	EARLY WARNING, SEALS OFF OBJ, LFS, RFS, ORP SECURITY	COMMO, NVGs, CLAYMORES, AT WPNS, BINO's, SIGNALS, MAP
SUPPORT	WSL, 3 GUN TEAMS COMPLETE +/-	SUPPRESSIVE FIRE ON KILL ZONE, SEC ON OBJ DURING RECON/REORG	MG's, COMMO, NVGs, AT WPNS, BINO's, SKEDCO, PEQ-2, SIGNALS., MAP
ASSAULT	2 SQUADS +/-	MAIN EFFORT, BLOCK, DESTROY, CANALIZE, CAPTURE PIR	NVGs, BINO's, CLAYMORE, AT, FLEX CUFFS, POLELESS LITTER, MARKINGS, SIGNALS

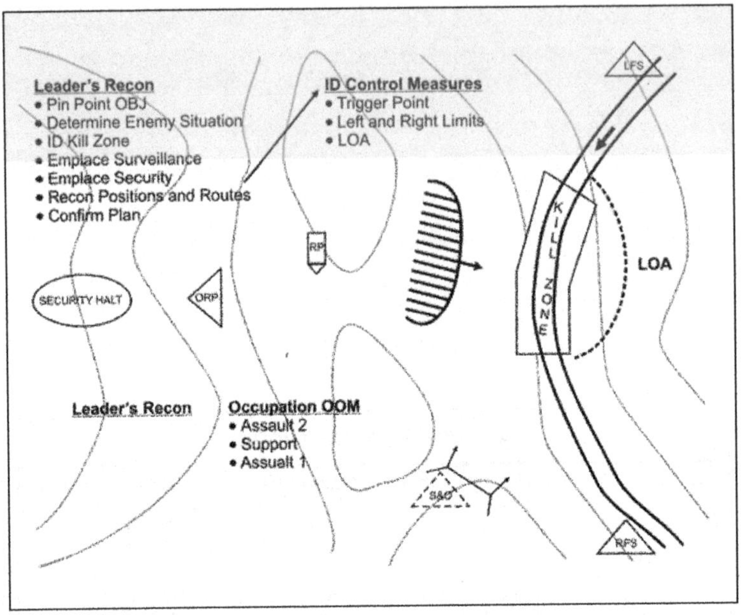

Leader's Recon
• Pin Point OBJ
• Determine Enemy Situation
• ID Kill Zone
• Emplace Surveillance
• Emplace Security
• Recon Positions and Routes
• Confirm Plan
ID Control Measures
• Trigger Point
• Left and Right Limits
• LOA
LFS
RP
KILL ZONE
LOA
SECURITY HALT
ORP
Leader's Recon
Occupation OOM
• Assault 2
• Support
• Assualt 1
S&O
RFS

AMBUSH BOARDS (RIGHT)
SOP'S

CONTINGENCIES

1. COMPROMISE
   a. ORP
   b. LEADERS RECON
   c. OCCUPATION
2. MASS CAL
3. COUNTERATTACK

SOP's

1. EPW SEARCH
2. AID LITTER
3. MASS CAL
4. MED EVAC
5. CCP

NOTES

Appendix B

QUICK REFERENCE CARDS

CASUALTY FEEDER CARD

NAME ____________________ ROS # __________

COMPANY ________ ALLEGERIES __________

LOCATION/EVENT ____________________

TIME ____________ TIME ____________

TEMP ____________ TEMP ____________

PULSE ____________ PULSE ____________

RESP ____________ RESP ____________

B/P ____________ B/P ____________

MENTAL STATUS ____________________

A & O x 1 2 3

NPA OPA ET Tube

SAO^2 ____________ GLUC ____________

IV ____ AC FOREARM HAND OTHER

1000-1000-500-500-500- NS RL D_5W

NOTES: ____________________

____

GTA 07-01-038
INFANTRY LEADERS' REFERENCE CARD
January 1995

HEADQUARTERS, DEPARTMENT OF THE ARMY
References: FM 3-3, FM 6-30, FM 6-121, FM 7-8, FM 7-10, FM 7-90
(Supersedes GTA 7-1-31)

TROOP-LEADING PROCEDURES

1. **Receive mission.**
2. **Issue warning order.**
3. **Make a tentative plan.**
4. **Start movement.**
5. **Reconnoiter.**
6. **Complete plan.**
7. **Issue plan.**
8. **Supervise.**

FIRE REQUEST (FM 6-30)

1. **Identification.**
2. **Warning order.**
3. **Target location.**
4. **Target description.**
5. **Method of engagement.**
6. **Method of control.**

ESTIMATE OF SITUATION (FM 7-10)

1. **Conduct a detailed mission analysis.**
2. **Analyze the situation and develop courses of action.**
3. **Analyze courses of action (wargame).**
4. **Compare courses of action.**
5. **Make a decision.**

OPERATION ORDER (FM 7-10)

Task Organization

1. Situation.
   a. Enemy.
   b. Friendly
   c. Attachments/detachments.
2. Mission
   - Who, what, where, when, and why.
   - Task and purpose.
3. Execution.
   a. Concept of operation.
      (1) Maneuver.
      (2) Fire support.
      (3) Engineer.
   b. Tasks to maneuver units.
   c. Tasks to combat support units.
   d. Coordinating instructions.
4. Service support.
   a. General.
   b. Materiel and services.
   c. Casualty evacuation.
   d. Miscellaneous.
5. Command and signal.
   a. Command.
   b. Signal.

SPOT REPORT

1. **Size.**
2. **Activity.**
3. **Location.**
4. **Unit/uniform.**
5. **Time.**
6. **Equipment.**

SHELL REPORT (FM 6-121)

1. **Observer identification.**
2. **Location (coded).**
3. **Azimuth to flash or sound.**
4. **Time (from and to).**
5. **Area shelled.**
6. **Nature of fire.**
7. **Type rounds received (artillery, mortars, etc.).**
8. **Damage (coded).**

WEAPONS (FM 7-8)

TYPE	MAX EFF RANGE (m)
M16A2	580 (pt) 800 (area) 200 (mov)
M203	150(pt) 350 (area)
M249	600 (pt) 800 (area)
M136 (AT4)	300
M47 (Dragon)	1,000 (sta) 100 (mov)
MK19	1500 (pt) 2,212 (area)
M3 RAAWS	700 (sta) 60 (mov)
M60 MG	1,100 (800 grazing)
.50 Caliber MG	1,800 (1,000 grazing)
TOW	3,000 (plng purposes)
TOW 2	3,750
105-mm	11,500
105-mm Tank	*2 to 2.5 km
120-mm Tank	*2 to 2.5 km
25-mm BFV	2,200
155-mm M109A3	18,100
M198	24,000
8-in Howitzer	22,800

WEAPONS (MORTAR) HE ONLY (FM 7-90)

	MIN	MAX
60-mm	70m	3,500m
81-mm (M252)	80m	5,800m
81-mm (M29A1)	73m	4,790m
4.2-inch	770m	6,840m
120-mm	200m	7,200m

FPFs (FM 7-90)

GUNS	MORT	WIDTH		DEPTH
2	60-mm	60m	x	30m
4	81 (M252)	150m	x	50m
4	81 (M29A1)	140m	x	40m
6	4.2-in	240m	x	40m
6	120-mm	360m	x	60m
Plt	155-mm	200m	x	50m
Btry	155-mm	400m	x	50m

Optimum engagement ranges

MEDEVAC REQUEST

1. Requesting unit identification.
2. Location.
3. Number of patients by type (litter or ambulatory).
4. Type of injuries.
5. Special equipment needed.
6. Tactical situation.

AIRCRAFT REQUEST

1. Identification.
2. Precedence/priority.
3. Target description.
4. Target location.
5. Target time/date.
6. Desired ordnance/results.
7. Final control.

DELIBERATE ATTACK CONSIDERATIONS

1. Reconnoiter, pinpoint objective/enemy positions/obstacles.
2. Determine weak points; designate supporting positions.
3. Assign platoon/squad objectives—identify the decisive point.
4. Determine main attack, supporting attack, reserve.
5. Assign breach, support, assault missions.
6. Designate fire control measures.
7. Coordinate indirect/direct fires and CAS to time of attack.
8. Control measures during attack.
9. Secure (ground and air).
10. Consolidate and reorganize.

NBC 1 OBSERVER'S INITIAL OR FOLLOW-UP REPORT

Instructions

1. Line items D and H are mandatory for NBC 1 reports.
2. Line items A, E, G, I, K, L, M, S, Y, and ZA are optional for NBC 1 reports.
3. Line items B, C, F, PAR, and PBR are reported if data is available.

Section I. Chemical or Biological Only

A. Strike serial number, if known (assigned by NBCE).
B. Position of observer.
C. Azimuth of attack from observer (state degrees or mils).
D. Date and time attack started (Zulu, local, or letter zone).
E. Time attack ended, if known.
F. Location of attack (UTM or place) (state actual or estimated).
G. Means of delivery, if known.
H. Type of agent and height of burst, if known.
I. Type and number of munitions or aircraft (state which).
K. Description of terrain (bare, scrubby vegetation, wooded, urban, or unknown).
S. Date and time contamination detected (Zulu, local, or letter zone).
Y. Representative downwind direction—4 digits (state degrees or mils), wind speed—3 digits (data kmph or knots).
ZA. Temperature (centigrade)—2 digits, cloud cover—1 digit, significant weather phenomena—1 digit, air stability—1 digit.
ZB. Remarks

Section II. Nuclear Only

A. Strike serial number, if known (assigned by NBCE).
B. Position of observer.
C. Azimuth of attack from observer (state degrees or mils and grid or magnetic).
D. Date and time attack started (Zulu, local, or letter zone).
F. Location of attack (UTM or place) (state actual or estimated).
G. Means of delivery, if known.
H. Type of burst (state air, surface, or unknown).
J. Flash-to-bang time (seconds).
K. Crater diameter (meters), if known.
L. Cloud width at H +5 minutes (degrees or mils).
M. Cloud angle(top or bottom) or cloud height(top or bottom) at H +10 minutes (state degrees, mils, meters, or feet).
PAR. Location of radioactive cloud outline (UTM).
PAB. Downwind direction of radioactive cloud (state degrees or mils).
ZB. Remarks.

DEFENSE PLANNING CONSIDERATIONS

1. Establish security (OP/patrols, PWs, MB),
2. Position key weapons:
   a. Coordinate with units on left and right
   b. Establish FPF or PDF for machine gun.
   c. Ensure mutual support between machine guns.
   d. Cover armor approaches with antiarmor systems.
   e. Establish fire control measures.
3. Prepare positions:
   a. Check sectors of fire.
   b. Check overhead cover and view positions from enemy vantage.
   c. Position in depth and achieve mutual support between positions.
   d. Select/prepare alternate and supplementary positions.
4. Integrate indirect fires, CAS, and obstacles with direct and indirect fire.
5. Check communications and establish emergency signals.
6. Designate ammunition, supply, PW, and casualty points.

IED / UXO

Procedures when IEDs are found

Security - Maintain 360 degrees. Scan close in and far out, up high and down low.

Always - Scan you immediate surroundings for more IEDs.

Move - Move away. Vary safe distances, but plan for 300m minimum safe distance and adapt to your METT-TC.

Attempt - To confirm suspected IEDs using optics while staying back as far as possible.

Cordon - Off the area. Direct people out of danger area. Do not allow anyone to enter except for EOD. Question, search, and detain suspects as defined by your Existing ROE.

Report - Your situation using the 9 line IED / UXO spot report format.

This could be your hand if you try to dispose of UXOs or IEDs. The enemy has developed anti-handling to catch you when you try defusing. Leave it to experts!

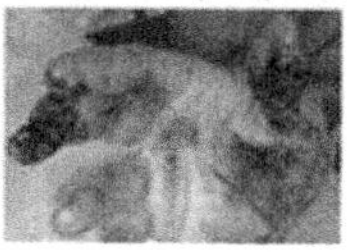

Call EOD
- Don't be a Hero!

IED / UXO Report

LINE 1. DATE-TIME-GROUP: When the item was discovered.

LINE 2. REPORT ACTIVITY AND LOCATION: Unit and grid location of the IED/UXO.

LINE 3. CONTACT METHOD: Radio frequency, call sign, POC, and telephone number.

LINE 4. TYPE OF ORDNANCE: Dropped, projected, placed, or thrown. Give the number of items, if more than one.

LINE 5. NBC CONTAMINATIONS: Be as specific as possible.

LINE 6. RESOURCES THREATENED: Equipment, facilities, or other assets that are threatened.

LINE 7. IMPACT ON MISSION: Short description of current tactical situation and how the IED/UXO affects the status of the mission.

LINE 8. PROTECTIVE MEASURES: Any measures taken to protect personnel and equipment.

LINE 9. RECOMMENDED PRIORITY: Immediate, Indirect, Minor, No Threat.

Priority

Immediate: Stops unit's maneuver and mission capability or threatens critical assets vital to the mission.

Indirect: Stops the unit's maneuver and mission capability or threatens critical assets important to the mission.

Minor: Reduces the unit's maneuver and mission capability or threatens non-critical assets of value.

No Threat: Has little or no effect on the unit's capabilities or assets.

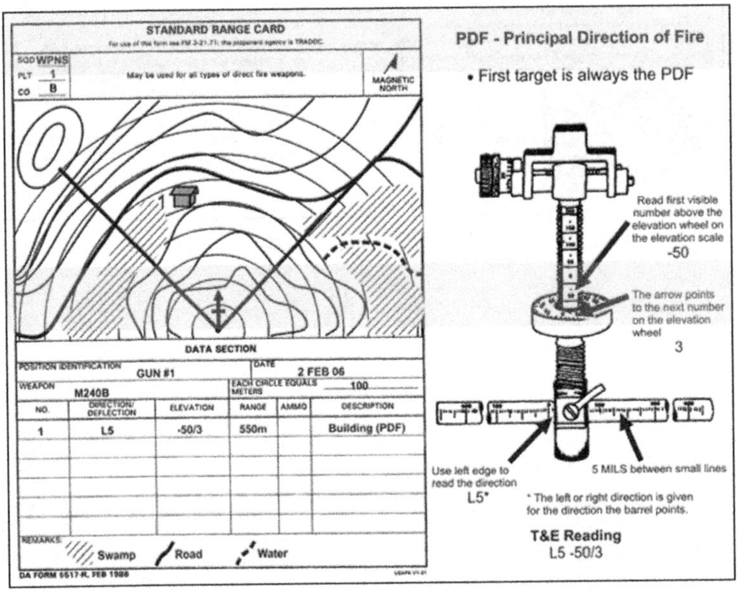
STANDARD RANGE CARD
For use of this form see FM 3-21.71; the proponent agency is TRADOC.
SQD WPNS
PLT 1
CO B
May be used for all types of direct fire weapons.
MAGNETIC NORTH
1
DATA SECTION
POSITION IDENTIFICATION GUN #1
DATE 2 FEB 06
WEAPON M240B
EACH CIRCLE EQUALS 100 METERS
NO.
DIRECTION/ DEFLECTION
ELEVATION
RANGE
AMMO
DESCRIPTION
1
L5
-50/3
550m
Building (PDF)
REMARKS:
Swamp
Road
Water
DA FORM 5517-R, FEB 1986
PDF - Principal Direction of Fire
• First target is always the PDF
Read first visible number above the elevation wheel on the elevation scale
-50
The arrow points to the next number on the elevation wheel
3
Use left edge to read the direction
L5*
5 MILS between small lines
* The left or right direction is given for the direction the barrel points.
T&E Reading
L5 -50/3

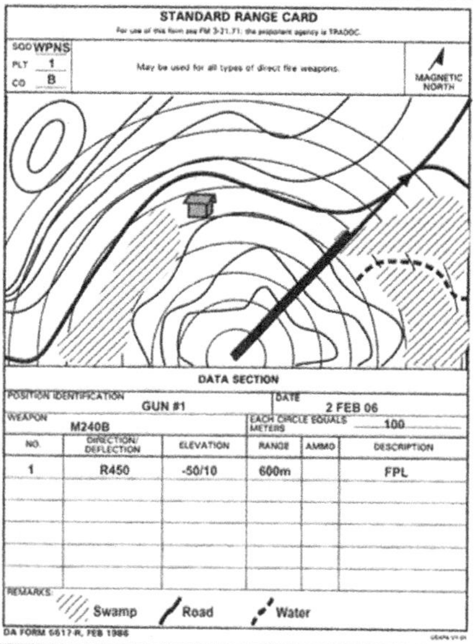

STANDARD RANGE CARD

For use of this form see FM 3-21.71; the proponent agency is TRADOC.

SQD WPNS
PLT 1
CO B

May be used for all types of direct fire weapons.

MAGNETIC NORTH

DATA SECTION

POSITION IDENTIFICATION: GUN #1
DATE: 2 FEB 06
WEAPON: M240B
EACH CIRCLE EQUALS ___ METERS: 100

NO.	DIRECTION/ DEFLECTION	ELEVATION	RANGE	AMMO	DESCRIPTION
1	R450	-50/10	600m		FPL

REMARKS: Swamp Road Water

DA FORM 5517-R, FEB 1986

FPL - Final Protective Line

- FPL is always target #1
- FPL will always be metal-to-metal
- Represented by thick line
  - Break in thick line for dead space out to 600 meters
  - The gap is equal to the width of the dead space

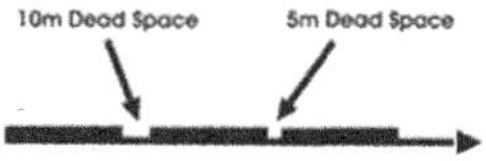

FPL - Final Protective Line

STANDARD RANGE CARD

For use of this form see FM 3-21.71; the proponent agency is TRADOC.

SQD ______
PLT ______
CO ______

May be used for all types of direct fire weapons.

MAGNETIC NORTH

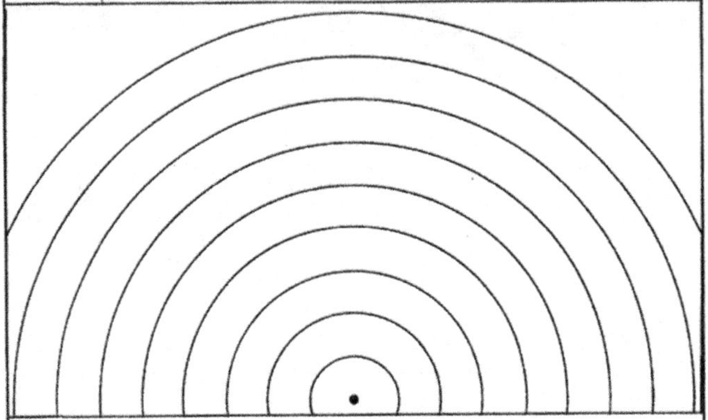

DATA SECTION

POSITION IDENTIFICATION | DATE

WEAPON | EACH CIRCLE EQUALS ______ METERS

NO.	DIRECTION/ DEFLECTION	ELEVATION	RANGE	AMMO	DESCRIPTION

REMARKS:

DA FORM 5517-R, FEB 1986

USAPA V1.01

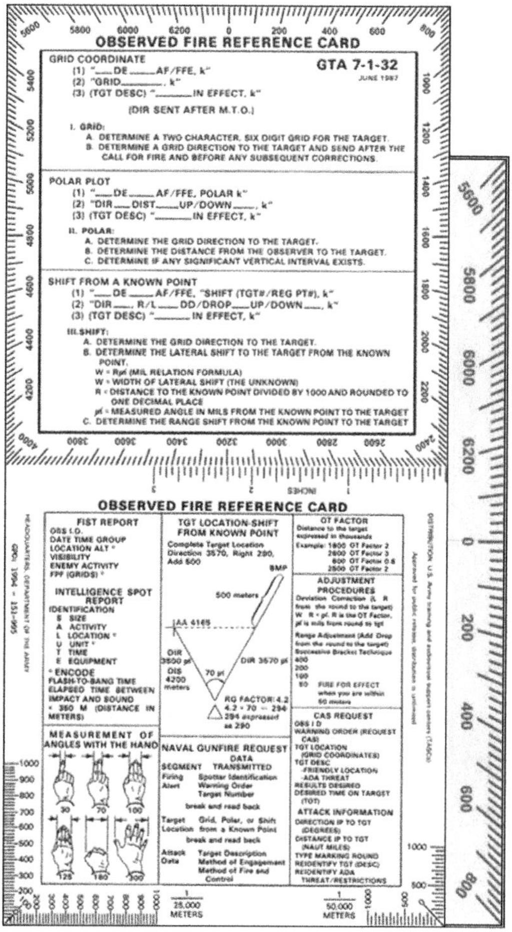

OBSERVED FIRE REFERENCE CARD

GTA 7-1-32
JUNE 1987

GRID COORDINATE
(1) "___DE____AF/FFE, k"
(2) "GRID_______, k"
(3) (TGT DESC) "______IN EFFECT, k"

(DIR SENT AFTER M.T.O.)

I. GRID:
A. DETERMINE A TWO CHARACTER, SIX DIGIT GRID FOR THE TARGET.
B. DETERMINE A GRID DIRECTION TO THE TARGET AND SEND AFTER THE CALL FOR FIRE AND BEFORE ANY SUBSEQUENT CORRECTIONS.

POLAR PLOT
(1) "___DE____AF/FFE, POLAR k"
(2) "DIR___DIST____UP/DOWN____, k"
(3) (TGT DESC) "______IN EFFECT, k"

II. POLAR:
A. DETERMINE THE GRID DIRECTION TO THE TARGET.
B. DETERMINE THE DISTANCE FROM THE OBSERVER TO THE TARGET.
C. DETERMINE IF ANY SIGNIFICANT VERTICAL INTERVAL EXISTS.

SHIFT FROM A KNOWN POINT
(1) "___DE____AF/FFE, "SHIFT (TGT#/REG PT#), k"
(2) "DIR____, R/L____DD/DROP____UP/DOWN____, k"
(3) (TGT DESC) "______IN EFFECT, k"

III. SHIFT:
A. DETERMINE THE GRID DIRECTION TO THE TARGET.
B. DETERMINE THE LATERAL SHIFT TO THE TARGET FROM THE KNOWN POINT.
W = Rμ (MIL RELATION FORMULA)
W = WIDTH OF LATERAL SHIFT (THE UNKNOWN)
R = DISTANCE TO THE KNOWN POINT DIVIDED BY 1000 AND ROUNDED TO ONE DECIMAL PLACE
μ = MEASURED ANGLE IN MILS FROM THE KNOWN POINT TO THE TARGET
C. DETERMINE THE RANGE SHIFT FROM THE KNOWN POINT TO THE TARGET

INCHES

OBSERVED FIRE REFERENCE CARD

FIST REPORT
OBS I.D.
DATE TIME GROUP
LOCATION ALT *
VISIBILITY
ENEMY ACTIVITY
FPF (GRIDS) *

INTELLIGENCE SPOT REPORT
IDENTIFICATION
S SIZE
A ACTIVITY
L LOCATION *
U UNIT *
T TIME
E EQUIPMENT
* ENCODE
FLASH-TO-BANG TIME
ELAPSED TIME BETWEEN IMPACT AND SOUND
× 350 M (DISTANCE IN METERS)

MEASUREMENT OF ANGLES WITH THE HAND
30 70 100
125 180 300

TGT LOCATION-SHIFT FROM KNOWN POINT
Complete Target Location
Direction 3570, Right 290, Add 500
BMP
500 meters
AA 4165
DIR 3500 μ
DIS 4200 meters
70 μ
DIR 3570 μ
RG FACTOR: 4.2
4.2 × 70 = 294
294 expressed as 290

NAVAL GUNFIRE REQUEST DATA
SEGMENT TRANSMITTED
Firing Alert — Spotter Identification, Warning Order, Target Number
break and read back
Target Location — Grid, Polar, or Shift from a Known Point
break and read back
Attack Data — Target Description, Method of Engagement, Method of Fire and Control

OT FACTOR
Distance to the target expressed in thousands
Example: 1800 OT Factor 2
2800 OT Factor 3
600 OT Factor 0.6
2500 OT Factor 2

ADJUSTMENT PROCEDURES
Deviation Correction (L R from the round to the target)
W = R × μ. R is the OT Factor, μ is mils from round to tgt
Range Adjustment (Add Drop from the round to the target)
Successive Bracket Technique
400
200
100
50 FIRE FOR EFFECT when you are within 50 meters

CAS REQUEST
OBS I D
WARNING ORDER (REQUEST CAS)
TGT LOCATION (GRID COORDINATES)
TGT DESC
-FRIENDLY LOCATION
-ADA THREAT
RESULTS DESIRED
DESIRED TIME ON TARGET (TOT)

ATTACK INFORMATION
DIRECTION IP TO TGT (DEGREES)
DISTANCE IP TO TGT (NAUT MILES)
TYPE MARKING ROUND
REIDENTIFY TGT (DESC)
REIDENTIFY ADA THREAT/RESTRICTIONS

1/25,000 METERS
1/50,000 METERS

HEADQUARTERS, DEPARTMENT OF THE ARMY
GPO: 1994 - 153-995

DISTRIBUTION: U.S. Army training and audiovisual support centers (TASCs)
Approved for public release; distribution is unlimited

9 Line MEDEVAC

LINE ITEM	EXPLANATION
1. Location of Pickup Site.	Encrypt grid coordinates. When using *DRYAD Numeral Cipher, the same SET line* will be used to encrypt grid zone letters and coordinates. To preclude misunderstanding, a statement is made that grid zone letters are included in the message (unless unit SOP specifies its use at all times).
2. Radio Frequency, Call Sign, Suffix.	Encrypt the frequency of the radio at the pickup site, *not* a relay frequency. The call sign (and suffix if used) of person to be contacted at the pickup site may be transmitted in the clear.
3. No. of Patients by Precedence.	Report only applicable info & encrypt brevity codes. A = Urgent, B = Urgent-Surg, C = Priority, D = Routine, E = Convenience. (If 2 or more categories reported in same request, insert the word "*break*" between each category).
4. Spec Equipment.	Encrypt applicable brevity codes. A = None, B = Hoist, C = Extraction equipment, D = Ventilator.
5. No. of Patients by Type.	Report only applicable information and encrypt brevity code. If requesting MEDEVAC for both types, insert the word "*break*" between the litter entry and ambulatory entry: L + # of Pnt - Litter; A + # of Pnt - Ambul (sitting).
6. Security Pickup Site (Wartime).	N = No enemy troops in area, P = Possibly enemy troops in area (approach with caution), E = Enemy troops in area (approach with caution), X = Enemy troops in area (armed escort required).
6. Number and type of Wound, Injury, Illness (Peacetime).	Specific information regarding patient wounds by type (gunshot or shrapnel). Report serious bleeding, along with patient blood type, if known.
7. Method of Marking Pickup Site.	Encrypt the brevity codes. *A* = Panels, *B* = Pyrotechnic signal, *C* = Smoke Signal, *D* = None, *E* = Other.
8. Patient Nationality and Status.	Number of patients in each category need not be transmitted. Encrypt only applicable brevity codes. A = US military, B = US civilian, C = Non-US mil, D = Non-US civilian, E = EPW.
9. NBC Contamination (Wartime).	Include this line only when applicable. Encrypt the applicable brevity codes. N = nuclear, B = biological, C = chemical.
9. Terrain Description (Peacetime).	Include details of terrain features in and around proposed landing site. If possible, describe the relationship of site to a prominent terrain feature (lake, mountain, tower).

Reference: FM 8-10-6, *Medical Evacuation in a Theater of Operations, pages 7-7 through 7-9.*